Welding

Level Two

SIXTH EDITION

Pearson NCCER National Center for
Construction Education
and Research

NCCER

President and Chief Executive Officer: Boyd Worsham
Vice President of Innovation and Advancement: Jennifer Wilkerson
Chief Learning Officer: Lisa Strite
Senior Manager, Curriculum Development: Chris Wilson
Production Manager: Graham Hack
Welding Project Manager: Dario VanHorne
Technical Writing Manager: Gary Ferguson
Technical Writer: Simone Flynn
Art Manager: Bree Rodriguez
Technical Illustrators: Judd Ivines, Liza Wailes
Production Artist: Chris Kersten
Permissions Specialists: Adam Black, Sherry Davis
Managing Editor: Hannah Murray
Lead Editor: Karina Kuchta
Editors: Zi Meng, Alexandria Willbond, James Singer III, Lauren Tygrest
Desktop Publishing Manager: Eric Caraballoso III
Desktop Publishing Coordinator: Daphney Milian
Production Specialist: Julie Watkins
Digital Content Manager: Kelly Beck
Digital Content Coordinator and Translation Manager: Yesenia Tejas
Digital Content Coordinator: Briana Rosa
Project Coordinator: Chelsi Santana

Pearson

Director of Association Partnerships: Tanja Eise
Program Manager: Vanessa Price
Senior Digital Content Producer: Shannon Stanton
Content Producer: Arup Kumar Ghosh
Employability Solutions Coordinator: Monica Perez
Cover Designer: Mary Siener
Rights and Permissions: Jenell Forschler

Composition: NCCER
Content Technologies: Gnostyx
Printer/Binder: Lakeside Book Company
Cover Printer: Lakeside Book Company
Typefaces: Palatino LT Pro and Helvetica Neue

Cover Image

Cover photo provided by: © Miller Electric Mfg. LLC

10 9 8 7 6 5 4 3 2 1   6 2024

Paperback
ISBN-13: 978-0-13-821616-0

Hardcover
ISBN-13: 978-0-13-821621-4

National Center for
Construction Education
and Research

PREFACE

To the Trainee

Welding stands as a paramount pillar in the structure of global industries, unifying pieces of our world in bonds of metal. The craft of welding operates as the backbone of numerous sectors, from construction to automotive. It's a realm of limitless potential and widespread applicability, where advancements continue to broaden its horizons and deepen its impacts.

Welding Level Two builds upon the foundational knowledge and skills acquired in *Welding Level One* and plunges deeper into the core of this ever-evolving field. This stage is meticulously crafted to bolster the proficiency and understanding of aspiring welders, granting them the ability to stride forward confidently in their welding journey and adapt to the diverse demands of this multifaceted trade.

Welding Level Two introduces advanced skills and knowledge base relevant to the craft. Mastering the intricate language of welding symbols and knowing how to read welding detail drawings are crucial for translating designs into tangible structures, ensuring precision and adherence to specifications. A deep understanding of metals' physical characteristics and mechanical properties equips a welder with the discernment to select and manipulate metals for a given project. Furthermore, expertise in preheating and postheating of metals ensures the longevity and quality of welds by mitigating potential defects and structural weaknesses. Together, these foundational skills set the stage for any welder to navigate the complexities of their craft confidently, positioning them for unparalleled success in the dynamic and lucrative world of welding.

New with *Welding Level Two*

The sixth edition of *Welding Level Two* has been reformatted to provide a better experience for both trainees and instructors. NCCER is proud to release this edition with our new design that links learning objectives to specific sections in each module. In addition to the design improvements, each module's content, images, and diagrams have been updated to reflect the most current safe welding practices.

Welding Level Two aligns with the current standards in American Welding Society's School Excelling through National Skills Education (SENSE) guidelines for entry welders. This means that, this program conforms to NCCER guidelines and can also be used to meet guidelines provided by AWS for entry welder training. For more information on the AWS SENSE program, contact AWS at 1-800-443-9353 or visit **www.aws.org**.

We wish you success as you progress through this training program. If you have any comments on how NCCER might improve upon this textbook, please complete the User Update form using the QR code on this page. NCCER appreciates and welcomes its customers' feedback. You may submit yours by emailing **support@nccer.org**. When doing so, please identify feedback on this title by listing *#WeldingL2* in the subject line.

Our website, **www.nccer.org**, has information on the latest product releases and training.

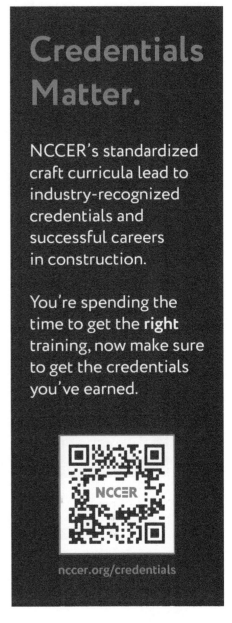

Credentials Matter.

NCCER's standardized craft curricula lead to industry-recognized credentials and successful careers in construction.

You're spending the time to get the **right** training, now make sure to get the credentials you've earned.

nccer.org/credentials

SCAN ME

NCCER Standardized Curricula

NCCER is a not-for-profit 501(c)(3) education foundation established in 1996 by the world's largest and most progressive construction companies and national construction associations. It was founded to address the severe workforce shortage facing the industry and to develop a standardized training process and curricula. Today, NCCER is supported by hundreds of leading construction and maintenance companies, manufacturers, and national associations. The NCCER Standardized Curricula was developed by NCCER in partnership with Pearson, the world's largest educational publisher.

Some features of the NCCER Standardized Curricula are as follows:

- An industry-proven record of success
- Curricula developed by the industry, for the industry
- National standardization providing portability of learned job skills and educational credits
- Compliance with the Office of Apprenticeship requirements for related classroom training (*CFR 29:29*)
- Well-illustrated, up-to-date, and practical information

NCCER maintains a secure online database that provides certificates, digital badges, transcripts, and wallet cards to individuals who successfully complete programs under an NCCER-accredited organization or through one of NCCER's self-paced, online programs. This system also allows individuals and employers to track and verify industry-recognized credentials and certifications in real time.

For information on NCCER's credentials, contact NCCER Customer Service at 1-888-622-3720 or visit **www.nccer.org**.

Digital Credentials

Show off your industry-recognized credentials online with NCCER's digital credentials!

NCCER is now providing online credentials. Transform your knowledge, skills, and achievements into digital credentials that you can share across social media platforms, send to your network, and add to your resume. For more information, visit **www.nccer.org**.

Cover Image

Cover image was taken for a Miller® photoshoot of a local Miller® supplier and customer using a new piece of welding equipment. They were pulsed gas metal arc welding (GMAW-P) a component for a Miller® Wirefeeder boom.

Miller® is about building things that matter. They lead the welding industry in building advanced, solution-focused products and meeting crucial needs for welding safety and health.

They're about the partnership and the work. Their products are designed with their users for manufacturing, fabrication, construction, aviation, motorsports, education, agriculture and marine applications.

DESIGN FEATURES

Content is organized and presented in a functional structure that allows you to access the information where you need it.

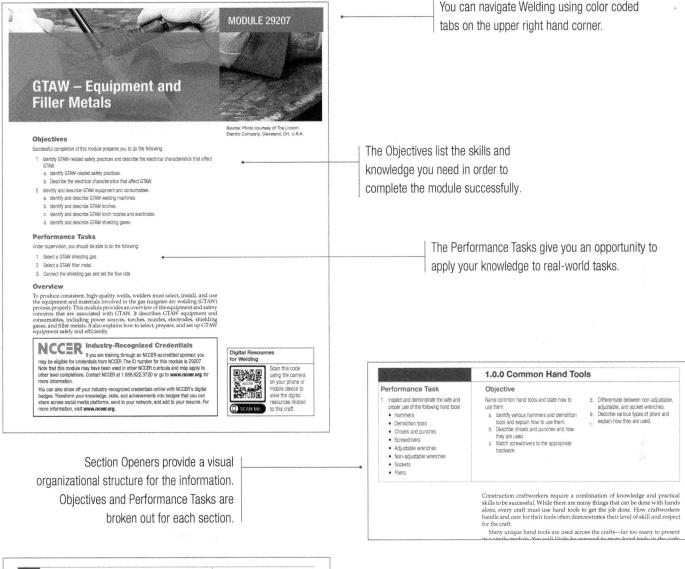

You can navigate Welding using color coded tabs on the upper right hand corner.

The Objectives list the skills and knowledge you need in order to complete the module successfully.

The Performance Tasks give you an opportunity to apply your knowledge to real-world tasks.

Section Openers provide a visual organizational structure for the information. Objectives and Performance Tasks are broken out for each section.

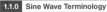

Trade Terms appear on the page adjacent to the text where they are first presented.

Step-by-step math equations help make the concepts clear and easy to grasp.

QR codes link directly to videos that highlight current content.

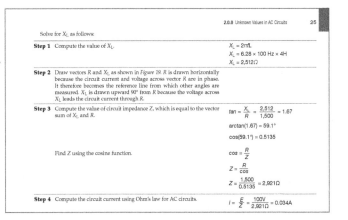

Important information is highlighted, illustrated, and presented to facilitate learning.

Placement of images near the text description and details such as callouts and labels help you absorb information.

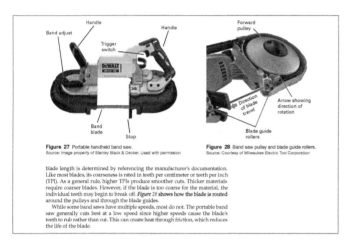

Figure 27 Portable handheld band saw.
Source: Image property of Stanley Black & Decker. Used with permission

Figure 28 Band saw pulley and blade guide rollers.
Source: Courtesy of Milwaukee Electric Tool Corporation

blade length is determined by referencing the manufacturer's documentation. Like most blades, its coarseness is rated in teeth per centimeter or teeth per inch (TPI). As a general rule, higher TPIs produce smoother cuts. Thicker materials require coarser blades. However, if the blade is too coarse for the material, the individual teeth may begin to break off. *Figure 28* shows how the blade is routed around the pulleys and through the blade guides.

While some band saws have multiple speeds, most do not. The portable band saw generally cuts best at a low speed since higher speeds cause the blade's teeth to rub rather than cut. This can create heat through friction, which reduces the life of the blade.

Preparing Drills with Keyless Chucks

Most cordless drills use a keyless chuck. While the steps for preparing a cordless drill are similar, there are some small differences. Follow the steps below when preparing to use drills with keyless chucks:

Step 1 Disconnect the drill from its power source by removing the battery pack before loading a bit.
Step 2 As shown in (*Figure 7A*), open the chuck by turning it counterclockwise until the jaws are wide enough to insert the bit shank.
Step 3 Insert the bit shank into the chuck opening (*Figure 7B*). Keeping the bit centered in the opening, turn the chuck by hand until the jaws grip the bit shank.
Step 4 Tighten the chuck securely with your hand so that the bit does not move (*Figure 7C*). You are now ready to use the cordless drill.

(A) Insert the Bit Shank.   **(B) Keep Bit Straight and Partially Tighten the Chuck.**   **(C) Tighten the Chuck Securely.**

Figure 7 Loading the bit on a keyless chuck.
Source: Cianbro Corporation

New boxes highlight safety and other important information. Warning boxes stress potentially dangerous situations, while Caution boxes alert to dangers that may cause damage to equipment. Note boxes provide additional information on a topic.

WARNING!

Dust, fumes, vapors, and fine machining particles of beryllium are cancer-causing agents. If inhaled, they can cause chronic beryllium disease. An independent air supply is required when working with beryllium.

CAUTION

Magnetic testing is sometimes unreliable because of alloys or changes that can occur during welding. Do not rely solely on magnetic testing.

NOTE

An example of an object and its corresponding three-view detail drawing is provided in *Appendix 29202A*.

Did You Know?
Lightning Rods

An interesting fact about grounding is that a lightning rod (air terminal) isn't meant to bring a bolt of lightning to ground. To do this, its conductors would have to be several feet (1 m or more) in diameter. The purpose of the rod is to dissipate the negative static charge that would cause the positive lightning charge to strike the house.

These boxed features provide additional information that enhances the text.

Induction Heaters

Because water cools more efficiently than air, water-cooled induction heating systems can be used for high-temperature preheating and stress-relieving applications. They have the ability to get to 1,450°F (788°C). This type of system is normally equipped with a temperature controller and temperature recorder, which are important components in stress-relieving applications.

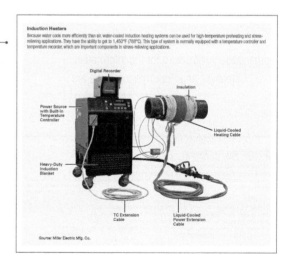

Source: Miller Electric Mfg. Co.

Around the World
Predicting Weather

Barometers measure atmospheric air pressure. Generally, an increased pressure indicates that the weather is pleasant. A reduced pressure indicates that the weather is cloudy and/or rainy.

Changes in barometric pressure are used to predict changes in weather conditions. For example, if it is raining outside, but the barometer is rising, it indicates that better weather is on the way. In the midst of a hurricane, the barometric pressure is an indicator of intensity. Extremely low pressures exist in powerful hurricanes. While traditional barometers work very well, sophisticated weather stations capture far more data and can transmit the information wirelessly.

Going Green
Residential Solar

As residential customers look for ways to reduce reliance on fossil fuels like coal and oil, they are turning to solar energy. Usually, they want to install photovoltaic (PV) panels on the roofs of their homes, or in their yards. A residential PV system usually includes the PV panels, a mounting structure that can support the panels, an inverter to convert the direct current (DC) electricity generated by solar photovoltaic modules into alternating current (AC) electricity, and a battery (or batteries) for storing the energy. Most states require a licensed electrician for solar energy system installation, but you should check the local requirements. (Source: **https://www.energy.gov/eere/solar/how-does-solar-work**)

Going Green looks at ways to preserve the environment, save energy, and make good choices regarding the health of the planet.

Cornerstone of Craftsmanship

John Lupacchino
Senior Design Engineer
Gaylor Electric, Inc.

How did you choose a career in the industry?
I knew that this was what I wanted to do since I was in 7th grade.

Who inspired you to enter the industry?
My grandfather, who was an electrician, inspired me to also become an electrician.

What types of training have you been through?
I went to vocational technical school for high school and studied electrical. Then I went on to complete an apprenticeship through the state of Connecticut.

How important is education and training in construction?
Training and education are extremely important in the construction industry. All craftworkers need to participate in continuing education in order to stay up with the advancements in technology.

How important are NCCER credentials to your career?
NCCER credentials are very important because they allow you to showcase your skills and abilities in a standardized way.

What kinds of work have you done in your career?
I have worked in all areas of electrical—residential, commercial, and industrial. Some of the types of facilities I have worked in include houses, stores, factories, warehouses, hospitals, prisons, steel mills, solar installations, cement plants, and many others.

Tell us about your current job.
In my present job as senior design engineer, I am responsible for estimating, designing, and managing electrical construction projects. I am also responsible for electrical code interpretation and compliance.

What do you enjoy most about your job?
My job is challenging and ever changing. It never gets boring. For me, it is great to be a part of building something that benefits others.

What factors have contributed most to your success?
I take advantage of the opportunities that come up, especially the training that is available. Applying my skills and putting in the effort required has definitely contributed to my success.

Would you suggest construction as a career to others? Why?
Yes! The construction industry has limitless opportunities. There will always be a need for building and maintaining facilities, which means there will always be a need for craftworkers.

What advice would you give to those new to the field?
Take advantage of any opportunities for training that you have. Show up to work on time with a "Get It Done" mentality. Do all you can to be the best you can be.

What is an interesting career-related fact or accomplishment?
I have been able to acquire licenses all over the country to enable my employer to work in various locations. I have also had the opportunity to be an instructor in the local ABC apprenticeship program for over 20 years.

How do you define craftsmanship?
Craftsmanship is the quality that comes from creating with passion, care, and attention to detail.

Cornerstone of Craftsmanship boxes feature career stories from people working in related fields.

Review questions at the end of each section and module allow you to measure your progress.

2.0.0 Section Review

1. A straight pull contains two raceways. One of the raceways has a trade size of 3" and one has a trade size of 2". The length of the box *must* be _____.
   a. 16"
   b. 24"
   c. 26"
   d. 32"
2. Where angle or U pulls are made, to determine the distance between raceway entries enclosing the same conductor, you need to multiply the trade size of the largest raceway by _____.
   a. two
   b. four
   c. six
   d. eight
3. If possible, pull boxes should be installed _____.
   a. high on the wall for security
   b. at a height/location that makes it easy to pull conductors
   c. behind wall coverings for a neater look
   d. with extra knockouts left open for air flow

Section Review questions can be found at the end of each section to test your knowledge of the content. Review Questions at the end of each module are provided to reinforce the knowledge you have gained.

Module 26207-23 Review Questions

1. In which *NEC®* article is cable tray installation primarily addressed?
   a. *NEC Article 300*
   b. *NEC Article 392*
   c. *NEC Article 517*
   d. *NEC Article 550*
2. Which type of tray is *best* suited for corrosive areas and atmospheres?
   a. Stainless steel cable tray
   b. Coated aluminum cable tray
   c. Nonmetallic cable tray
   d. Basket tray
3. When a cable tray has a solid bottom, it is referred to as _____.
   a. basket tray
   b. ladder tray
   c. trough
   d. raceway
4. Cable tray is generally manufactured in _____.
   a. 2' and 4' lengths
   b. 6' and 8' lengths
   c. 6' and 12' lengths
   d. 12' and 24' lengths

Additional information can be found in Appendixes at the back of the book.

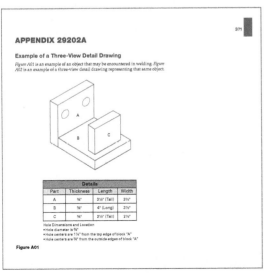

371

APPENDIX 29202A

Example of a Three-View Detail Drawing

Figure A01 is an example of an object that may be encountered in welding. *Figure A02* is an example of a three-view detail drawing representing that same object.

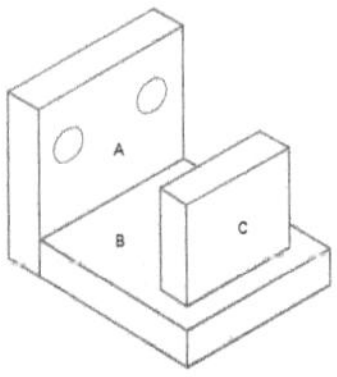

Details			
Part	Thickness	Length	Width
A	¾"	3½" (Tall)	3½"
B	¾"	4" (Long)	3½"
C	¾"	2½" (Tall)	2½"

Hole Dimensions and Location
- Hole diameter is ¾"
- Hole centers are 1¼" from the top edge of block "A"
- Hole centers are ¾" from the outside edges of block "A"

Figure A01

Some modules have corresponding Appendixes that provide supplementary information or activities to enhance your understanding of the material.

NCCERCONNECT

This interactive online course is a unique web-based supplement that provides a range of visual, auditory, and interactive elements to enhance training. Also included is a full eText.

Visit **www.nccerconnect.com** for more information!

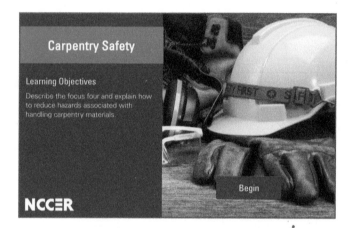

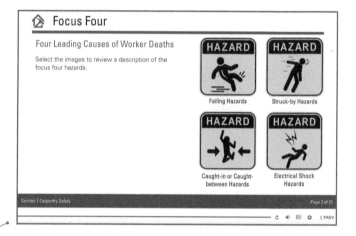

Use the interactive self-guided learning lesson to understand key concepts and terms needed for a career in construction.

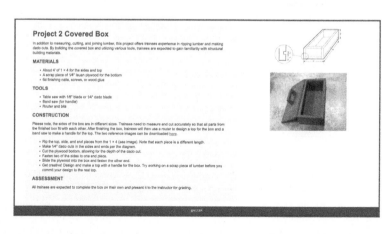

ACKNOWLEDGMENTS

This curriculum was revised as a result of the vision and leadership of the following sponsors:

All Things Metal

Bechtel

College of Southern Nevada

Kentucky Welding Institute

Lee College

Mesa Community College

Miller Electric

NW Florida State College

Raycap

United Group Services

Windham School District

Yates

This curriculum would not exist without the dedication and unselfish energy of those volunteers who served on the Authoring Team. A sincere thanks is extended to the following:

Jim Casey

Ben Pletcher

John Elliott

Matthew Aschoff

Donald "Greg" MacLiver

Brian Dennis

Jimmy Perry

Richard Samanich

Scottie Smith

Don Herron

Ashley Applegate

Curtis Casey

Nick Peterson

A final note: This book is the result of a collabortive effort involving the production, editorial, and development staff at Pearson Education, Inc., and NCCER. Thanks to all of the dedicated people involved in the many stages of this project.

NCCER PARTNERS

To see a full list of NCCER Partners, please visit:

www.nccer.org/about-us/partners.

You can also scan this code using the camera on your phone or mobile device to view these partnering organizations.

CONTENTS

Module 29201   Welding Symbols

Module 29202   Reading Welding Detail Drawings

Module 29203 Physical Characteristics and Mechanical Properties of Metals

Module 29204 Preheating and Postheating of Metals

Module 29205 GMAW and FCAW – Equipment and Filler Metals

Module 29209 GMAW – Plate

Module 29210   FCAW – Plate

Module 29207 GTAW – Equipment and Filler Metals

Module 29208 GTAW – Plate

Welding Appendixes

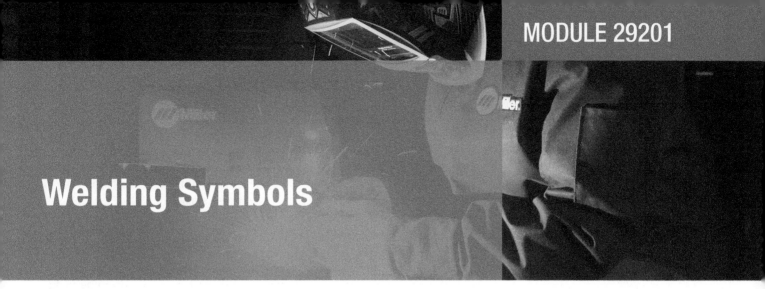

Welding Symbols

Objective

Successful completion of this module prepares you to do the following:

1. Identify and interpret welding symbols and their structure.
   a. Describe the structure and placement of welding symbols and identify basic symbols.
   b. Identify and interpret size and dimension markings for common types of welds.
   c. Identify and interpret various supplemental symbols.
   d. Identify and interpret less common welding symbols.

Performance Task

Under supervision, you should be able to do the following:

1. Identify and interpret welding symbols on an instructor-provided drawing.

Overview

Project drawings and specifications contain many symbols that communicate critical information about the welds to be used at various locations throughout the project. Welders must learn this symbolic language so they can properly interpret the symbols and make welds that meet design specifications. This module introduces a wide range of welding symbols, describes how they are structured, and explains the basic rules for applying the information that the symbols convey.

NCCER Industry-Recognized Credentials

If you are training through an NCCER-accredited sponsor, you may be eligible for credentials from NCCER. The ID number for this module is 29201. Note that this module may have been used in other NCCER curricula and may apply to other level completions. Contact NCCER at 1.888.622.3720 or go to **www.nccer.org** for more information.

You can also show off your industry-recognized credentials online with NCCER's digital credentials. Transform your knowledge, skills, and achievements into credentials that you can share across social media platforms, send to your network, and add to your resume. For more information, visit **www.nccer.org**.

Digital Resources for Welding

Scan this code using the camera on your phone or mobile device to view the digital resources related to this craft.

1.0.0 Introduction to Welding Symbols

Performance Task

1. Identify and interpret welding symbols on an instructor-provided drawing.

Objective

Identify and interpret welding symbols and their structure.

a. Describe the structure and placement of welding symbols and identify basic symbols.
b. Identify and interpret size and dimension markings for common types of welds.
c. Identify and interpret various supplemental symbols.
d. Identify and interpret less common welding symbols.

It is important for welders to understand the meaning of welding symbols used in drawings. They convey critical information about the size of the welds, the number of welds, the location of the welds, and the type of welds to be used. Ultimately, these symbols convey a design that the welder is responsible for welding.

1.1.0 Welding Symbol Construction

Reference line: The horizontal line in the center of the welding symbol from which all elements of the welding symbol are referenced. The reference line is one of the most important elements of the welding symbol.

Arrow line: The line drawn at an angle from the reference line (either end or both ends) to an arrowhead at the location of the weld.

Welding symbol: A graphical representation of the specifications for producing a welded joint; includes a reference line and arrow line and many also include a weld symbol.

The base for all welding symbols is the horizontal **reference line**, which has an arrow at one end. The **arrow line**, which can be on either side of the reference line, points to the location to which the welding symbol applies (*Figure 1*). The **welding symbol** describes the type of weld, its size, and its surface finish.

Welding symbols are used on drawings, project specifications, and Welding Procedure Specifications (WPS) to convey the design specifications for welds. A series of symbols is used to indicate the joint configuration and weld type, location, size, and length of weld required. Welders must be able to properly interpret welding symbols to ensure that the welds they make will meet the design specifications.

Most companies use the symbols that have been standardized by the American Welding Society (AWS) in *AWS A2.4, Standard Symbols for Welding, Brazing, and Nondestructive Examination*. Most companies also have a site quality standard that provides guidelines and examples of how welding symbols are to be used. Always refer to the site-specific quality standard when interpreting welding symbols.

The opposite end of the reference line, called the *tail*, is used for information that aids in making the weld but does not have its own special place on the symbol (*Figure 2*). For instance, the tail may be used to indicate the welding and cutting processes, the reference to a note, or the welding procedure or electrode type to be used. When a reference is not required, the tail is omitted, as shown in *Figure 1*.

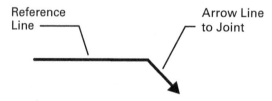

Figure 1 Horizontal reference line.

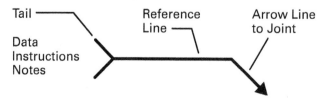

Figure 2 Tail-side information.

In the case of a T-joint, the arrow points to the two sides of the joint, and welding may be accomplished on either side. The welding symbol distinguishes between the two sides of a joint by using the arrow line and the spaces above and below the reference line. The side of the joint to which the arrow points is called the *arrow side*, and its weld is made in accordance with the instructions below the reference line. The opposite side of the joint is referred to as the *other side*, and its weld is made in accordance with the instructions found above the reference line. *Figure 3* shows examples of these symbols.

Regardless of which end of the reference line the arrow line is on, information on the reference line is always read from left to right. *Figure 4* shows a welding symbol base and the numerous elements of a welding symbol.

The **weld symbols** used to identify the type of weld to be made signify a basic type of weld or joint preparation (*Figure 5*). It is important to note the difference between welding symbols and weld symbols. A weld symbol is a graphic character connected to the reference line of a welding symbol, specifying the weld type. A welding symbol graphically represents the specifications for producing a welded joint; it includes any weld symbols as well as the additional components. All welding symbols include a reference line and arrow line, at a minimum. Many also include a weld symbol.

Weld symbols: Graphic characters connected to the reference line of a welding symbol specifying the weld type.

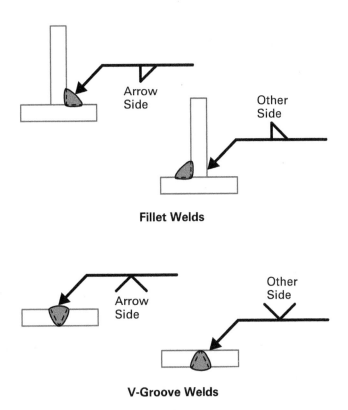

Fillet Welds

V-Groove Welds

Figure 3 Arrow-side and other-side significance.

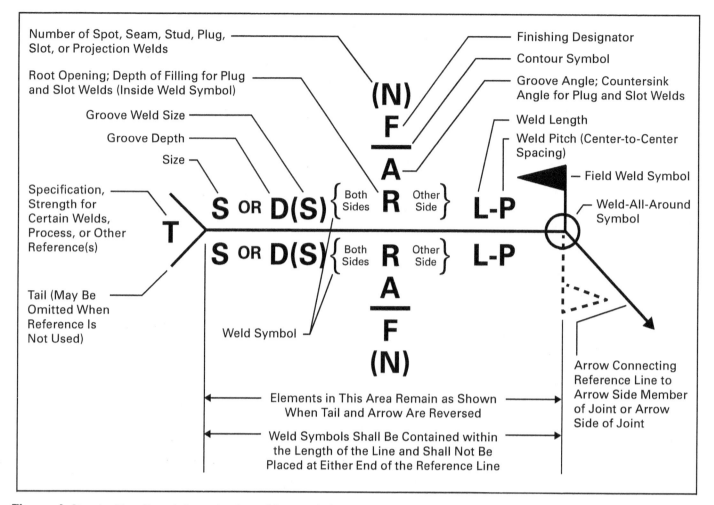

Figure 4 Standard location of elements of a welding symbol.
Source: AWS A2.4:2020, Figure 4.3, Reproduced with permission from the American Welding Society (AWS), Miami, FL, USA

Groove Welds							
Square	Scarf	Bevel	Flare V	Flare Bevel	V	U	J

Other Welds							
Fillet	Plug or Slot	Stud	Spot or Projection	Seam	Back or Backing	Surfacing	Edge

Figure 5 Basic weld symbols.

1.1.1 Symbols for Welds

The type of weld used is determined by the way the members to be welded are positioned to form the joint. The five basic types of weld joints are the butt, corner, lap, tee, and edge joints (*Figure 6*).

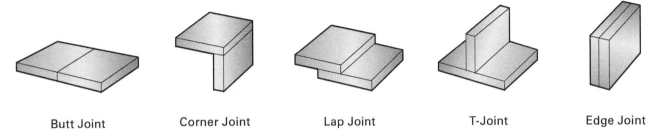

Butt Joint   Corner Joint   Lap Joint   T-Joint   Edge Joint

Figure 6 Basic types of weld joints.

1.1.2 Location of Weld Symbols

When the weld symbol is on both the top and bottom of the reference line, the information applies to both sides of the joint or surface to which the arrow is pointing. Symbols appear on both sides of the reference line when a double-fillet or double groove weld is specified. *Figure 7* shows symbols for arrow-side, other-side, and both sides for fillet and V-groove welds.

The shape of the groove weld symbol indicates how the groove is to be prepared. In the case of the bevel groove and J-groove weld symbols, only one of the two members to be welded is prepared. The arrow indicates which member is to be prepared by breaking toward that member. The broken segment that includes the arrowhead is the one that indicates the member that receives the preparation (*Figure 8*). Notice also that the vertical leg of the weld symbol is shown drawn to the left of the slanted/curved leg(s). Regardless of whether the symbol is for a fillet, bevel groove, J-groove, or flare bevel groove weld, the vertical leg is always drawn to the left.

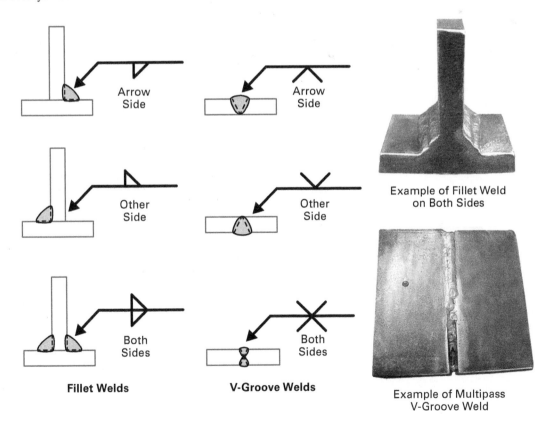

Figure 7 Fillet and V-groove welds.

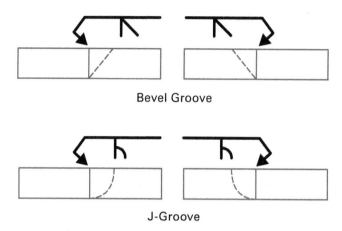

Bevel Groove

J-Groove

Note: Dashed lines indicate preparation
specified by weld symbol.

Figure 8 Bevel groove and J-groove weld symbols.

Think About It

Breaking the Arrow

Are the symbol representations shown in *Figure 8* the only ways a breaking arrow can be shown?

1.1.3 Combining Weld Symbols

A typical combination weld is made by adding a fillet weld to a single or double bevel groove weld in a T-joint. The fillet weld symbol is always placed on top of the groove weld symbol, just as it would be on the actual weld. When more than one type of weld is to be made on the same joint, it is necessary to combine weld symbols as shown in *Figure 9*. Note, a broken arrow is not needed when there is only one member that can receive a single bevel preparation, as in the case of this T-joint. When there are different edge shapes in a joint, combinations of J-, flare bevel, or bevel groove weld symbols are permissible. Adjusting the weld symbol in this way will represent the different edge preparations present in the weld.

1.1.4 Multiple Reference Lines

When more than one welding operation must be performed in a certain sequence, two or more reference lines may be used to indicate the sequence. *Figure 10* illustrates the typical sequences of operation when multiple reference lines are used. In each sequence, the first operation to be performed will be closest to the arrow. Subsequent operations may be shown sequentially on other reference lines. The last operation to be performed will be the one farthest from the arrow.

Multiple reference lines may also be used to show data supplementary to welding symbol information that is on the reference line nearest the arrow. As shown in *Figure 11*, test information may be shown on a second or third line, away from the arrow.

When required, the weld-all-around symbol, shown in *Figure 12* (A), must be placed at the junction of the arrow line and reference line for each operation to which it applies. The field weld symbol shown in *Figure 12* (B) may also be used in this manner. Supplemental symbols are discussed in more detail later in this module.

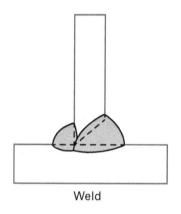

Weld

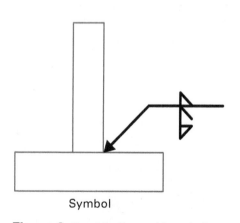

Symbol

Figure 9 Combination weld symbol.

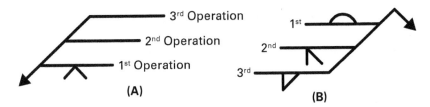

Figure 10 Multiple reference lines.

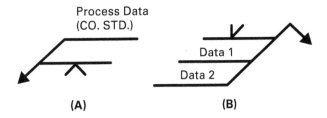

Figure 11 Supplementary data.

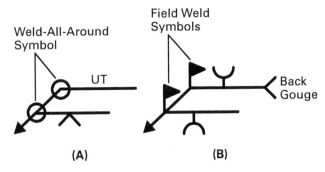

Figure 12 Typical placement of supplemental symbols.

Nondestructive Examination Symbols

Nondestructive examination symbols may appear on the same drawings as (and look very similar to) welding symbols. Inspectors use nondestructive examination symbols to examine joints once welding is complete. The nondestructive methods of inspection are defined by letter codes on the reference line. For instance, VT indicates visual testing, UT indicates ultrasonic testing, RT indicates radiographic testing, and PT indicates penetrant testing. *AWS A2.4* requires that only a horizontal reference line, arrow, and examination method(s) are required elements for nondestructive examination symbols. Additional elements may be included to convey specific nondestructive examination information.

NDT symbols and weld symbols may be combined by using two reference lines. In contrast to the arrow for welding symbols, the arrow for NDT always points to the weld that is to be tested, and never to a weld on the other side of the joint. If a double fillet weld were to be tested, for example, this would be shown with a separate arrow on each side of the joint.

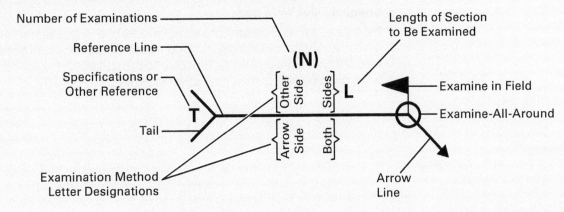

1.2.0   Sizing and Dimensioning Welds

Unless defined in a drawing note, the size data for a weld is always shown to the left of the symbol to which it applies. For example, the size of the fillet weld shown in *Figure 13* is ¼". The length of the weld is always shown to the right of the symbol to which the length applies. An example of this is seen in *Figure 14*, where the length is 6".

Dimensions can be given as fractions, decimals, or metric measurements. The following sections explain how to size and dimension the following weld types:

- Fillet
- Groove
- Plug
- Slot

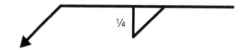

Figure 13 Sizing welds.

Figure 14 Dimensioning welds.

1.2.1 Sizing Fillet Welds

A fillet weld is a weld made in a corner. It is roughly triangular in shape and joins two welding surfaces that are perpendicular to each other, such as in a lap, T-, or corner joint. The size of a fillet weld is the length of the legs (sides) of the largest triangle that can be inscribed within the fillet weld cross section. *Figure 15* shows the profiles of convex and concave welds.

The lengths of the legs on a typical fillet weld are equal, so only one dimension is given. Equal-leg fillet welds are sized by placing the size to the left of the fillet weld symbol or as defined in a drawing note.

If the fillet weld has unequal legs, the size of each leg is given in parentheses or brackets to the left of the fillet weld symbol with a drawing or explanation of leg orientation. An example of this is shown in *Figure 16* where the unequal legs are ¼" (6.4 mm) and ½" (12.7 mm).

Unequal Fillet Weld Legs

The only practical use of unequal fillet weld legs is on metal of unequal thickness. Unequal legs do not make the weld stronger; the thinner material prevails. An unequal fillet leg weld helps distribute loads and eliminates incomplete fusion discontinuities that would be caused by making a small weld on heavy material without preheat.

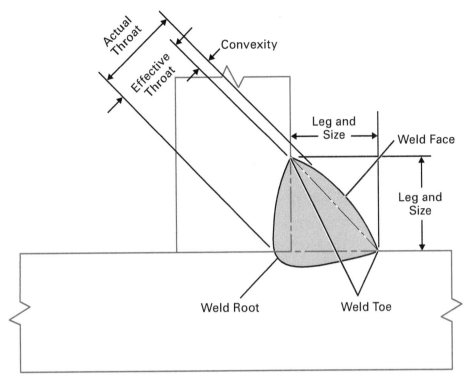

Convex Fillet Weld

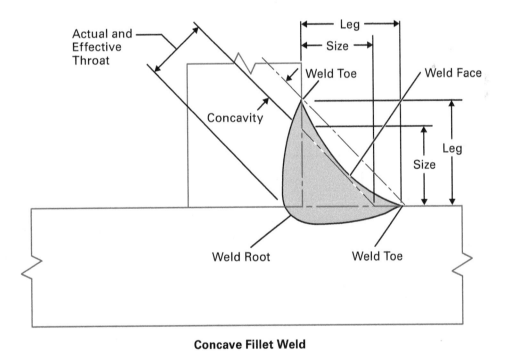

Concave Fillet Weld

Figure 15 Profiles of fillet welds.

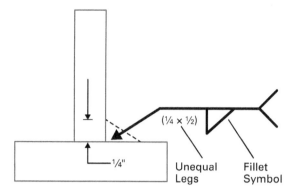

Figure 16 Sizing fillet welds.

1.2.2 Dimensioning Fillet Welds

The length of a fillet weld is shown to the right of the fillet weld symbol. If the fillet weld is to be an intermittent fillet weld, the length (length of each segment) and the **pitch** (center-to-center distance between welds) are shown with a dash. The dash appears between the length, which is always shown first, and the pitch, which is always shown second. *Figure 17* shows a length of 2" (5.1 cm) and a pitch of 6" (15.2 cm). If fillet welds are required on both sides of the joint, they can be back-to-back or staggered. If the welds are back-to-back, as shown in *Figure 17* (A), the fillet weld symbols are aligned evenly on both sides of the reference line, as shown in *Figure 17* (B). If the welds are to be staggered, as shown in *Figure 17* (C), the fillet weld symbols above and below the reference line are offset, as shown in *Figure 17* (D).

When there is no dimension to the right of the fillet weld symbol, the weld extends the full length of the joint.

Pitch: The center-to-center distance between welds.

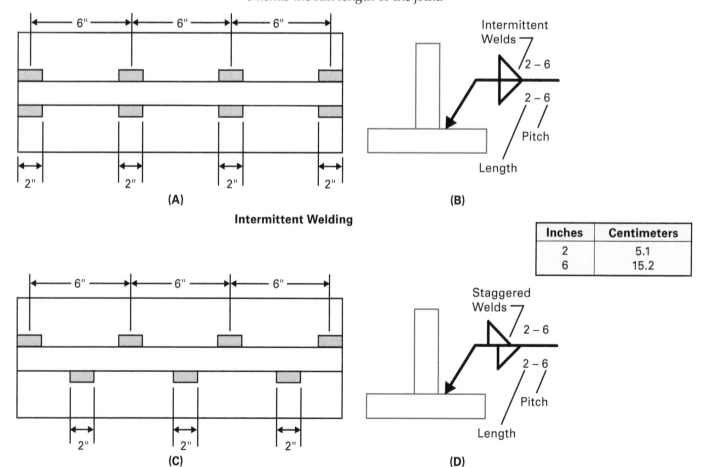

Inches	Centimeters
2	5.1
6	15.2

Figure 17 Dimensioning fillet welds.

1.2.3 Sizing Groove Welds

A groove weld is a weld that is made in the groove formed by two beveled pieces of metal (*Figure 18*). Several factors must be considered when sizing groove welds. These factors apply to all groove weld types.

Groove preparation is the depth to which the groove extends into the base metal. When the groove weld extends all the way through the joint, as shown in *Figure 19* (A), no preparation size needs to be shown. When the groove weld extends only part of the way through the joint, as shown in *Figure 19* (B), the depth of the groove is shown to the left of the symbol. *Figure 19* (B) shows sizing a groove with a depth of $\frac{1}{4}$" (6.4 mm).

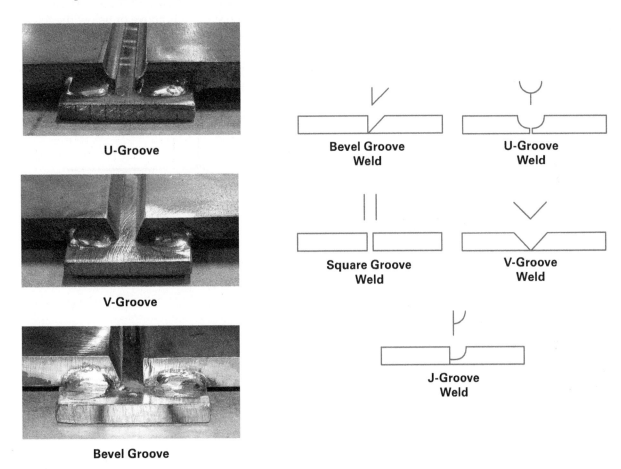

U-Groove

V-Groove

Bevel Groove

Bevel Groove Weld

U-Groove Weld

Square Groove Weld

V-Groove Weld

J-Groove Weld

Figure 18 Types of groove welds.
Source: Specialty Welding & Fabricating of New York, Inc.

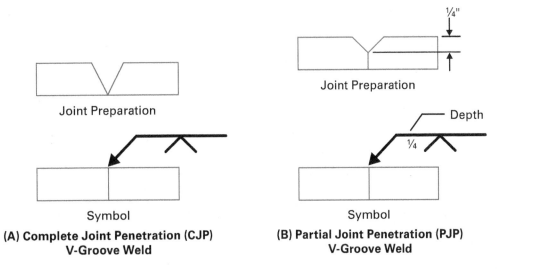

Joint Preparation

Symbol

(A) Complete Joint Penetration (CJP) V-Groove Weld

$\frac{1}{4}$"

Joint Preparation

Depth

$\frac{1}{4}$

Symbol

(B) Partial Joint Penetration (PJP) V-Groove Weld

Figure 19 Sizing groove welds.

For partial joint penetration (PJP) groove welds, both the depth of the groove and weld size (joint penetration) are shown, as in *Figure 20*. The depth of the groove is shown first, and the weld size is shown next, in parentheses. In this example, the groove depth is 0.25" (6.4 mm) and the weld size is 0.385" (9.8 mm). For complete joint penetration (CJP) groove welds, the weld metal extends through the joint thickness.

If a root opening is required, the root opening dimension of a groove weld will be shown inside the symbol. In *Figure 21*, the root opening is $\frac{1}{8}$" (3.6 mm) and is shown on only the arrow-side symbol because the opening is common to both grooves.

The angle for the type of groove is shown above or below the welding symbol, depending on whether the symbol is above or below the reference line. In *Figure 22*, the groove angle is shown as 45° and is shown only by the arrow-side symbol because the groove is the same on both sides. The bevel angle for groove preparation on each workpiece is 22.5°.

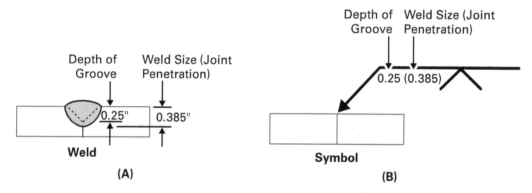

Figure 20 Sizing partial joint preparation (PJP) welds.

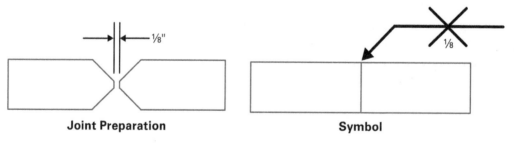

Figure 21 Sizing the root opening.

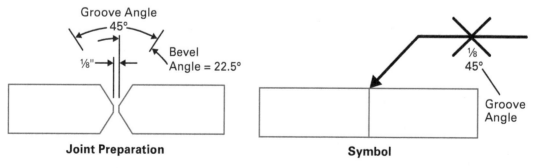

Figure 22 Specifying the groove angle.

1.2.4 Dimensioning Groove Welds

The length of a groove weld is shown to the right of the groove weld symbol. If there is no dimension on the right of the groove weld symbol, the weld extends the full length of the joint.

1.2.5 Sizing and Dimensioning Plug Welds

A plug weld is a weld made in a circular hole in one member of a joint, fusing that member to another member. *AWS A2.4* requires the size of a plug weld, which is the diameter of the hole at the faying surface, to be shown to the left of the welding symbol and preceded by the diameter symbol, Ø. The number of plug (and slot) welds can be shown below the symbol. If the plug is not completely filled, the depth of the fill is given inside the plug weld symbol, as shown in *Figure 23* (A). If the plug weld is to be countersunk, the angle of the **countersink** is shown below or above the weld symbol, depending on which side of the reference line the symbol is on, as shown in *Figure 23* (B). The pitch, shown in *Figure 23* (C), represents the center-to-center spacing of the plugs shown in *Figure 23* (B) to the right of the plug weld symbol. *Figure 23* shows a composite view of sizing and dimensioning plug welds.

Countersink: A hole with tapered sides and a wider opening that allows a flat-head fastener to seat flush to the surface of the material in which the hole is made.

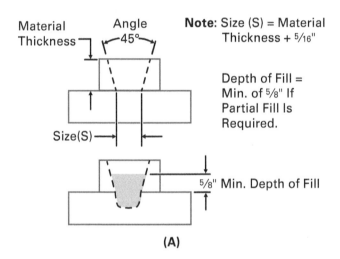

(A)

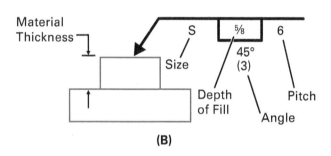

(B)

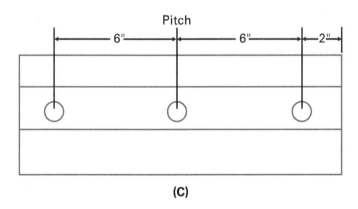

(C)

Typical Plug Weld

Inches	Millimeters
5/16	7.9
5/8	15.9
2	50.8
6	152.4

Figure 23 Sizing and dimensioning plug welds.
Source: Specialty Welding & Fabricating of New York, Inc.

1.2.6 Sizing and Dimensioning Slot Welds

The only time depth of fill is used with a slot weld symbol is when the slots are partially filled. The partial depth of fill is shown in the symbol (*Figure 24*). All other information, such as length, width, spacing, angle of countersink, and location of slots, is shown in a special detail on the print. Note, a slot welding symbol will usually have the width of the slot called out on the left of the weld symbol as the weld size, taken at the faying surface, and the length of the slot called out on the right side of weld symbol, followed by a dash and the pitch.

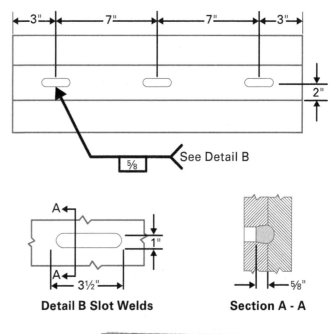

Detail B Slot Welds   **Section A - A**

Typical Slot Weld

Inches	Millimeters
⅝	15.9
1	25.4
2	50.8
3	76.2

Figure 24 Sizing slot welds.

1.3.0 Supplemental Symbols

The following supplemental symbols are used to convey special instructions that apply to the welding symbol:

- Weld-all-around
- Field weld
- Contour finish (flush, convex, concave)

Figure 25 shows these and other supplemental symbols.

1.3.1 Weld-All-Around

When a weld is to extend completely around the joint, such as when welding a channel to a base plate, a small circle is shown where the arrow line meets the reference line. *Figure 25* and *Figure 26* both show examples of the symbol placement. *Figure 26* also shows the intended result of a typical weld-all-around symbol.

Weld-All-Around	Field Weld	Melt-Through	Consumable Insert (Square)	Backing and Spacer (Rectangular)	Flush or Flat	Convex	Concave
				Backing / Spacer			

Inches	Millimeters
3½	88.9
7	177.8

Figure 25 Supplemental symbols.

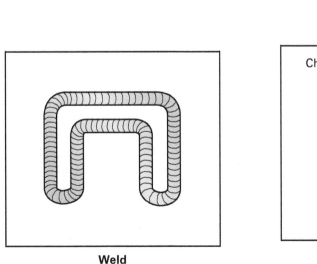

Weld

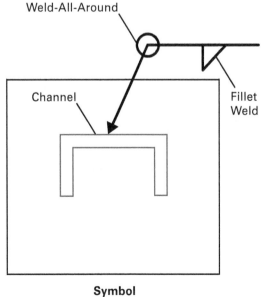

Symbol

Figure 26 Weld-all-around symbol.

1.3.2 Field Weld

When the weld indicated by the welding symbol is to be made at a location other than that of the initial construction, a small flag is placed where the arrow line meets the reference line. The flag can be pointing either left or right. It can also point either up or down. *Figure 27* shows two examples of field welding symbols.

1.3.3 Contour Finish

When the face of the weld must have a finished shape that is not its normal as-welded condition, a finish symbol that indicates a flush, flat, convex, or concave surface is placed adjacent to the welding symbol. The method used to finish the weld face is specified by a letter, such as M for machining (*Figure 28*). When welds are required to be mechanically finished but the method of finishing is unspecified, the designator "U" shall be added to the appropriate contour symbol.

Table 1 shows the related finishing methods with their letter symbols.

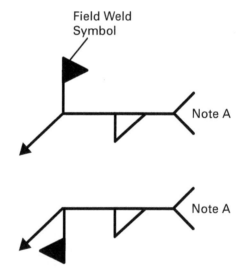

Figure 27 Field welding symbols.

Think About It

Unspecified Contour Finishes

Refer to *Table 1*. What might an unspecified (U) contour finish mean?

TABLE 1 Finishing Methods

Method	Letter
Chipping	C
Grinding	G
Hammering	H
Machining	M
Rolling	R
Unspecified	U

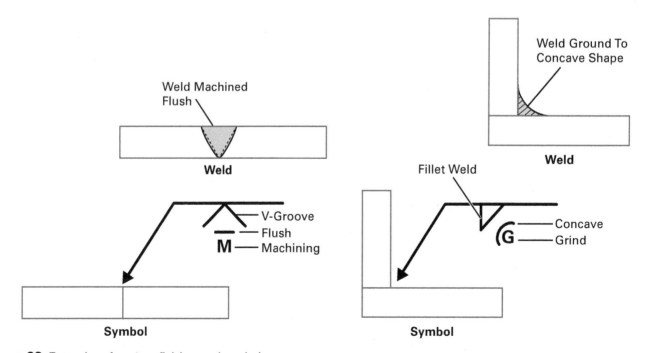

Figure 28 Examples of contour finishes and symbols.

1.4.0   Other Symbols

Other, less common weld symbols are shown in *Figure 29*. They include the following:

- Backing and spacer
- Back or backing
- Melt-through
- Surfacing
- Edge
- Spot
- Seam
- AWS abbreviations

If a consumable insert is required, details are provided with the symbol.

The Letters TYP

The letters TYP (meaning typical) are placed in the tail of the welding symbol to indicate multiple joints of the same configuration. This eliminates duplication of the same welding symbol on the drawing.

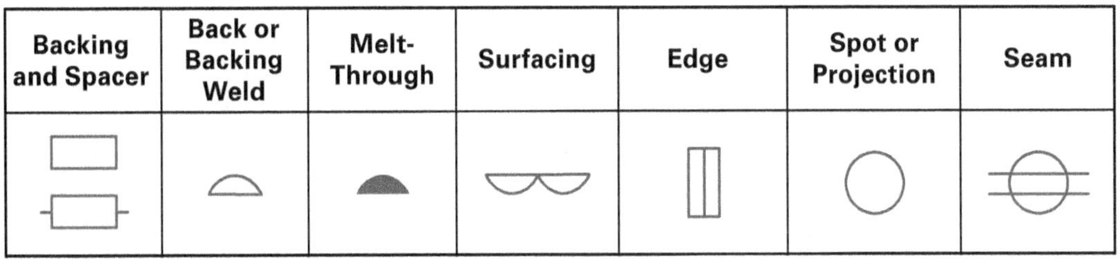

Backing and Spacer	Back or Backing Weld	Melt-Through	Surfacing	Edge	Spot or Projection	Seam

Figure 29 Other symbols.

1.4.1   Backing and Spacer Symbols

Backing is used at the root of the weld to control melt-through. The backing can be metal strips or shapes, flux-coated tape, fiberglass-coated tape, or ceramic tape. The backing symbol is placed opposite a groove weld symbol on the reference line. The material, size, and shape of the backing are specified in the tail of the welding symbol. Spacers are metal strips or shapes that are placed at the root of the joint to control burn-through. They are used most often on double groove joint preparations. The spacer symbol is centered on the reference line. The joint weld symbol is modified to project from the corners of the spacer symbol. The material and size of the spacer are noted in the tail of the welding symbol. *Figure 30* includes illustrations of typical backing and spacer symbols.

1.4.2   Back or Backing Weld Symbols

A back weld is made after the groove weld. A backing weld is made before the groove weld. The same symbol is used to indicate either a back or a backing weld (*Figure 31*). The back or the backing weld symbol is placed opposite a groove weld symbol on the reference line. A note in the tail of the welding symbol indicates which weld is to be made.

1.4.3   Melt-Through Symbols

When complete penetration is required from one side of a joint and root reinforcement is desired, a melt-through symbol (*Figure 32*) can be used to specify the amount of melt-through. The symbol is placed opposite a groove weld symbol on the reference line. A melt-through symbol is similar in shape to the semicircle of a back or a backing symbol. However, the semicircle for a melt-through symbol is solidly filled in, instead of remaining blank. The amount of melt-through required is placed to the left of the melt-through symbol.

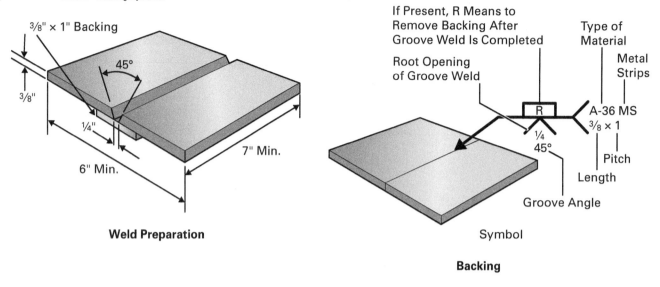

Weld Preparation

Symbol

Backing

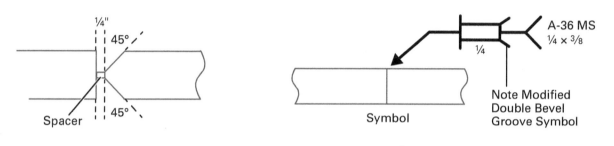

Weld Preparation

Symbol

Spacer

Inches	Millimeters
¼	6.4
⅜	9.5
1	25.4
6	152.4
7	177.8

Figure 30 Backing and spacer symbols.

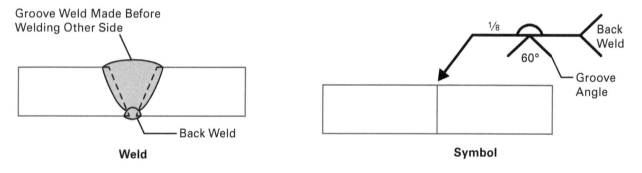

Weld

Symbol

Figure 31 Back or backing weld symbol.

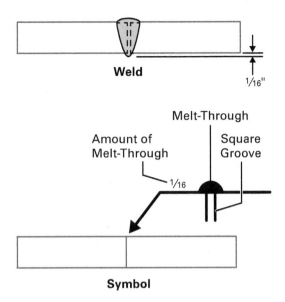

Figure 32 Melt-through symbol.

An Alternate Welding Symbol

The welding symbol shown in *Figure 31* could also be presented as a multiple line symbol. Using a multiple line symbol eliminates the need for the tail information showing the back weld. The multiple line symbol indicates that the first operation is the groove weld and the second operation, by definition, can only be a back weld rather than a backing weld.

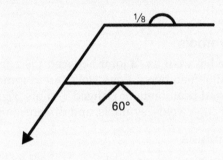

1.4.4 Surfacing Weld Symbols

The AWS defines surfacing as "the application by welding, brazing, or thermal spraying of a layer, or layers, of material to a surface to obtain specified properties or dimensions, as opposed to joining." Examples of surfacing welds are shown in *Figure 33*. A surfacing weld symbol is used when a surface is to be built up with weld. The depth of buildup is indicated to the left of the symbol.

When less than the entire surface is to be built up, the surface to be built up is dimensioned on the blueprint in a detail drawing.

NOTE

A surfacing weld is not applied to an actual joint, and as such, is only shown with a weld symbol on the arrow side. Since the definition for other-side welds is the opposite side of the weld joint from the arrow, the symbol only makes sense when used on the arrow side.

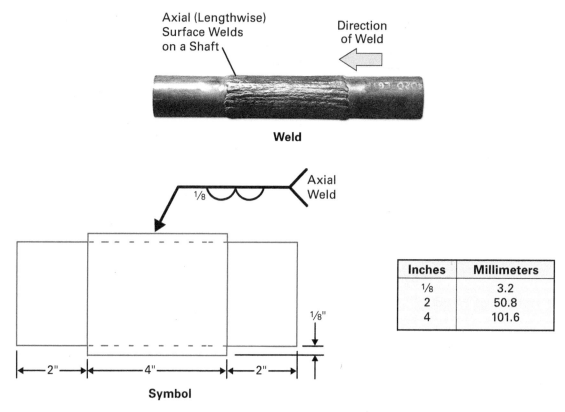

Weld

Symbol

Inches	Millimeters
1/8	3.2
2	50.8
4	101.6

Figure 33 Surfacing welds and symbols.

1.4.5 Edge Weld Symbols

The AWS defines an edge weld as "a joint between the edges of two or more parallel or nearly parallel members." Edge welds are commonly used on sheet metal and can be flanged butt joints or flanged T-joints. *Figure 34A* and *Figure 34B* shows a variety of edge welds, symbols, and dimensioning.

1.4.6 Spot Weld Symbols

Spot welding (*Figure 35*) is a process used to join sheet metal using a series of small spots. There are two types: resistance spot welding (RSW) and arc spot welding. RSW uses electrical resistance heating and clamping pressure to fuse sheet metal panels. Arc spot welding involves either gas metal arc welding (GMAW) or gas tungsten arc welding (GTAW). The process used to make the spot weld is specified in the tail. The size of a spot weld is indicated to the left of the symbol. The pitch is indicated to the right of the symbol. The number of spot welds is defined in parentheses beneath the symbol. For arrow-side spot welds, the circle is placed on the bottom of the reference line. When the spot weld is placed on the other side, the circle is placed on top of the reference line. When there is no side significance (such as with a resistance spot), the circle is centered on the reference line. The drawings will define the starting or ending locations for the spot welds.

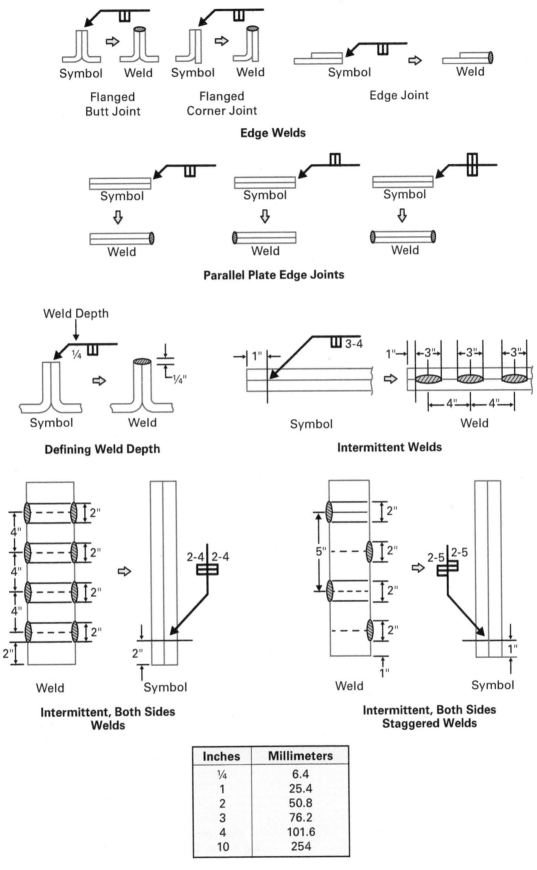

Figure 34A Edge welds and symbols (1 of 2).

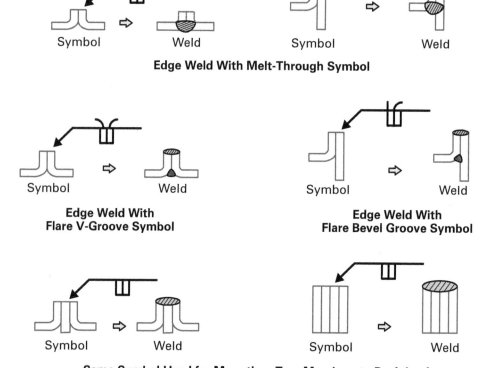

Figure 34B Edge welds and symbols (2 of 2).

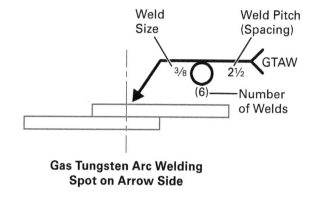

Figure 35 Spot weld symbols.

Projection Welding

Projection welding (PW) is a type of spot welding in which a number of welds are performed simultaneously with one welding machine. Because of tooling costs, projection welding is primarily used in high-production applications.

Prior to welding, one of the members to be joined is passed through a machine that presses projections at specified locations on the member. This member is joined to another member in a welding machine by pressing the two members together and applying current through the projections. The resistance at the points of contact of the projections causes the members to heat, melt, and fuse together while under pressure. This results in welded members that are fused closely together. The side without the projections has minimal surface marking as a result of the weld.

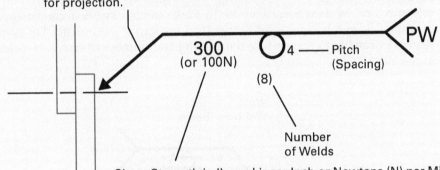

Symbol location on side of reference line for arrow indicates member to be pressed for projection.

300 (or 100N)

4 — Pitch (Spacing)

(8)

PW

Projection welding (PW) process reference must be present or weld is classified as a spot weld.

Number of Welds

Shear Strength in lb per Linear Inch or Newtons (N) per Millimeter (Diameter of projection weld may be used instead of strength.)

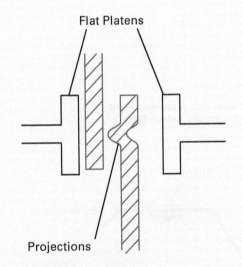

Flat Platens

Projections

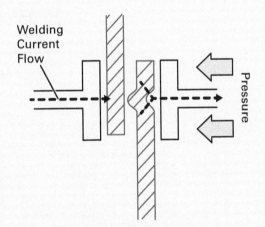

Welding Current Flow

Pressure

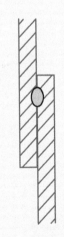

Step 1 One member is pressed to create projections and both members are placed between flat platens in welding machine.

Step 2 Members heat and melt at projections under pressure.

Step 3 Completed Weld

1.4.7 Seam Weld Symbols

Seam welding is used to join two pieces of sheet metal in a continuous seam or in a series of overlapping spot welds. Several processes can be used to make a seam weld. The most common are GMAW, GTAW, and resistance seam welding (RSEW). The process used to make the seam weld is specified in the tail. The length and pitch are given on the right of the symbol for intermittent welds. For welds that are less than the full length of the joined members, the drawing details the starting point and the length of the weld.

For arrow-side seam welds, the circle is placed on the bottom of the reference line, as shown in *Figure 36* (A). When the seam weld is placed on the other side, the circle is placed on top of the reference line. When there is no side significance, such as with a resistance weld seam, the circle is centered on the reference line, as shown in *Figure 36* (B). Flush contour symbols can be applied to seam weld symbols for seams other than resistance weld seams. For parallel seams, a drawing note detailing the spacing and number of seams is included in the tail.

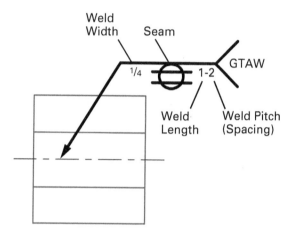

**(A) Gas Tungsten Arc Weld
Seam On Arrow Side**

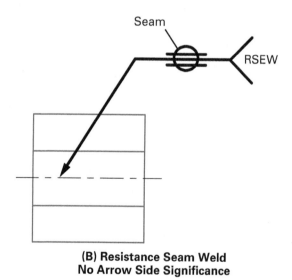

**(B) Resistance Seam Weld
No Arrow Side Significance**

Figure 36 Seam weld symbols.

Around the World

AWS and ISO Welding Symbol Standards

AWS A2.4, Standard Symbols for Welding, Brazing, and Nondestructive Examination is the welding symbols standard used in the United States. AWS welding symbols comply with the requirements of the American National Standards Institute (ANSI) and are designated ANSI/AWS. The AWS system of welding symbols has long been used in the oil industry, which is a global enterprise. Today, the AWS system is used by approximately half of the world's welding industry.

The other widely used welding symbols standard is *ISO 2553:2019, Welded, Brazed, and Soldered Joints — Symbolic Representation on Drawings*. The ISO system was designed by the International Organization for Standardization (ISO). The *ISO 2553:2019* update includes two systems for welding symbols: System A and System B. The AWS and the ISO System B are very similar. For example, the reference line and arrow system in the ISO System B are the same as in the AWS system. There are, however, differences that welders need to be aware of in order to interpret the symbols correctly. The most significant difference, as shown in the first image, is that the ISO System A uses a dashed identification line that is not used in the AWS system. The dashed identification line, which may be drawn above or below the solid reference line, is used to indicate the other side of the joint. Information that applies to the arrow side of the joint is placed on the solid reference line. Information that applies to the other side of the joint is placed on the dashed line.

The second image uses fillet welding symbols as an example to further illustrate typical differences between AWS and ISO welding symbols.

Some countries that have extensive international trade relationships have actually used one system of welding symbols in their own country and symbols from the other system as needed to satisfy the requirements of their overseas customers. With the publication of *ISO 2553:2019*, the ISO sought to provide a set of standardized welding symbols that recognizes both the AWS and the ISO welding symbol systems and can be applied on a worldwide basis.

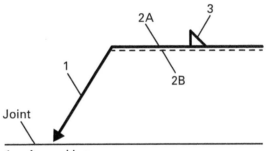

1. Arrow Line
2A. Reference Line (Continuous Line)
2B. Identification Line (Dashed Line)
3. Welding Symbol

ISO System A Welding Symbols

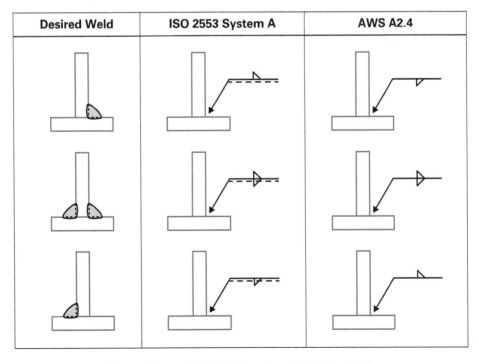

Comparison of ISO System A and AWS Symbols

1.4.8 AWS Abbreviations for Welding and Cutting Processes

Table 2, *Table 3*, and *Table 4* list AWS abbreviations for various welding and cutting processes. Abbreviations from all three tables can be used in tail references, or in single-line or multiple-line welding symbols.

TABLE 2 AWS Abbreviations for Welding Processes

Letter Designations for Welding Processes

AAW	Air Acetylene Welding	GTAW	Gas Tungsten Arc Welding
AB	Adhesive Bonding	GTAW-P	Gas Tungsten Arc Welding–Pulsed Arc
AHW	Atomic Hydrogen Welding	HPW	Hot Pressure Welding
AW	Arc Welding	IB	Induction Brazing
B	Brazing	IRB	Infrared Brazing
BB	Block Brazing	IW	Induction Welding
BMAW	Bare Metal Arc Welding	LBW	Laser Beam Welding
CAW	Carbon Arc Welding	OAW	Oxyacetylene Welding
CAW-G	Gas Carbon Arc Welding	OHW	Oxyhydrogen Welding
CAW-T	Twin Carbon Arc Welding	PAW	Plasma Arc Welding
CEW	Coextrusion Welding	PEW	Percussion Welding
CW	Cold Welding	PGW	Pressure Gas Welding
DP	Dip Brazing	PW	Projection Welding
DFB	Diffusion Brazing	RB	Resistance Brazing
DFW	Diffusion Welding	ROW	Roll Welding
EBW	Electronic Beam Welding	RSEW	Resistance Seam Welding
EBW-HV	Electronic Beam Welding–High Vacuum	RSEW-HF	Resistance Seam Welding–High Frequency
EBW-MV	Electronic Beam Welding–Medium Vacuum	RSEW-I	Resistance Seam Welding–Induction
EBW-NV	Electronic Beam Welding–Nonvacuum	RSW	Resistance Spot Welding
EGW	Electrogas Welding	RW	Resistance Welding
ESW	Electroslag Welding	SAW	Submerged Arc Welding
EXW	Explosion Welding	SAW-S	Series Submerged Arc Welding
FB	Furnace Brazing	SMAW	Shielded Metal Arc Welding
FCAW	Flux Cored Arc Welding	SSW	Solid-State Welding
FLB	Flow Brazing	SW	Stud Arc Welding
FLOW	Flow Welding	TB	Torch Brazing
FOW	Forge Welding	TCAB	Twin Carbon Arc Brazing
FRW	Friction Welding	TW	Thermite Welding

TABLE 3 AWS Abbreviations for Cutting Processes

Letter Designations for Cutting Processes

AC	Arc Cutting	LOC	Oxygen Lance Cutting
AOC	Oxygen Arc Cutting	MAC	Metal Arc Cutting
CAC	Carbon Arc Cutting	OC	Oxygen Cutting
CAC-A	Air Carbon Arc Cutting	OC-F	Flux Cutting
EBC	Electron Beam Cutting	OFC	Oxyfuel Gas Cutting
GMAC	Gas Metal Arc Cutting	OFC-A	Oxyacetylene Cutting
GTAC	Gas Tungsten Arc Cutting	OFC-H	Oxyhydrogen Cutting
LBC	Laser Beam Cutting	OFC-N	Oxynatural Gas Cutting
LBC-A	Laser Beam Cutting–Air	OFC-P	Oxypropane Cutting
LBC-EV	Laser Beam Cutting–Evaporative	OC-P	Metal Powder Cutting
LBC-IG	Laser Beam Cutting–Inert Gas	PAC	Plasma Arc Cutting
LBC-O	Laser Beam Cutting–Oxygen	SMAC	Shielded Metal Arc Cutting

TABLE 4 AWS Abbreviations for Applying Welding Processes

Letter Designations for Applying Welding Processes

AU	Automatic
ME	Mechanized
MA	Manual
SA	Semi-Automatic

1.0.0 Section Review

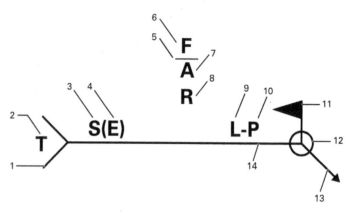

Figure SR01

1. In the welding symbol shown in *Figure SR01,* what is the symbol indicated by the number 4?
   a. Arrow line
   b. Groove weld size
   c. Field weld symbol
   d. Specification or process

2. A weld made in a corner is a _____.
   a. slot weld
   b. fillet weld
   c. plug weld
   d. groove weld

3. Which one of the symbols in *Figure SR02* contains a supplemental symbol?
   a. Symbol A
   b. Symbol B
   c. Symbol C
   d. Symbol D

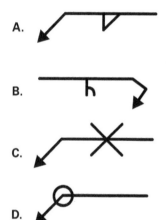

Figure SR02

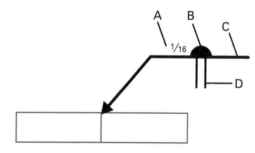

Figure SR03

4. The symbol in *Figure SR03* that indicates that complete penetration from one side of a joint and root reinforcement are required is _____.
   a. Symbol A
   b. Symbol B
   c. Symbol C
   d. Symbol D

1. The welding symbol describes the type of weld, its size, and its _____.
   a. surface finish
   b. tail references
   c. cutting process
   d. depth of field

2. The side of the joint or surface that is opposite the side to which the arrow of a reference line is pointing is called the _____.
   a. arrow side
   b. other side
   c. tail joint
   d. countersink

3. Regardless of which end of the reference line the arrow line is on, information on the reference line is always read from _____.
   a. top to bottom
   b. bottom to top
   c. left to right
   d. right to left

4. In the welding symbol shown in *Figure RQ01*, the operation to be conducted first is _____.
   a. Operation A
   b. Operation B
   c. Operation C
   d. not critical

Figure RQ01

5. In the welding symbol shown in *Figure RQ02*, ½ represents the _____.
   a. size of the weld
   b. width of the weld
   c. length of the weld
   d. depth of preparation

Figure RQ02

6. In relation to the fillet weld symbol, the fillet weld size is placed _____.
   a. above the symbol
   b. below the symbol
   c. to the left of the symbol
   d. to the right the symbol

7. In the welding symbol shown in *Figure RQ03*, the number 9 represents the weld _____.
   a. pitch
   b. depth
   c. width
   d. length

Figure RQ03

8. In relation to the plug weld symbol, the pitch between the plug welds is shown _____.
   a. above the weld symbol
   b. below the weld symbol
   c. to the left of the weld symbol
   d. to the right of the weld symbol

9. The field weld symbol is a(n) _____.
   a. number
   b. letter
   c. arrow
   d. flag

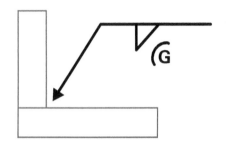

Figure RQ04

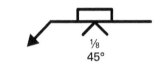

Figure RQ05

Figure RQ06

Figure RQ07

10. When the face of the weld must have a finished shape, a finish symbol is placed adjacent to the welding symbol, along with a letter indicating _____.
   a. size of the weld
   b. method of finish
   c. smoothness of the finish
   d. width of the finished area

11. The finish of the weld face in *Figure RQ04* is to be achieved by _____.
   a. filleting
   b. grinding
   c. chipping
   d. reaming

12. When backing is to be used at the root of the weld to control melt-through, the backing symbol is placed opposite which weld symbol on the reference line?
   a. tail
   b. field
   c. spacer
   d. groove

13. In the welding symbol shown in *Figure RQ05*, the numbers inside the symbol represent the _____.
   a. root opening and groove angle
   b. groove angle and depth
   c. root opening and pitch
   d. pitch and groove angle

14. The type of weld symbol shown above the reference line in *Figure RQ06* is a _____.
   a. spot weld
   b. back weld
   c. surfacing weld
   d. melt-through

15. The type of weld represented by the welding symbol in *Figure RQ07* is a _____.
   a. spot weld
   b. backing weld
   c. surfacing weld
   d. melt-through weld

Answers to odd-numbered questions are found in the Review Question Answer Keys at the back of this book.

Answers to Section Review Questions

Answer	Section Reference	Objective
Section 1.0.0		
1. b	1.0.0	1a
2. b	1.2.1	1b
3. d	1.3.0	1c
4. b	1.4.3	1d

User Update

Did you find an error? Submit a correction by visiting **https://www.nccer.org/olf** or by scanning the QR code using your mobile device.

Reading Welding Detail Drawings

Source: © Miller Electric Mfg. LLC

Objectives

Successful completion of this module prepares you to do the following:

1. Describe welding detail drawings and identify basic drawing elements and features.
   a. Describe the object views used to depict welding details.
   b. Identify basic drawing elements related to welding detail drawings.
2. Identify and explain how to interpret dimensional information, notes, and a bill of materials.
   a. Identify and explain how to interpret dimensional information.
   b. Identify and explain how to interpret notes and a bill of materials.

Performance Task

Under supervision, you should be able to do the following:

1. Draw or sketch a welding drawing based on an instructor-provided image or object.

Overview

Reading and interpreting the information contained on construction drawings is an essential skill for welders. Incorrectly interpreting this information can result in producing work that fails to meet design specifications, building codes, and safety standards. This module focuses on how to read and interpret assembly and detail drawings. It introduces the various object views and basic drawing elements used in welding detail drawings. It also explains how to interpret dimensional information, notes, and bills of materials.

NCCER **Industry-Recognized Credentials**

If you are training through an NCCER-accredited sponsor, you may be eligible for credentials from NCCER. The ID number for this module is 29202. Note that this module may have been used in other NCCER curricula and may apply to other level completions. Contact NCCER at 1.888.622.3720 or go to **www.nccer.org** for more information.

You can also show off your industry-recognized credentials online with NCCER's digital credentials. Transform your knowledge, skills, and achievements into credentials that you can share across social media platforms, send to your network, and add to your resume. For more information, visit **www.nccer.org**.

Digital Resources for Welding

Scan this code using the camera on your phone or mobile device to view the digital resources related to this craft.

SCAN ME

1.0.0 Basic Welding Detail Drawings

Performance Task

1. Draw or sketch a welding drawing based on an instructor-provided image or object.

Objective

Describe welding detail drawings and identify basic drawing elements and features.

a. Describe the object views used to depict welding details.

b. Identify basic drawing elements related to welding detail drawings.

Engineers, designers, and drafters use drawings, also called prints or blueprints, to convey and receive information needed for work on a job. There are many categories of drawings, including civil, structural, mechanical, piping, electrical, and instrumentation. Welders must be able to correctly interpret drawings so the design specifications and safety factors shown in the drawings will be included in the finished product.

Altering or failing to comply with a drawing could result in a violation of federal or state standards or building codes, which could lead to fines and other penalties. More importantly, a deviation from the drawings could create an unsafe structure or product, which might result in death osr injury. To avoid these problems, welders must be able to read and interpret drawings accurately. This module explains how to read and interpret assembly and detail drawings.

Each category contains several types of drawings used for specific purposes. The most common drawing used by welders is the welding detail drawing, which generally depicts the weldment that must be fabricated and the pieces to be joined by welding. The parts are drawn assembled, as they will appear after welding, except the welds are not shown. Welding symbols show how the pieces are to be joined. *Figure 1* is an example of a small bracket that may need to be fabricated on the job.

Figure 2A and *Figure 2B* show how the same bracket would appear on a welding detail drawing.

NOTE

An example of an object and its corresponding three-view detail drawing is provided in *Appendix 29202A*.

Blueprints

Blueprints got their name from an obsolete chemical process used to develop chemically treated print paper, which turned paper blue. The lines produced on the paper were white against the blue background. Drawings, although often still referred to as blueprints, are now made using other methods and are typically black on white.

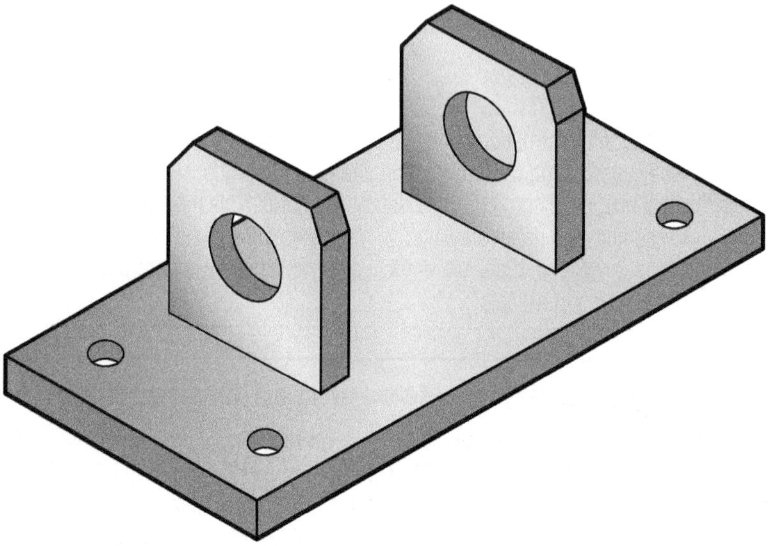

Figure 1 Bracket.

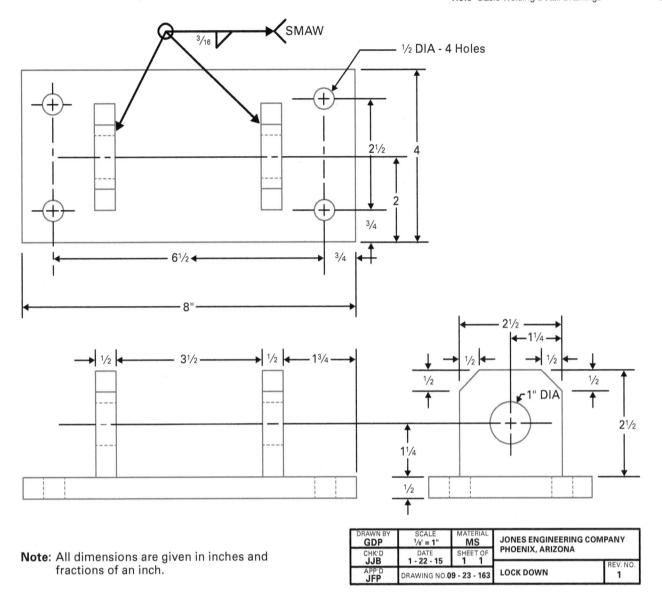

SMAW

½ DIA - 4 Holes

Note: All dimensions are given in inches and fractions of an inch.

DRAWN BY GDP	SCALE ⅛' = 1"	MATERIAL MS	JONES ENGINEERING COMPANY PHOENIX, ARIZONA	
CHK'D JJB	DATE 1 - 22 - 15	SHEET OF 1 1		
APP'D JFP	DRAWING NO.09 - 23 - 163		LOCK DOWN	REV. NO. 1

(A) Imperial Measurements

Figure 2A Welding detail drawing (1 of 2).

When the object to be fabricated is too complicated for one drawing, the object will be shown on an assembly drawing as well as one or more detail drawings. The assembly drawing (*Figure 3*) shows the completed and assembled object. The main purpose of the assembly drawing is to show how the parts are assembled in relation to one another. Detail drawings of each part are then provided (*Figure 4*). The detail drawings contain all the information necessary for the welder to make the part.

NOTE

Letters and leader lines identify the parts of the object in an assembly drawing. These same letters are then used to label the parts in the corresponding detail drawing.

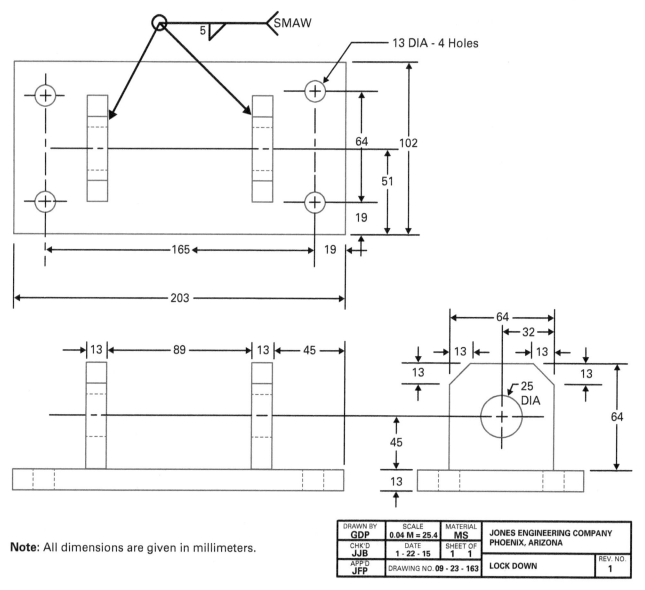

Note: All dimensions are given in millimeters.

DRAWN BY **GDP**	SCALE **0.04 M = 25.4**	MATERIAL **MS**	**JONES ENGINEERING COMPANY** **PHOENIX, ARIZONA**	
CHK'D **JJB**	DATE **1 - 22 - 15**	SHEET OF **1 1**		
APP'D **JFP**	DRAWING NO. **09 - 23 - 163**		**LOCK DOWN**	REV. NO. **1**

(B) Metric Measurements

Figure 2B Welding detail drawing (2 of 2).

1.1.0 Object Views

Drawings must clearly show the shape of an object. To accomplish this goal, most drawings will show more than one view of the object. The most commonly used views include the following:

- Isometric
- Multiview
- Section

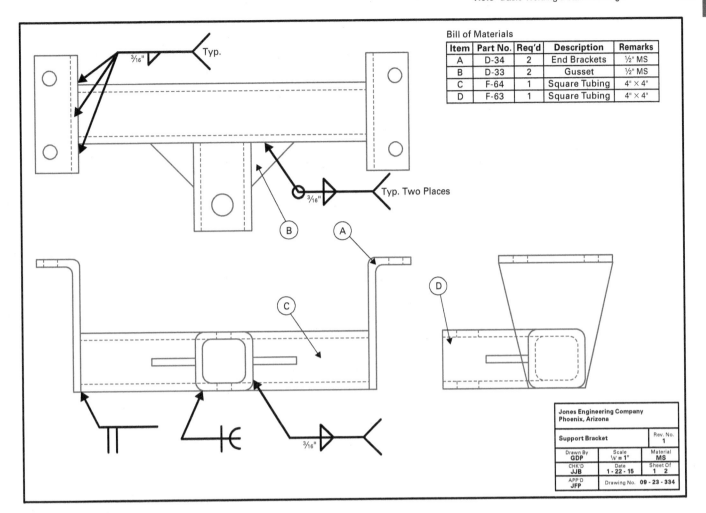

Figure 3 Assembly drawing.

1.1.1 Isometric View

The **isometric view** shows a three-dimensional picture of an object (*Figure 5*). The isometric, or pictorial, view is an excellent way to show what an object looks like but is a difficult view to use when providing details such as dimensions, hole locations, sizes, or angles. For this reason, the isometric view is generally not used on detail drawings.

Isometric view: A drawing depicting a three-dimensional view of an object.

1.1.2 Multiview

Multiviews are developed by a process called **orthographic projection**, and they are commonly used on detail drawings. Multiviews include several views, each from a different direction, developed and shown on the same drawing. The views are designed as if the object had been placed inside a transparent box (*Figure 6*). Each view is then expanded by projecting points at right angles from the object onto a surface of the transparent box.

> **NOTE**
>
> To help visualize a multiview, think of a cardboard box. If the sides of this cardboard box were folded open, a flat, multiview drawing would result, as shown in *Figure 6*.

Orthographic projection: A method of developing multiple flat views of an object by projecting lines out from the object at right angles from the face.

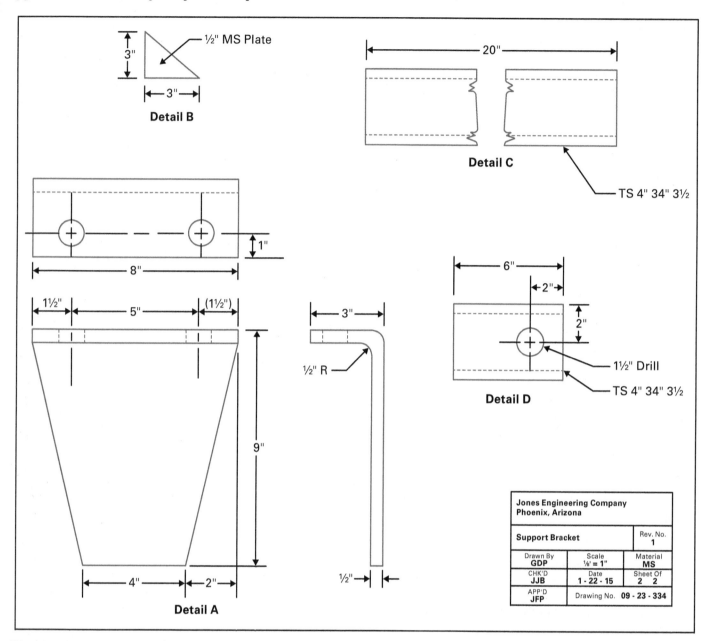

Figure 4 Detail drawing.

Each view in an orthographic projection has a set position in relation to the other views. In addition, the views are aligned vertically and horizontally. All sides of an object can be shown, although the views used most frequently are the front, top, and right side. The front view is selected as the view that best represents the object's shape. All other views are positioned in reference to the front view. *Figure 7* shows the relationship of possible views used with orthographic projection.

Figure 8 shows the relationship of views on a welding detail drawing.

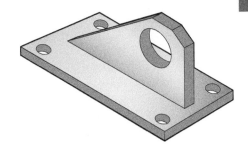

Figure 5 Isometric view.

Isometric Views

An isometric drawing of an object shows the object at a 30° angle. This causes the object to appear on paper similar to the way the real, three-dimensional object appears.

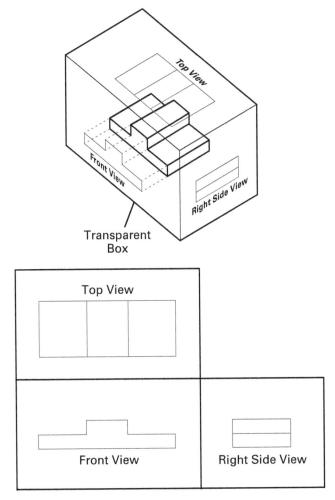

Figure 6 Orthographic projection.

> **NOTE**
>
> Standards for the projection and placement of orthographic views vary between countries. In the United States, the American National Standards Institute (ANSI) recommends using third-angle projection for orthographic views. Most other countries follow the International Organization for Standardization (ISO) standard, which calls for first-angle projection.
>
> Understanding these different drawing conventions is important because confusion over orthographic views can cause problems on worksites. This could result in unnecessary field revision of shop-fabricated piping. To prevent confusion, look at the projection symbol on the drawing to identify what system is being used.

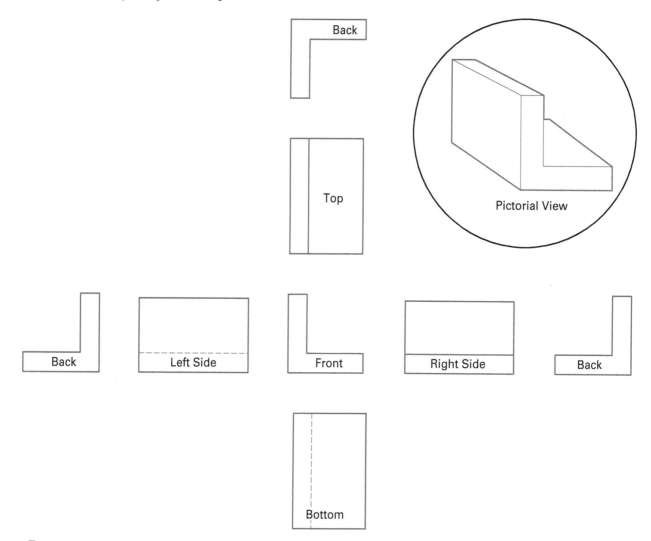

Figure 7 Relationship of possible views.

1.1.3 Section

A full section, or cross section, is a view of an object used to show internal details that would not normally be visible (*Figure 9*). A section line is used on one of the complete views to indicate where the part has been cut or sectioned. The arrowheads on the section line point in the direction from which the sectioned object is to be viewed. Because there can be more than one section view, each is identified by placing a letter next to the arrowheads. The sectioned view is then labeled with the appropriate letters.

A drawing shows a more detailed view of a part by showing different sections of the same part. Each of these sections is identified by pairs of letters located near the points of the cutting plane line.

1.2.0   Drawing Components

Welding drawings have several components, including lines, material symbols, solid round breaks, pipe or tubing breaks, and revolved sections.

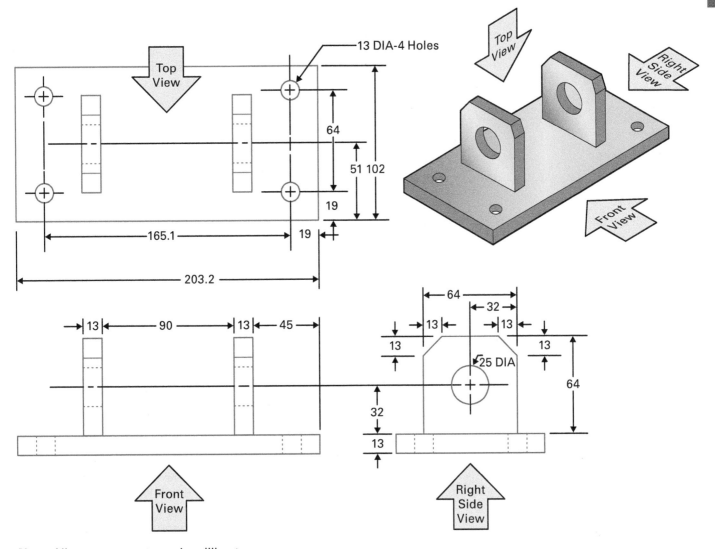

Note: All measurements are in millimeters.

Figure 8 Relationship of views.

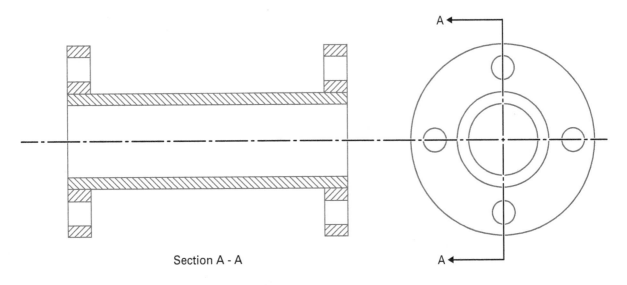

Section A - A

Inches	Millimeters
6½	165
8	204

Figure 9 Section view.

1.2.1 Lines

Many different types of lines are used in drawings, each with its own special meaning. The lines are drawn to a standard that specifies thickness (thick, medium, or thin) and the way the lines are interpreted or broken. The more commonly used lines are shown in *Figure 10* and are described as follows:

- *Object line* — A thick, solid line used to show all visible edges of a part; also called a *visible line*.

- *Hidden line* — A medium, broken line made up of evenly spaced dashes; used to show edges that are not visible.

- *Center line* — A thin, broken line made up of short and long dashes; used to show the center of holes or other objects.

- *Dimension line* — A thin line with arrowheads that shows the extent of a dimension; the line is broken by the dimension number.

- *Extension line* — A thin line used with dimension lines that extends from the point on the object to which the dimension applies.

- *Leader line* — A thin, horizontal line with an angled arrowhead at one end; used with a note or dimension. The arrowhead points directly to the surface of the note or dimension to which the line applies.

- *Cutting plane line* — A heavy line that can be made up of either a series of one long and two short dashes or a series of short dashes; used to show where an imaginary cut has been made through an object. Short lines with arrowheads are on both ends of the line. The arrowheads point in the direction in which the section should be viewed.

- *Short break line* — A thick, solid, irregular line drawn freehand to indicate that part of the object has been removed or broken away; used to conserve space or show internal detail.

- *Long break line* — A thin, solid line with zigzags inserted at several places; used to show where a section of an object has been removed to conserve space on the drawing.

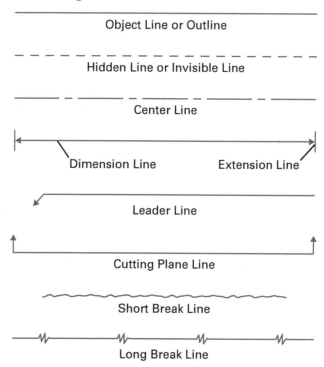

Figure 10 Common drawing lines.

1.2.2 Material Fill Symbols

Material symbols are a series of thin lines, either solid or a combination of solid and broken, arranged in a pattern to represent various types of materials. The symbols are used to identify internal surfaces that have been exposed by a cutting plane line or a short cutaway view. *Figure 11* shows common material symbols used to represent various metals.

1.2.3 Bar Stock and Pipe Breaks

A bar stock or pipe break is represented by thick, curved lines that loop back on themselves (*Figure 12*). The approach is the same regardless of the shape of the stock. The loop is filled with a material symbol, such as cross-hatching, to indicate an internal surface. Breaks are used to show where a section of an object has been removed to conserve space on the drawing.

Material Symbols

Do not select materials based on material symbols shown on a drawing. These symbols only represent the described material and do not provide specific information about the material. For example, two or three symbols represent steel, but there are many different types of steel, each with specific characteristics. Always use the drawing notes, material list, or bill of materials to determine the exact materials required.

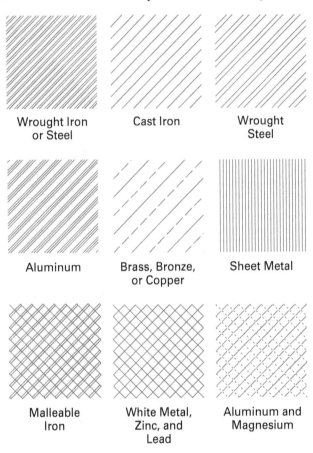

Wrought Iron or Steel

Cast Iron

Wrought Steel

Aluminum

Brass, Bronze, or Copper

Sheet Metal

Malleable Iron

White Metal, Zinc, and Lead

Aluminum and Magnesium

Figure 11 Material symbols.

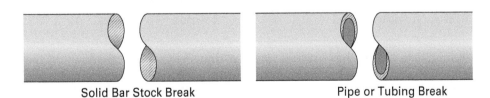

Solid Bar Stock Break

Pipe or Tubing Break

Figure 12 Bar stock and pipe breaks.

1.2.4 Revolved Sections

Revolved sections are similar to solid round breaks and pipe or tubing breaks except that the shape of the object that is broken is shown in the break as a section view (*Figure 13*).

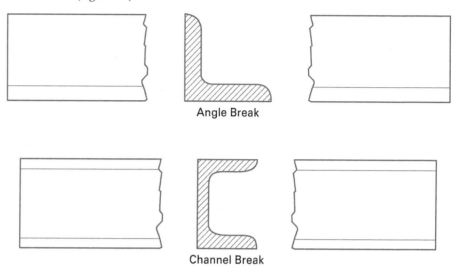

Angle Break

Channel Break

Figure 13 Revolved sections.

1.0.0 Section Review

1. The *most* common drawing used by welders is the _____.
   a. section view
   b. isometric view
   c. assembly drawing
   d. welding detail drawing

2. In a drawing, a solid round break and a pipe or tubing break show where a section of an object has been _____.
   a. broken
   b. repaired
   c. inserted
   d. removed

2.0.0 Dimensions, Notes, and Materials

Performance Task

1. Draw or sketch a welding drawing based on an instructor-provided image or object.

Objective

Identify and explain how to interpret dimensional information, notes, and a bill of materials.

a. Identify and explain how to interpret dimensional information.

b. Identify and explain how to interpret notes and a bill of materials.

Welding detail drawings include precise dimensional information for the various pieces that make up an object. Additional details about the construction of the object are often provided in notes and a bill of materials. To fabricate an object that meets design specifications, welders must be able to read and correctly interpret the dimensional information, general or local notes, and the bill of materials.

2.1.0 Dimensions

Dimensions identify sizes. Drawings must have dimensions if the object shown is to be fabricated. The drawing must include all dimensions in correct relationship to one another to give a complete size description of the object. The two basic methods for dimensioning a part are conventional dimensioning and baseline dimensioning.

In conventional dimensioning (*Figure 14*), the overall dimensions of length, width, and height are given. Dimensions A and B are not given because they are easily found by adding and/or subtracting the given dimensions. If they were given, dimensions A and B would be identified in parentheses as reference dimensions. A reference dimension, such as the $1\frac{3}{4}$" dimension in parentheses on the drawing, is obtained by adding or subtracting other dimensions found on the drawing. If the overall size of a dimensioned part is not given, then those dimensions that make up the total must be added to find the overall size. Reference dimensions help speed up the reading of drawings by allowing for quick visualization of dimensions.

> **Think About It**
>
> **Dimension Calculation**
>
> What is the reference dimension of a section whose overall part dimension is $4\frac{7}{8}$"? The part consists of three sections. Two of the sections are $1\frac{7}{8}$" and $1\frac{1}{8}$", respectively.

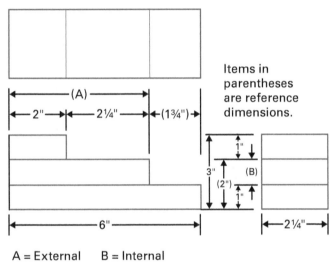

A = External   B = Internal

Figure 14 Conventional dimensioning.

In baseline dimensioning (*Figure 15*), all dimensions originate from baselines. When imperial measurements are used, baseline dimensions are usually expressed as three-place decimals for accuracy. Metric baseline dimensions are also normally expressed as three-place decimals, depending upon the chosen unit of measure.

> **NOTE**
>
> Baseline dimensions on metric drawings are never fractional because, unlike imperial measurements, metric measurements are always shown as decimals. When working with drawings that have baseline dimensions, check to make sure you know on which measurement system (metric or imperial) the drawing is based.

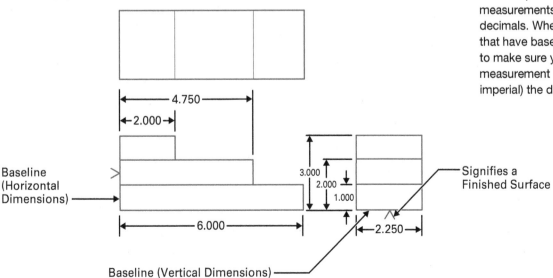

Figure 15 Baseline dimensioning.

Tolerances: The allowances over or under the specified dimension, value, or other parameter.

Scale: A method of drawing an object larger or smaller than its actual size while keeping the proportions the same.

The baselines are typically machined surfaces indicated by the symbol V. Baselines may also originate from center lines or a critical edge. The advantage of baseline dimensioning is that the approach eliminates the cumulative error that often occurs with conventional dimensioning. Baseline dimensioning does not require calculations to find a dimension. Most welding drawings that require hand assembly and welding do not use three-place decimal dimensioning because of the difficulty in maintaining tight **tolerances**.

2.1.1 Scale

Most drawings are made in **scale**. This means the drawing can be larger, the same size, or smaller than the actual object so long as the proportions in the drawing are the same as the proportions of the actual object. Most objects shown in welding detail drawings are too large to fit on a standard drawing, so they are drawn smaller than the actual object's size. The scale used on the drawing is given in the title block in the form of an equation. Many different scales can be used; for example:

- $\frac{1}{4}$" = 1"
- $\frac{1}{8}$" = 1"
- $\frac{1}{4}$" = 1'

The figure to the left of the equal sign in a scale equation represents the measure used for the drawing, while the figure to the right of the equal sign corresponds to the measure used for the object. For example, if the scale $\frac{1}{4}$" = 1" is used, an actual surface 3" long would be represented on the drawing by a line $\frac{3}{4}$" long ($\frac{1}{4}$" for each actual inch). If the scale $\frac{1}{4}$" = 1' is used, an actual surface 4' long would be shown on the drawing by a line 1" long. *Figure 16* shows how scale is used to reduce the size of a drawing.

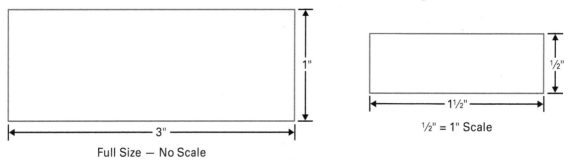

Full Size — No Scale

Figure 16 Drawing in scale.

On drawings that are dimensioned in metric units, the scale is stated as a ratio. The actual size would be stated as 1:1. A reduction ratio would be stated as 1:2, 1:5, 1:10, etc. An increasing ratio would be stated as 2:1, 5:1, etc. The drawing title block will usually specify the applicable scale unless more than one scale is used on the drawing.

Drawing Measurements

Be sure to use only the dimensions provided on the drawing. If a dimension is needed but is not provided, add or subtract the dimensions that were furnished to obtain the needed dimension. Avoid measuring directly from a drawing. Measuring directly from a drawing is called scaling a drawing and will almost always result in errors. If a welder cannot determine a measurement by adding or subtracting given dimensions, they should check with their supervisor.

2.1.2 Size and Location Dimensions

The most common dimensions are size and location. Size dimensions provide the measurements for the overall size of the object. Location dimensions provide measurements from one feature of an object to another part, hole, break, or component. Examples of both dimension types are shown in *Figure 17*.

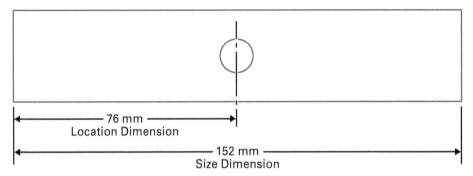

Figure 17 Size and location dimensions.

2.1.3 Hole Dimensions

Drilled hole dimensions are shown with a leader line and a note (*Figure 18*). The note gives the hole size and the number of holes to be drilled if more than one. If the hole is to be drilled only partway through the object, called a **blind hole**, then the depth will also be given. If the depth is not given, then the hole goes completely through the object and is called a **through hole**.

Blind hole: A drilled hole that does not go all the way through the object.

Through hole: A drilled hole that goes completely through an object.

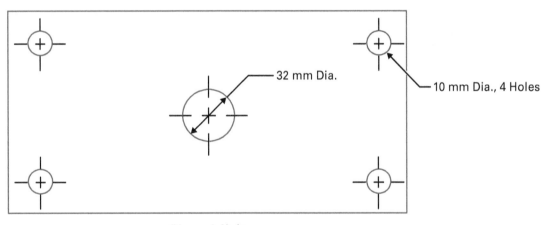

Through Holes

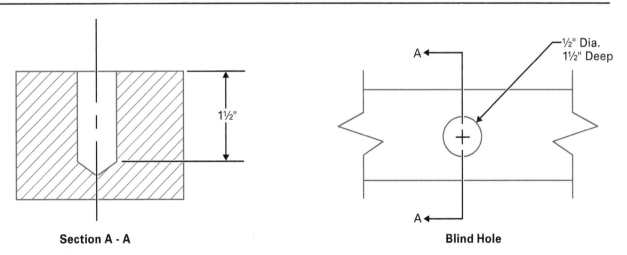

Section A - A   **Blind Hole**

Figure 18 Dimensioning holes.

Counterbore: To enlarge a portion of a drilled hole so that a socket head fastener will seat flush or below the surface.

It is also possible to **counterbore** or countersink holes. A hole that is countersunk is tapered from the surface to the drilled hole size and is used for flat-head fasteners. A hole that is counterbored has a square shoulder used to recess socket head fasteners. The required dimensions for countersinking or counterboring holes are added to the note (*Figure 19*).

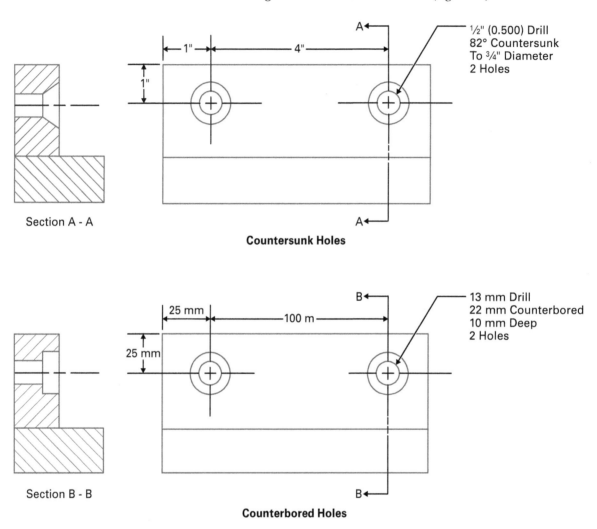

Figure 19 Dimensioning counterbored and countersunk holes.

Structural Steel and Pipe Sizes

Be aware that structural steel and pipe sizes, if included on drawings, are usually approximate sizes only. Structural steel is classified as hot rolled or cold rolled. Hot rolled varieties are somewhat oversized and have rounded corners and edges. The sizes of cold rolled types are more precise, and the corners and edges are square. Size specifications for some structural shapes, including S-beams (formerly called I-beams), H-beams, and wide flange (W) beams, are nominal.

Pipe is sometimes sized through the application of a schedule number as part of a specification. The eleven schedules commonly used are 5, 10, 20, 30, 40, 50, 60, 80, 100, 120, and 160. The schedule numbers, which represent the strength of the pipe, refer to wall thickness. Pipe schedule tables provide the nominal wall thickness, weight, and outside diameter for each size and type of pipe.

2.1.4 Angle and Bevel Dimensions

Angles and bevels are common on welding detail drawings because they are used with weld joint preparation. They can be specified with a leader line and a note or by using extension and dimension lines (*Figure 20*). The size can be given as angular, linear, or a combination of both dimensions.

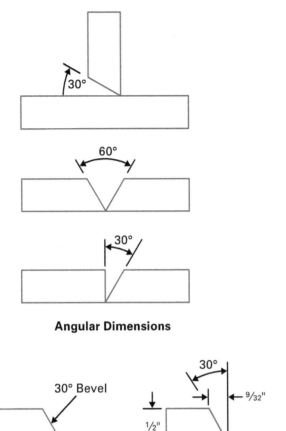

Figure 20 Specifying angles and bevels.

2.1.5 Radius and Arc Dimensions

When the ends, sides, or corners of an object are to be rounded, a radius or arc is shown and dimensioned (*Figure 21*). A combination of center, radius, extension, and dimension lines is typically used. To obtain the arc required, the radius line may start outside the physical shape of the object.

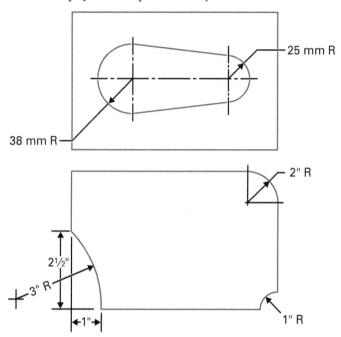

Figure 21 Dimensioning radius and arcs.

2.1.6 Tolerances

Tolerance is an important part of dimensioning because it identifies the variance in size that is allowable on a part. Depending on its intended use, a part can have a large or small tolerance. If a part is fabricated outside the allowable tolerance, the part will have to be reworked or scrapped at an additional cost. On the other hand, if a part with a large tolerance is fabricated to very fine tolerances, this will generally add to the cost since parts with large tolerances typically can be fabricated faster than parts with smaller tolerances. For these reasons, it is important to identify the tolerances on a drawing and use this information for fabrication.

Tolerances can be given following the dimension as a note or in a table as part of the title block. They are generally given as a plus (+) or minus (–) quantity. For example, if the dimension 10" was followed by the tolerance + or $-\frac{1}{8}$", the object could be $9\frac{7}{8}$" through $10\frac{1}{8}$".

The plus and minus tolerances for a size can be different. For example, a dimension of 12", $+\frac{1}{8}$" and –0", indicates that the object can be from 12" through $12\frac{1}{8}$". *Figure 22* shows three ways of indicating tolerances.

Think About It

Cut Tolerance

What are the minimum and maximum allowable sizes of an 18" cut that has to be made with a tolerance of $\pm 6\frac{1}{4}$"?

ς—10⅛" ± ⅛"—ς ς—60 mm ± 2 mm—ς

ς—8.750" ± 0.005"—ς ς—30° ± 2°—ς

Tolerance Following a Dimension

Note: Unless Otherwise Specified,
Tolerances Are as Follows:

Fractional Dimension	± ⅛"
Metric Dimensions	± 2 mm
Decimal Dimensions	± 0.005"
Angular Dimensions	± 2°

Tolerances as a Note

Unless Otherwise Specified Dimensions Are in Inches	Jones Engineering Company Phoenix, Arizona		
	Support Bracket		Rev. No. 1
Tolerances:	Drawn By GDP	Scale ⅛' = 1"	Material MS
Fractions ± ⅛	CHK'D JJB	Date 1 - 22 - 15	Sheet Of 1 2
Decimals ± 0.010	APP'D JFP	Drawing No. 09 - 23 - 334	
Angles ± 3°			

Tolerances as Part of the Title Block

Figure 22 Dimensioning tolerances.

2.2.0 Notes and Bill of Materials

Additional information that can be found on many drawings includes notes and a bill of materials. These forms of documentation provide data that are not given elsewhere on the drawing.

2.2.1 Notes

A note is an explanatory or critical comment concerning the details of construction. Two types of notes are found on drawings: general notes and local, or specific, notes (*Figure 23*). A general note applies to the drawing in its entirety and is placed in an area away from the views so that it will be noticed. On large jobs, general notes are placed on separate drawings. On smaller jobs, general notes are placed on detail drawings where they refer to the entire drawing. General notes are grouped together and numbered.

General Notes — Steel

1. Design, fabrication, and erection of structural steel shall be in accordance with the American Institute of Steel Construction (AISC) *Steel Construction Manual*, Thirteenth Edition (2007) unless otherwise modified on the drawings or in the specifications.

2. Material shall meet the requirements of the following specifications unless otherwise noted:
Structural Steel	—	ASTM A36 - 08
High-Strength Bolts	—	ASTM A325 - 07a, ⅞" Dia.
Weld Steel	—	AWS D1.1 - 08, Class E70

3. Shop and field paint shall be as called for in the specifications.

4. Steel to be encased in concrete shall not be painted.

5. Hot dipped galvanized surfaces that have been damaged by welding, cutting, burning, shearing, or other damage incurred during transit or erection shall be repaired with ZRC Cold Galvanizing Compound as manufactured or approved equivalent. Repair shall be in strict accordance with the manufacturer's recommendations.

6. All shop connections shall be welded or made with high-strength bolts unless noted specifically.

7. Erectors shall provide all temporary shoring and bracing needed for stability until structure is completed.

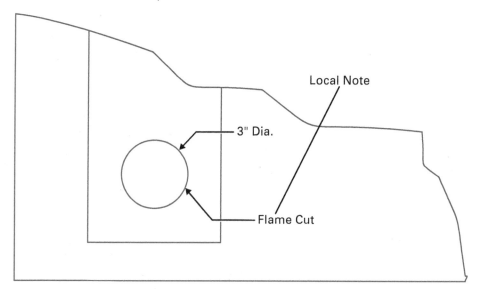

Figure 23 General notes and local notes.

Items typically found in general notes include the following:

*   References to specifications, procedures, or other notes
*   Identification of work responsibility
*   Surface finish requirements (smoothness, type of finish)
*   Special handling requirements
*   Material requirements such as specification, grade, and class
*   Special assembly requirements

A local note applies to a particular part of an object. The local note is placed near one of the views that represent the part. Local notes are used for the following types of information:

- Thread fastener type and size
- Location of special items (parts to be supplied by others)
- Reference to details on another drawing
- **Elevation**
- Special treatments or finishes
- Drilled hole size

Elevation: A height measurement above sea level or other identified point; also, the vertical sides of an object in an orthographic view.

A specification is a note that provides a detailed description of requirements for the type and size of the parts or material to be used. For instance, it may specify the type and size of the electrode and/or the welding process to be used. In *Figure 24*, the letter A refers to Specification A. In this case, Specification A in the tail of the arrow indicates that No. 30, $\frac{1}{8}$" copper rod is to be used for the weld. Specifications are usually located near the views to which they refer. If there are many specifications, they are attached on a separate sheet of paper and referenced to the drawing.

Specification A = No. 30, ⅛" Copper Rod

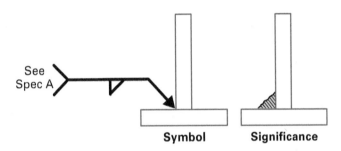

Figure 24 Application of a specification.

2.2.2 Bill of Materials

A bill of materials, also called a list of materials, parts list, or schedule of parts, is a list of all the parts required to fabricate an object. It generally identifies each part with a number or letter, sometimes called a **mark**. The bill of materials includes a description of the part, the part number, and the quantity needed for each assembly. Some bills of materials may also include size and dimensions of the part. For example, size and dimension are often included on bills of materials when the parts being described are structural shapes. *Figure 25* shows a portion of a drawing that includes a bill of materials. Not every drawing, however, will have a bill of materials. For larger projects, the bill of materials alone may require several drawing sheets.

Mark: A system of letters and numbers used with structural steel to identify the type of member and where it is placed.

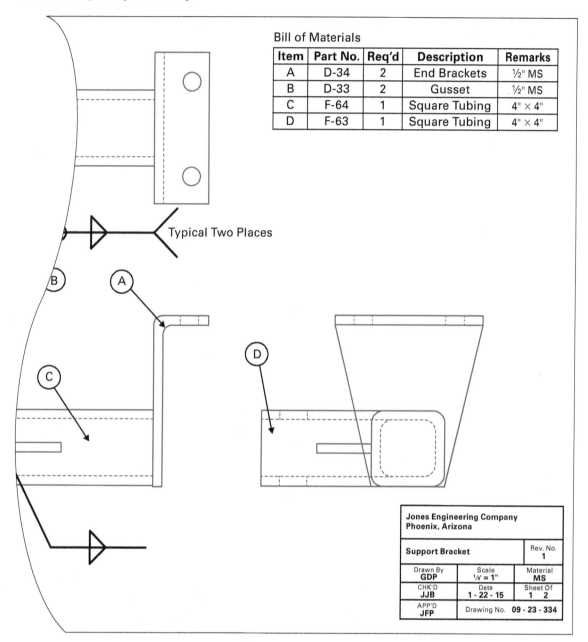

Bill of Materials

Item	Part No.	Req'd	Description	Remarks
A	D-34	2	End Brackets	½" MS
B	D-33	2	Gusset	½" MS
C	F-64	1	Square Tubing	4" × 4"
D	F-63	1	Square Tubing	4" × 4"

Typical Two Places

Jones Engineering Company Phoenix, Arizona		
Support Bracket		Rev. No. 1
Drawn By GDP	Scale ⅛' = 1"	Material MS
CHK'D JJB	Date 1 - 22 - 15	Sheet Of 1 2
APP'D JFP	Drawing No. 09 - 23 - 334	

Figure 25 Bill of materials.

2.0.0 Section Review

1. A hole that will only be drilled part of the way through an object is called a _____.
   a. counterbored hole
   b. countersunk hole
   c. blind hole
   d. through hole

2. Items such as identification of work responsibility and special assembly requirements are given in _____.
   a. classification schedules
   b. specification notes
   c. welding parts lists
   d. general notes

1. What is included in a welding detail drawing to show how pieces are to be joined in a weld?
   a. A welding equation
   b. A welding sign
   c. A welding groove
   d. A welding symbol

2. When an object to be fabricated is too complicated for one drawing, the complete and assembled object may be shown on one or more detail drawings plus a(n) _____.
   a. general note
   b. bill of materials
   c. structural print
   d. assembly drawing

3. On a typical drawing, a thick, solid line used to show all visible edges of a part is called a(n) _____.
   a. object line
   b. center mark
   c. leader line
   d. hidden section

4. A series of thin lines that are either solid, or solid and broken, and are arranged in a pattern to indicate various types of substances, such as metals, is called a(n) _____.
   a. material symbol
   b. cut-away projection
   c. cutting plane
   d. extension grid

5. On a welding detail drawing, drilled hole dimensions are shown with a note and _____.
   a. dotted line
   b. leader line
   c. dashed line
   d. section line

6. Angles and bevels are common on welding detail drawings because they are used with _____.
   a. fillet welds
   b. groove welds
   c. oxyfuel cuts
   d. weld joint preparation

7. The tolerances used for the dimensions of a drawing can be given (shown) following the dimension, as a note, or in the _____.
   a. title block
   b. scale factor
   c. drawing title
   d. object drawing itself

8. What is shown on a bill of materials to identify each part needed to fabricate an object?
   a. A weld symbol
   b. A decimal
   c. A mark
   d. A dollar amount

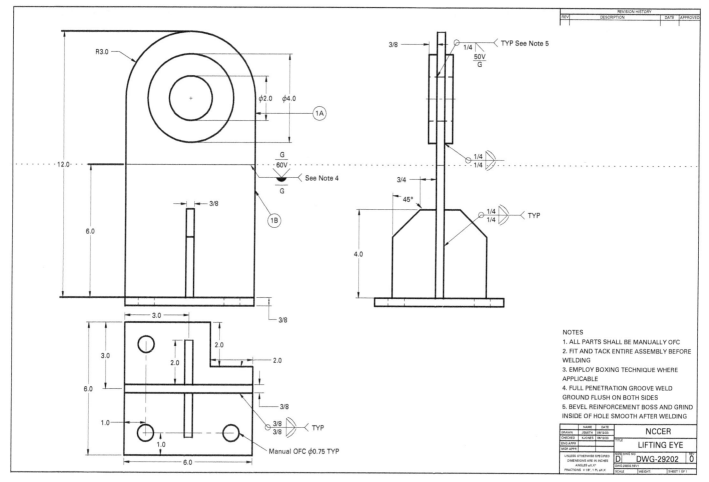

Figure RQ01

9. How tall is part 1A of the object shown in *Figure RQ01*?
    a. 2"
    b. 4"
    c. 6"
    d. 12"

10. What degree of angle is used for the bevel groove weld joining 1A and 1B in *Figure RQ01*?
    a. 50°
    b. 60°
    c. 30°
    d. 20°

Answers to odd-numbered questions are found in the Review Question Answer Keys at the back of this book.

Answers to Section Review Questions

Answer	Section Reference	Objective
Section 1.0.0		
1. d	1.0.0	1a
2. d	1.2.3	1b
Section 2.0.0		
1. c	2.1.3	2a
2. d	2.2.1	2b

User Update

SCAN ME

Did you find an error? Submit a correction by visiting **https://www.nccer.org/olf** or by scanning the QR code using your mobile device.

Physical Characteristics and Mechanical Properties of Metals

Objectives

Successful completion of this module prepares you to do the following:

1. Describe the composition and classification systems for a variety of welding base metals.
   a. Describe the composition and classification system for ferrous metals.
   b. Describe the composition and classification system for low-alloy steel.
   c. Describe the composition and classification system for stainless steel.
   d. Describe the composition and classification system for nonferrous metals.
2. Describe the physical and mechanical characteristics of metals and explain how to identify base metals.
   a. Describe the physical characteristics of different metals.
   b. Describe the mechanical properties of different metals.
   c. Explain how to identify base metals in field conditions.
   d. Describe metallurgy-related considerations for welding.
3. Identify the common structural steel shapes of metal.
   a. Identify the most common structural steel shapes.
   b. Identify different structural beam shapes.
   c. Identify pipe and tubing types.
   d. Identify other common metal forms, including rebar.

Performance Tasks

This is a knowledge-based module. There are no Performance Tasks.

Overview

In order for welders to accurately identify base metals and select the proper filler metals, they must have a thorough understanding of the composition, classification, physical characteristics, and mechanical properties of common metals and alloys. They must be familiar with metallurgical factors that need to be considered during welding activities. They must also be able to recognize standard commercial steel shapes used for fabrication and construction.

1.0.0 Metal Composition and Classification

Performance Tasks

There are no Performance Tasks in this section.

Objective

Describe the composition and classification systems for a variety of welding base metals.

a. Describe the composition and classification system for ferrous metals.
b. Describe the composition and classification system for low-alloy steel.
c. Describe the composition and classification system for stainless steel.
d. Describe the composition and classification system for nonferrous metals.

Metallic: Having characteristics of metal, such as ductility, malleability, luster, and heat and electrical conductivity.

Alloy: A metal that has had other elements added to it that substantially change its mechanical properties.

Ferrous: Relating to iron or an alloy that is mostly iron; a metal containing iron.

Nonferrous: Relating to a metal, such as aluminum, copper, or brass, that lacks sufficient quantities of iron to be affected by its properties.

Wrought: Formed or shaped by hammering or rolling.

Malleable: Able to be hammered or pressed into another shape without breaking.

Ductile: Able to be bent or shaped without breaking.

Casting: The metal object produced by pouring molten metal into a mold.

All living and nonliving things are made up of chemical elements that can be altered by changes in temperature. A steak left on a grill will eventually burn. A piece of wood thrown into a fire will also burn. Metals are no different. When metals are heated to a high enough temperature, some will melt while others will burn.

Metals, like other substances, are a collection of chemical elements bound together by natural forces or by human-made production processes. The chemical elements in some metals make them stronger than other metals. Those same chemical elements allow some metals to withstand stress or force better than others.

The composition of a common welding base metal varies widely, from a metal made of essentially one **metallic** element, to an **alloy** (mixture) of metallic and nonmetallic components. Metals are classified into two basic groups: **ferrous** metals, which are composed mainly of iron, and **nonferrous** metals, which contain very little or no iron. Ferrous metals include all steel, cast iron, **wrought** iron, **malleable** iron, and **ductile** (nodular) iron. Nonferrous metals and their alloys include the light metals (aluminum, magnesium, titanium), the heavy metals (copper, nickel, lead, tin, zinc), and the precious metals (platinum, gold, silver).

1.1.0 Ferrous Metal

Ferrous metals contain mostly iron and some carbon. The amount of carbon ranges from almost none in commercially pure iron to about 1.4%. Steel contains less than 1.7% carbon. With above 1.4% carbon, steel develops brittle properties like those of cast iron. Carbon steel, the largest group of ferrous-based metals, contains much smaller amounts of carbon than cast iron. As the carbon content increases in carbon steel, the steel becomes harder and more wear-resistant. The carbon content of a particular carbon steel is the largest factor in determining its use. A steel **casting** usually contains a higher percentage of carbon than a rolled plate and other rolled shapes. The casting foundry can usually provide the alloy formula.

Steels also contain impurities and trace amounts of elements such as manganese, silicon, phosphorus, copper, and lead. Sometimes additional amounts of these elements are intentionally added to steel to change its characteristics, such as to increase its **forgeability** or to decrease a tendency toward forming **blowholes**.

Forgeability: The ability to be heated and then beaten or hammered into shape.

Blowholes: The vents that form when air or gases escape.

Carbon and Alloy Steels

Steel is an alloy of iron and typically contains less than 1% carbon. All steels also contain varying amounts of other elements such as manganese, silicon, phosphorus, sulfur, and oxygen. In addition, some standard alloy steels can also contain elements such as nickel, chromium, and molybdenum. There are currently about 3,500 different grades of steel with many different properties. Most of the steels produced today are carbon and alloy steels, which make up 95% of the construction and fabrication metals used worldwide.

Source: Photo Courtesy of U.S. Army

Steels, including carbon steels, low-alloy carbon steels, alloy steels, and stainless steels, are classified by various systems, such as specification number and grade or manufacturer's trade name and number. Steel classification systems include those developed by the **American Iron and Steel Institute (AISI)**, the **ASTM International**, and the **Unified Numbering System (UNS)**, as well as manufacturer's trade names and identification numbers. Stainless steels have their own classification systems. They are sometimes referred by the percentages of their chromium and nickel content, such as 18/8, 25/20, and 18/10, but this system has largely been replaced by the AISI stainless steel classification system. Common classification systems for steels are explained in the following sections.

American Iron and Steel Institute (AISI): An industry organization responsible for preparing standards for steel and steel alloys that are based upon the code system of SAE International.

ASTM International: Formerly known as American Society for Testing and Materials, an organization that developed a code system for identifying and labeling steels.

Unified numbering system (UNS): An industry-accepted numbering system for identifying ferrous metals and alloys.

SAE International: Formerly named the Society of Automotive Engineers, an automotive society responsible for creating an early system for classifying carbon steels.

1.1.1 Carbon Steel Classification

The principal classification system for carbon steels is the AISI numerical designation of Standard Carbon and Alloy Steels. It was originally a four-digit system developed by **SAE International** for carbon steels commonly used in structural shapes, plate, strip, sheet, and welded tubing. Five digits are now used for some alloys. The classification system may be referred to as the *AISI, SAE,* or *AISI/SAE system.*

The AISI and SAE designations are essentially the same, except that the AISI system sometimes uses a letter prefix to indicate the manufacturing process that produced the steel. The following are examples of letter designations:

- *A* — Open-hearth steel
- *B* — Acid Bessemer carbon steel
- *C* — Basic open-hearth carbon steel
- *D* — Acid open-hearth carbon steel
- *E* — Electric furnace steel

Steel Alloying Elements

Some alloying elements used with steel and the reasons they are added include the following:

- Manganese — Promotes forgeability
- Silicon — Prevents blowholes
- Phosphorus — Improves machinability
- Copper — Inhibits corrosion
- Lead — Improves machinability

The absence of a letter prefix indicates basic open-hearth or acid Bessemer carbon steel. The prefixing letter (if any) designates the manufacturing process. The first two numerical digits represent the series (type and class) of steel. The third and fourth (and sometimes fifth) numerical digits specify the approximate percentage of carbon content. *Table 1* shows the AISI/SAE system for designating carbon and alloy steel types and classes and includes the AISI/SAE number.

TABLE 1   AISI/SAE System for Designating Carbon and Alloy Steel Types and Classes

AISI/SAE Number	Carbon Steel Types and Classes
1XXX	**Carbon steels**
10XX	Plain (nonresulfurized) carbon steel grades
11XX	Free machining, resulfurized (screw stock)
12XX	Free machining, resulfurized, rephosphorized
13XX	Manganese 1.75%
15XX	High-manganese (1.00% to 1.65%) carburizing steels
2XXX	**Nickel steels**
23XX	3.50% nickel
25XX	5.00% nickel
3XXX	**Nickel-chromium steels**
31XX	1.25% nickel, 0.65% or 0.80% chromium
32XX	1.75% nickel, 1.07% chromium
33XX	3.50% nickel, 1.50% or 1.57% chromium
34XX	3.00% nickel, 0.77% chromium
4XXX	**Molybdenum steels**
40XX	0.20% or 0.25% molybdenum
41XX	0.50%, 0.80%, or 0.95% chromium, 0.12%, 0.20%, 0.25%, or 0.30% molybdenum
43XX	1.82% nickel, 0.50% or 0.80% chromium, 0.25% molybdenum
46XX	0.85% or 1.82% nickel, 0.20% or 0.25% molybdenum
47XX	1.05% nickel, 0.45% chromium, 0.20% or 0.35% molybdenum
48XX	3.50% nickel, 0.25% molybdenum
5XXX	**Chromium steels**
50XX	0.27%, 0.40%, 0.50%, or 0.65% chromium
51XX	0.80%, 0.87%, 0.92%, 0.95%, 1.00%, or 1.05% chromium
5XXXX	1.00% carbon, 0.50%, 1.02%, or 1.45% chromium
6XXX	**Chromium-vanadium steels**
61XX	0.60%, 0.80%, or 0.95% chromium, 0.10% or 0.15% (minimum) vanadium
86XX	0.55% nickel, 0.50% chromium, 0.20% molybdenum
92XX	0.65%, 0.82%, or 0.85% manganese, 1.40% or 2.00% silicon, 0.00% or 0.65% chromium
93XX	3.25% nickel, 1.20% chromium, 0.12% molybdenum
94XX	0.45% nickel, 0.40% chromium, 0.12% molybdenum
97XX	0.55% nickel, 0.20% chromium, 0.20% molybdenum
98XX	1.00% nickel, 0.80% chromium, 0.25% molybdenum
XXBXX	Boron steels
XXLXX	Leaded steels

The following examples demonstrate the use of the AISI/SAE numbers for carbon steels:

- AISI number C1020:
  - C = Indicates basic open-hearth carbon steel
  - 10 = Carbon steel, nonresulfurized
  - 20 = Contains approximately 0.20% carbon
- AISI number E2512:
  - E = Indicates electric furnace steel
  - 25 = Designates steel alloyed with approximately 5% nickel
  - 12 = Designates steel containing approximately 0.12% carbon
- AISI number E52100:
  - E = Indicates electric furnace steel
  - 5 = Contains approximately 0.50%, 1.00%, or 1.45% chromium, designated by the next digit—0, 1, or 2, respectively
  - 2 = Designates approximately 1.45% chromium
  - 100 = Designates approximately 1% carbon

The common group classification of carbon steels is primarily based on carbon content. Plain-carbon steels (AISI series 10XX through 11XX) are basically iron-carbon alloys. High-strength low-alloy (HSLA) carbon steels (AISI series 13XX through 98XX) have small amounts of alloying elements to improve strength, hardness, and toughness or to increase resistance to corrosion, heat, and environmental damage. The plain-carbon steels are commonly grouped as follows:

- *Low carbon* — 0.10% to 0.15% carbon, 0.25% to 1.50% manganese
- *Mild carbon* — 0.15% to 0.30% carbon, 0.60% to 0.70% manganese
- *Medium carbon* — 0.30% to 0.50% carbon, 0.60% to 1.65% manganese
- *High carbon* — 0.50% to 1.00% carbon, 0.30% to 1.00% manganese

The HSLA carbon steels are alloyed with one or more elements, including manganese, nickel, chromium, or molybdenum, to provide higher strength, better toughness, improved weldability, and in some cases, greater resistance to corrosion. As the percentage of these elements increases, changes in the weldability of the low-alloy steels make electrode selection and welding procedures more critical. HSLA steels include the following:

- Low-nickel steels
- Low-nickel chrome steels
- Low-manganese steels
- Low-alloy chromium steels
- Weathering steels

Another class of steel in the AISI series is sulfur steel. It is not an alloy steel but is still included in the series.

1.1.2 ASTM Steel Specification System

ASTM International develops and publishes specifications for use in material production and testing. Welding codes and procedures typically use the ASTM designation to identify metal types because it includes specifications for mechanical and physical characteristics of the material as well as chemical composition. Piping and tubing are the largest group of materials fabricated by welding covered in ASTM specifications.

UNS

The UNS used to identify metals was developed jointly by ASTM International and SAE International as a way of relating the different numbering systems for metals and alloys. The UNS aims to eliminate the confusion caused by using more than one identification number to identify the same material or using the same identification number to identify two different materials. It is anticipated that the UNS will eventually replace the AISI and other systems.

Each UNS number has a letter prefix followed by five digits. Often the prefix letter relates to the family of metals. For example, UNS numbers A00001 through A99999 refer to aluminum and aluminum alloys, and UNS numbers C00001 through C99999 signify copper and copper alloys. When possible, the digits in the UNS sequence contain the same numbering sequences used in the other systems. For example, the UNS number G10200 corresponds to the AISI/SAE number 1020 for carbon steel. Most engineering and materials reference books contain tables that cross-reference AISI and SAE numbers for metals to the corresponding UNS numbers.

American Society of Mechanical Engineers (ASME): An educational and technical organization founded for the practice of mechanical and multidisciplinary engineering.

NOTE

The complete collection of ASTM standards consists of 15 sections comprising 65 volumes, with a separate index. The table is provided only to illustrate sample ASTM numbers.

Mill test report (MTR): A quality assurance report provided to a customer by a metal manufacturer that shows the chemical content and testing results of the metal being purchased; also called a *certified mill test report (CMTR)*.

Prefix letters are a general part of each specification and provide a general idea of the specification content. For example, A is used for ferrous metals, B for nonferrous metals, and E for miscellaneous subjects, such as examination and testing.

Sometimes an S is added before the prefix letter. This indicates that the **American Society of Mechanical Engineers (ASME)**, publishers of the *Boiler and Pressure Vessel Code*, has adopted that specification.

Apart from a few mild-steel grades, such as SA-36 (mild-steel plate) and SA-53 (mild-steel pipe), ASTM specifications are three-digit numbers. Most specifications also have various grades, classes, or types. *Table 2* gives examples of some of the ASTM specifications for alloy steels.

When companies buy steel, the manufacturer will provide the customer with a certificate indicating the quality of the steel. This document is often called a **mill test report (MTR)** or a *certified mill test report (CMTR)*. The report mainly shows the chemical content and testing results of the metal that was purchased (*Figure 1*). Such information is necessary in case the structure being built from the material fails. For liability reasons, this information must be tracked and kept on record.

TABLE 2 ASTM Specification Standards

ASTM Number	Grade	Tensile Strength (psi)	Alloying Elements (%)
A209	Grade T1	55,000	0.50 Mo
	Grade T1a	60,000	0.50 Mo
	Grade T1b	53,000	0.50 Mo
A213	Grade T17	60,000	1.00 Cr 0.15 V
A334	Grade 8	100,000	9.00 Ni
A335	Grade P1	55,000	0.50 Mo
	Grade P2	55,000	0.60 Cr 0.50 Mo
	Grade P11	60,000	1.25 Cr 0.55 Mo 0.75 Si
	Grade P12	60,000	0.95 Cr 0.55 Mo
	Grade P15	60,000	1.40 Si 0.55 Mo
A517	Grade A	115,000	0.65 Cr 0.25 Mo 0.60 Si
	Grade B	115,000	0.50 Cr 0.20 Mo 0.05 V
	Grade P	115,000	1.25 Ni 1.0 Cr 0.50 Mo

Notes:
C—Carbon; Cr—Chromium; Mo—Molybdenum; Ni—Nickel; Si—Silicon; V—Vanadium

Wide flange (W) beam: A metal I-beam that has an extra wide flange at the top and bottom.

Included in the mill test report are the ASTM specifications, size, grade, and length of the product, which in this case is a **wide flange (W) beam**. The report also shows the heat number, chemical analysis, and physical properties of the steel. These markings are also found on the materials covered by the report. *Figure 2* shows how the markings look on an actual beam.

AM/NS Calvert LLC
1 AM/NS Way
Calvert, Al. , AL 36513 USA

COPY PRINT

Mill Certificate

Order - Item 2490A-10	Certificate Number 109514866	Delivery No A3218022-10	Ship Date 04/12/2023	Page 1 of 1

Customer No: 102	Cust PO: 6ZA-344C6

Customer Part No: RMS-001-340XL

Customer Sold to:	Customer Ship to:	Contact - Quality Manager AM/NS Calvert LLC 1 AM/NS Way CALVERT AL 36513 USA Email: xx.yy@ArcelorMittal.com Ph : 1-251-xxx-xxxx

Steel Grade / Customer Specification
Hot Roll Black Coil Conv to A36 / 0.1770 " X 48.0000 " ACCORDING TO A1011-18a {Light < 0.230"(6.0 mm)}-Hot Roll Base

Type of Product/Surface
Hot Roll Black Dry Exposed CTL FOR LASER APPLICATION

TEST METHOD   Melted in Brazil   Manufactured in USA

ASTM

MATERIAL DESCRIPTION

	ORDERED	Heat No.	Coil No.	Weight Net LB	Weight Gross LB
(mm)	4.496	C10074	109514866	46,032.000	46,032.000
(in)	0.1770				

CHEMICAL COMPOSITION OF THE LADLE *

Heat No.	C	Si	Mn	P	S	Al	Cr	Cu	Mo	N
C10074	0.1970	0.01	0.92	0.020	0.009	0.035	0.02	0.011	0.001	0.0048

Ni	Nb	Ti	B	V	Ca
0.008	0.000	0.000	0.0000	0.000	0.0002

TENSILE TEST

Test Direction	Yield Strength	Tensile Strength	% Total Elong.
T	51 ksi	70 ksi	32

AM/NS Calvert LLC certify that the material herein described has been manufactured, sampled, tested and inspected in accordance with the contract requirements and is fully in compliance. Test certificates are prepared in accordance with procedures outlined in DIN EN 10204:2005 Type 3.1

* - This test is not covered by our current A2LA accreditation

Director
Quality Management, AM/NS Calvert

Rev.

Figure 1 Example of a mill test report.
Source: AM/NS Calvert

Figure 2 Mill markings on a beam.

The markings show who made the steel, where it was made, its original dimensions, and its heat number or batch number. Heat or batch numbers are tracking numbers that the steel producer assigns to a particular ladle of steel. When a product, such as a beam or plate, is made from that ladle of steel, the tracking number is placed on that product in case of recall or problems with the chemistry. The following are explanations of the letters and numbers shown on the beam in *Figure 2*.

- *SDI* — Typically the name of the manufacturer.
- *USA* — Country of origin where the steel was melted.
- *A046196* — The heat number or batch number, which can be used to match the material to a material test report. Heat numbers are normally unique to each manufacturer, and they are usually meaningful only to the producer. The number provides a way to trace the product. The material test report lists the mechanical test results and chemistry. These can be checked against the material specification to ensure that the material test report results meet the requirements.
- *W16×36* — This beam has a depth of approximately 16" (actual depth 15⅞" or about 403 mm) and weighs 36 pounds per foot of length (lb/ft) or 53.6 kilograms per meter (kg/m). If you looked up *W16×36* in the *Steel Construction Manual* published by the American Institute of Steel Construction (AISC), you would find the beam's flange width and thickness, **web** thickness, beam depth, and many other design properties.

Plate steel also has mill markings, such as batch or heat numbers (*Figure 3*). The information at the top of the plate indicates that the plate steel was not heat-treated after being rolled. The *S7-09438* heat number is the batch number. The size information is given. Grade A36 indicates that the plate was manufactured in accordance with the ASTM specifications for A36 steel plate. The other information at the bottom of the plate shows who made the plate steel and who bought the plate steel.

Web: The plate joining the flanges of a girder, joist, or rail.

NOTE

One piece of information that is not included on the beam in *Figure 2* is the ASTM specification to which the beam was made (for example, A36, A441, A572, A588, etc.).

Figure 3 Mill markings on plate steel.

There is no set format for labeling steel, so steel producers have their own way of labeling their products. The mill markings on the beams and plates shown in *Figure 2* and *Figure 3* were stenciled or sprayed onto the steel. Other manufacturers attach paper labels to their products, as shown in *Figure 4*.

The mill markings on most steel will not be as clearly visible as those shown in these examples. Steel is often stored outside in the weather and may be exposed to other rough conditions or handling that can damage the markings.

When preparations are being made for a welding job, the job planner determines which steel to use. On some jobs, lay-down yard personnel may bring the materials to the work area. However, the fabricators (fitters and welders) must be able to ensure that the delivered steel is the same as that indicated on the welding procedure specification (WPS). *Figure 5A* and *Figure 5B* show a WPS with standard or English measurements. *Figure 6* shows a WPS with metric values.

When only a part of the original piece of steel is needed, welding personnel must transfer the heat number of the original piece to the piece they are cutting from the main piece. The transfer of numbers must be done before the part is cut away from the original piece. Transferring the heat number ensures that the removed piece of steel can be tracked when it is welded into place. Follow site-specific guidelines for transferring numbers.

On high-profile jobs, the process of tracking these numbers can be very involved because of liability concerns. Many jobs today require the welders to closely track the materials being used. The owner representatives (usually the quality control personnel) can use these heat or batch numbers to trace where the raw materials came from, the day and shift on which the material was produced, and who was on duty at the time. Most plants will not allow material to be used if the heat number has been ground off or rusted over. From a contractor's point of view, material that lacks a heat number should never be used in any critical application. If a failure occurred, and the installer could not provide a valid heat number, the contractor/installer would assume full responsibility, even though the steel was produced by someone else.

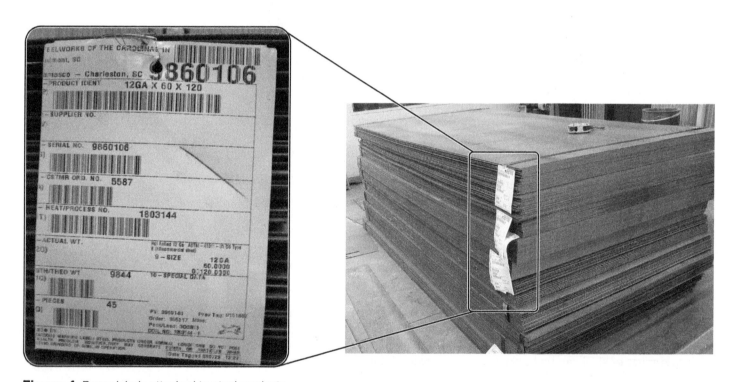

Figure 4 Paper labels attached to steel products.

Zachry Industrial, Inc.

ASME IX – Welding Procedure Specification (WPS)

WPS record no.	IGMA-11-AF	Rev.	3	Qualified to	ASME Section IX	Date	2/14/2019
Company	ZACHRY			Supporting PQR(s)	IGMA - 11 PAGE B - Rev. 1		
Scope				Reference documents			
Applicability	Groove, Fillet, With PWHT			Project			
Joint sketches	In production drawings. In engineering specs. V or U Groove Root Opening = 1/4" (+1/8")(3/16")						

Base Materials/Joints

						Material Dimensions	As welded		With PWHT	
							Min	Max	Min	Max
Qualified material	Carbon steel (P1)	P no.	1	Grp no.		Thickness CJP (in.)			0.188	8
Welded to	Carbon steel (P1)	P no.	1	Grp no.		Thickness PJP (in.)			0.188	8
Backing	When required	P no.	1	Grp no.		Thickness toughness (in.)			-	-
Retainers	None					Fillet welds (in.)			no min.	no max.
Other						Diameter (in.)			no min.	no max.

Filler Materials

	SFA	Classification	F no.	A no.	Trade name/chem. analysis	Deposit Thickness	As welded		With PWHT	
							Min	Max	Min	Max
GTAW	5.18	ER70-S	6	1	NR				no min.	0.75
SMAW	5.1	E7018	4	1	NR				no min.	8
Consumable insert							(None)			

Electrode coating	Low H potassium, Fe powder	Active flux	Without
Product form GTAW	Bare (solid)		
Other			

Processes

Process		GTAW				SMAW	
Type		Manual				Manual	
Autogenous		No				-	
Filler size (in.)	1/16	3/32	1/8	3/32	1/8	5/32	
Layer	1, 2	1, 2	1, 2	Remainder	Remainder	Remainder	
Max pass thickness (in.)		-			Less than 0.5		

Position

Position	All	All
Progression	Uphill	Uphill

Preheat

Preheat temp min. (°F)	50	50
Interpass temp max. (°F)	600	600

Gas

Shielding gas type		Argon (A5.32 SG-A)		-		
Shielding flow rate (cu.ft/hr)	15 - 35	15 - 35	15 - 35	-		
Trailing gas type		None		-		
Backing gas type		None		-		

Electrical

Heat input calculation		N/A			N/A	
Welding current (A)	50 - 150	75 - 180	100 - 250	50 - 120	70 - 150	90 - 200
Arc voltage (V)	NR	NR	NR	NR	NR	NR
Electrode run out length (in.)		-				
Travel speed (in/min)	Manual	Manual	Manual	Manual	Manual	Manual
Heat input (KJ/in.)	N/A	N/A	N/A	N/A	N/A	N/A
Current type/polarity	DCEN	DCEN	DCEN	DCEP	DCEP	DCEP
Tungsten type	SFA 5.12 EWTh-2			-		
Tungsten size (in.)	1/16, 3/32, 1/8			-		
Pulse welding details	Not used			-		

Other Information

String/Weave	Stringer or Weave	Stringer or Weave
Orifice/Gas cup size (in.)	#4 #12	-
Multi/Single pass/side	Multiple passes	Multiple passes
Initial/Interpass cleaning	Brushing and Grinding	Brushing and Grinding
Backgouging method	Grinding if required.	Grinding if required.
Groove preparation method	Mechanical	Mechanical
Peening	Not used	Not used

Figure 5A Sample WPS with standard measurements (1 of 2).
Source: Zachry Group

Zachry Industrial, Inc.

ASME IX – Welding Procedure Specification (WPS)

WPS record no.	IGMA-11-AF	Rev.	3	Date	2/14/2019	Project		
Company	ZACHRY	Qualified to	ASME Section IX			Supporting PQR(s)	IGMA - 11 PAGE B - Rev. 1	

PWHT

Temperature (°F)	1150	Time (hrs)		1hr/in. 15min min.	Type		Stress relief
Heating rate (°F/hr)	Per design code	Heating method		Not recorded			
Cooling rate (°F/hr)	Per design code	Cooling method		Not recorded			

Notes

1. A potential health risk exists from inhaling grinding dust from EWTh-2 tungsten electrodes. As an alternative, EWCe-2 and/or EWLa-2 tungsten electrodes are acceptable.

Preheat rules:

200 (°F) for thickness over 1 (in.) with C over 0.30%.

50 (°F) for thickness not over 1 (in.) with C over 0.30%, and for thickness over 1 (in.) with C not over 0.30%.

Welding Engineer

Jason D. Praster
2/14/2019

Figure 5B Sample WPS with standard measurements (2 of 2).
Source: Zachry Group

Weld Procedure Number	30 P1 TIG 01 Issue A
Qualifying Welding Procedure (WPAR)	WPT17/A

Manufacturer: National Fabs Ltd 25 Lane End Birkenshaw Leeds Location: Workshop Welding Process: Manual TIG Joint Type: Single Sided Butt Weld	Method Of Preparation and Cleaning: Machine and Degrease Parent Metal Specification: Grade 304L Stainless Steel Parent Metal Thickness 3 to 8mm Wall Pipe Outside Diameter 25 to 100mm Welding Position: All Positions Welding Progression: Upwards

Joint Design   **Welding Sequences**

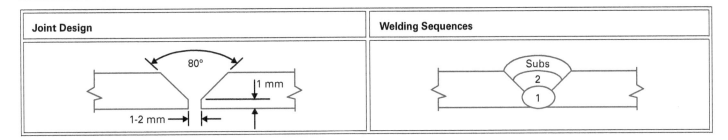

Run	Process	Size of Filler Metal	Current A	Voltage V	Type of Current/Polarity	Wire Feed Speed	Travel Speed	Heat Input
1 2 And Subs	TIG TIG	1.2 mm 1.6 mm	70 - 90 80 - 140	N/A	DC– DC–	N/A	N/A	N/A

Welding Consumables:- Type, Designation Trade Name: Any Special Baking or Drying: Gas Flux: Gas Flow Rate - Shield: - Backing: Tungsten Electrode Type/ Size: Details of Back Gouging/Backing: Preheat Temperature: Interpass temperature: Post Weld Heat Treatment Time, temperature, method: Heating and Cooling Rates*:	BS 2901 Part 2 : 308S92 No Argon 99.99% Purity 8 - 12 LPM 5 LPM 2% Thoriated 2.4mm Dia Gas Backing 5°C Min 200°C Max Not Required	**Production Sequence** 1. Clean weld and 25mm borders to bright metal using approved solvent. 2. Position items to be welded ensuring good fit up and apply purge. 3. Tack weld parts together using TIG, tacks to at least 5mm min length. 4. Deposit root run using 1.2mm dia. wire. 5. Inspect root run internally. 6. Complete weld using 1.6mm dia wire using stringer beads as required. 7. 100% Visual inspection of completed weld

Revision History

Date	Issue	Changes	Authorization
26/11/2005	A	First changes	Jack Straw *Jack Straw*

Figure 6 Sample WPS with metric measurements.

1.2.0 Low-Alloy Steel Groups

Low-alloy steels fall into the following general groups:

- High-strength low-alloy steels
- Quench-and-tempered steels
- Heat-treatable low-alloy steels
- Chromium-molybdenum steels

1.2.1 High-Strength Low-Alloy Steels

High-strength low-alloy (HSLA) steels are designed to meet specific mechanical requirements per ASTM specifications. The principal difference between plain-carbon steels and HSLA steels is the addition of alloys such as manganese, chromium, nickel, copper, molybdenum, and vanadium. Typically, these alloys total less than 2%. The carbon content of HSLA steels is about the same as that in low-carbon and mild steels (0.06% to 0.28%). Compared to mild steels, HSLA steels are designed to provide a combination of higher strength, better corrosion resistance, and improved notch toughness. The weldability of HSLA steel is like that of mild steel.

HSLA steels can be divided into two general groups: Group A and Group B. Steels in Group A are designed for high strength, while steels in Group B are designed for high strength, improved atmospheric corrosion resistance, and notch toughness. Typical steels in Group A are A441, A572, and A633, which are used for structural applications such as pipelines, buildings, bridges, machinery, and railroad equipment. Other steels in Group A are A225 and A737, which are used for pressure vessels. Typical steels in Group B are A242 and A588, which are used for high-strength structural applications that may include unpainted applications. Group B steels specifically intended for low-temperature applications (the nickel content is as high as 9%) are A203 and A353.

HSLA steels can be welded using the common welding processes, although successful welding requires consideration of preheating, interpass temperature, and control of hydrogen, including the use of low-hydrogen electrodes when applicable. This information can be found on the WPS. Typically, only the root pass and possibly the hot pass are made with the gas tungsten arc welding (GTAW) process on HSLA steels. Fill passes and cover passes are often made with faster processes such as shielded metal arc welding (SMAW) or gas metal arc welding (GMAW). One advantage of using the GTAW process is that hydrogen is easier to control because fluxes are not used. This advantage is also true for GMAW, another solid wire process with no other fluxes. Minimum preheat and interpass temperatures are determined based on the alloy type and the thickness of the thinner part being welded.

Quench: To rapidly cool a hot metal using air, water, or oil.

Notch toughness: The impact load required to break a test piece of weld metal measured by a notch toughness test (Charpy V-notch test). This test may be performed on metal below room temperature, at which point it may be more brittle.

CAUTION

The welding recommendations provided are general guidelines only. When welding HSLA steel, the site WPS or site quality standards must be followed. Failure to follow proper standards can result in rejected and/or defective welds.

CAUTION

Steels that are wet or below 32°F (0°C) should always be preheated to at least 70°F (21°C) and maintained at that temperature to remove moisture and prevent its formation. Failure to preheat metals may result in rejected and/or defective welds. Always follow the governing WPS for any preheat requirements.

Interpass temperature: In a multipass weld, the temperature of the base metal at the time the next weld pass is started; it is usually the same temperature that is used for preheating.

Hot Tapping

Steel pipes are commonly hot tapped to install a bypass around a repair area or to add a branch line. With hot tapping, threaded or welded taps (called *threadolets* and *weldolets*, respectively) are welded to the outside of the pipe while material or liquid is still flowing. Special tools can then be used to work through the installed fitting, penetrate the pipe, and install a shutoff valve without the loss of the pipe contents. In other words, the complete connection is made without draining the pipe contents. The pipe remains hot.

Low-alloy pipes cannot be hot tapped because the material flowing through the pipe will cool the pipe rapidly and prevent proper preheat, interpass, and postheat temperature maintenance. This can result in defective welds, as well as pipe failure during welding. A site engineer should be consulted before any hot tap is made due to concerns related to the pipe contents. Hot taps can be extremely dangerous, and the site safety department should always be involved.

TABLE 3 Minimum Preheat and Interpass Temperatures for Typical HSLA Steels Welded with Low-Hydrogen Electrodes

ASTM Steel	Thickness Range		Minimum Temperature
	(in)	(mm)	
A242 A441 A572 Gr 42, 50 A588	Up to 0.75	Up to 19	32°F (0°C)
	0.81 to 1.50	21 to 38	50°F (10°C)
	1.56 to 2.50	40 to 64	150°F (66°C)
	Over 2.50	Over 64	225°F (107°C)
A572, Gr 60, 65 A633, Gr E	Up to 0.75	Up to 19	50°F (10°C)
	0.81 to 1.50	21 to 38	150°F (66°C)
	1.56 to 2.50	40 to 64	225°F (107°C)
	Over 2.50	Over 64	300°F (140°C)

Table 3 provides minimum preheat and interpass temperatures for typical HSLA steels and also includes the ASTM steel numbers and grades.

Filler metal for welding HSLA steel depends on the alloy and the application. Mild-steel filler metal can be used when joint strength is not critical or when alloy steel is being joined to mild steel. Alloy steel should be used when the strength of the joint must equal or approach that of the base metal. High-alloy stainless steel is used for special applications including welding dissimilar metals or welding high-nickel steel, such as A353.

Typically, HSLA steels are not postweld heat-treated. However, this may be required to improve **ductility** or to maintain the dimensions of the weldment during machining.

1.2.2 Quench-and-Tempered Steels

Quench-and-tempered (Q&T) steels are furnished in the heat-treated condition, which generally consists of **austenitizing**, quenching, and **tempering**. Some Q&T steels fall into the carbon steel classification, some into the low-alloy steel classification, and others into the alloy classification of the AISI/SAE system. Q&T steels combine high yield and **tensile strength** with good notch toughness, ductility, corrosion resistance, or weldability, depending on their intended use. Q&T steels have less than 0.25% carbon, with yield strengths of 50,000 psi to 180,000 psi (≈344,738 kPa to 1,241,056 kPa). Most Q&T steels are furnished as plate, although some are available as seamless pipe.

Q&T steels are used where high strength and minimum weight are essential. Many of the Q&T steels are covered by ASTM classifications. A common group of Q&T steels known as *HY steels* are covered by military specifications (MIL). The most common HY steel is HY-80, which has a yield stress of 80,000 psi (≈551,581 kPa). HY-100, HY-130, HY-150, and HY-180 are also available. Types and typical uses of Q&T steels include the following:

- *A514 and A517* — Used for earth-moving equipment, pressure vessels, bridges, TV towers, and ships.
- *A533 Grade B* — Used for nuclear pressure vessels.
- *A543 Type B* — Used in nuclear reactor vessels, ships, and submarines. Classes 1 and 2 steels are similar to HY-80 and HY-100.
- *A553* — Used for cryogenic applications. It contains 8% to 9% nickel for high strength at extremely low temperatures and toughness at service temperatures of -320°F (-196°C).
- *A678* — Used for high-strength structural applications.

Ductility: The characteristic of metal that allows it to be stretched, drawn, or hammered without breaking.

Austenitizing: Heating a steel up to or above the transformation temperature range to achieve partial or complete transformation to austenite grain structure.

Tempering: The process of reheating quench-hardened or normalized steel to a temperature below the transformation range and then cooling it at any rate desired.

Tensile strength: The maximum stress or force that a material can withstand without breaking.

CAUTION

When welding Q&T steels, pay close attention to the welding procedures concerning preheat and interpass temperature recommendations to prevent loss of strength or cracking.

To prevent cracking, most Q&T steels can be welded by common welding processes using preheat and interpass temperatures as shown in *Table 4*. These temperatures vary depending on the type of steel being welded. Because Q&T steels obtain their strength from heat treatment, slow cooling can destroy the benefits of the original heat treatment. Slow cooling is generally caused by the following:

- Excess preheat
- Exceeding the recommended interpass temperature
- High heat input, such as high amperage with slow travel speed and the use of weave beads instead of stringers

On the other hand, cooling too quickly can cause a hard, brittle zone that may crack. Fast cooling is caused by the following:

- Insufficient preheat
- Insufficient interpass temperature
- Insufficient heat input during welding

When welding Q&T steels, always use low-hydrogen techniques and stringer beads to minimize heat input. Typically, the GTAW process is used only on thin sections or for the first one or two passes.

Most Q&T steels are designed to be used in the as-welded condition, with no postheat. Sometimes thick sections of A533 or A537 steels may require postheat treatment. If postheat treatment is required, use extreme care and follow the procedures exactly to prevent adversely affecting the base metal.

TABLE 4 Minimum Preheat and Interpass Temperatures for Q&T Steels

Thickness Range		Minimum Temperature				
(in)	(mm)	A514/A517	A533	A537	A543	A678
Up to 0.50	Up to 13	50°F (10°C)	50°F (10°C)	50°F (10°C)	100°F (38°C)	50°F (10°C)
0.56 to 0.75	14 to 19	50°F (10°C)	100°F (38°C)	50°F (10°C)	125°F (52°C)	100°F (38°C)
0.81 to 1.00	21 to 25	125°F (52°C)	100°F (38°C)	50°F (10°C)	150°F (66°C)	100°F (38°C)
1.10 to 1.50	27 to 38	125°F (52°C)	200°F (93°C)	100°F (38°C)	200°F (93°C)	150°F (66°C)
1.60 to 2.00	40 to 51	175°F (79°C)	200°F (93°C)	150°F (66°C)	200°F (93°C)	150°F (66°C)
2.10 to 2.50	53 to 64	175°F (79°C)	300°F (149°C)	150°F (66°C)	300°F (149°C)	150°F (66°C)
Over 2.50	Over 64	225°F (107°C)	300°F (149°C)	225°F (107°C)	300°F (149°C)	—

1.2.3 Heat-Treatable Low-Alloy Steels

Hardenability: A characteristic of a metal that enables it to become hard, usually through heat treatment.

Heat-treatable low-alloy (HTLA) steels are alloy steels that contain about 0.25% to 0.45% carbon and small amounts of chromium, nickel, and/or molybdenum to enhance **hardenability**. They are generally welded in the annealed condition, and then the entire weldment is heat-treated. The relatively high carbon content of HTLA steels allows them to be heat-treated to a very high strength and hardness. Because HTLA steels must be heat-treated after welding, they are typically used for smaller parts that can be placed in tempering ovens.

HTLA steels are produced to contain very low levels of impurities. During welding, low-hydrogen techniques are used, and extreme care must be taken to keep the weld joint and filler metal clean and free of impurities. For this reason, GTAW is the preferred process for welding HTLA steels. Heat input is also very important, and welding procedures that consider current, voltage, travel speed, and preheat temperatures must be followed.

Postweld heat treatment (PWHT): The process of heat-treating the base metal and weld after the weld has been made.

HTLA steels generally require higher preheat and interpass temperatures than other types of alloy steels to prevent hardening of the weld zone and hot cracks. They also require **postweld heat treatment (PWHT)**. *Table 5* shows minimum preheat and interpass temperatures for HTLA steels and includes the AISI steel numbers and thickness ranges.

TABLE 5 Minimum Preheat and Interpass Temperatures for HTLA Steels

AISI Steel	Thickness Range		Minimum Temperature
	(in)	(mm)	
4027	Up to 0.50	Up to 13	50°F (10°C)
	0.60 to 1.0	15 to 25	250°F (121°C)
	1.1 to 2.0	27 to 51	150°F (66°C)
4037	Up to 0.50	Up to 13	100°F (38°C)
	0.60 to 1.0	15 to 25	200°F (93°C)
	1.1 to 2.0	27 to 51	300°F (149°C)
4130 5140	Up to 0.50	Up to 13	300°F (149°C)
	0.60 to 1.0	15 to 25	400°F (204°C)
	1.1 to 2.0	27 to 51	500°F (260°C)
4135 4140	Up to 0.50	Up to 13	350°F (177°C)
	0.60 to 1.0	15 to 25	450°F (232°C)
	1.1 to 2.0	27 to 51	500°F (260°C)
4340 8630	Up to 0.5	Up to 13	200°F (93°C)
	0.6 to 1.0	15 to 25	250°F (121°C)
	1.1 to 2.0	27 to 51	300°F (149°C)
8640	Up to 0.50	Up to 13	200°F (93°C)
	0.60 to 1.0	15 to 25	300°F (149°C)
	1.1 to 2.0	27 to 51	350°F (177°C)

HTLA steels should receive postweld heat treatment immediately after they are welded. The type of postweld heat treatment depends on the type of steel and the preheat and interpass temperature maintained during welding. Allowing the weldment to cool after welding without some type of heat treatment generally results in cracks. Immediately after it is welded, HTLA steel should be stress relieved following strict guidelines depending on the alloy and the preheat and interpass temperatures used.

1.2.4 Chromium-Molybdenum Steels

Chromium-molybdenum (Cr-Mo) steels contain 0.5% to 9% chromium and 0.5% to 1% molybdenum. The carbon content is generally less than 0.20%. Cr-Mo steels, which are sometimes called *chrome-moly* or *heat-resisting alloy steels*, are widely used in the petroleum industry and in steam-generating plants for elevated-temperature applications. The chromium provides improved resistance to oxidation and corrosion, and the molybdenum increases strength at elevated temperatures. Cr-Mo steels are also used in the aircraft and racing industries as tubing for highly stressed parts (*Figure 7*). Cr-Mo steels that are in the AISI/SAE 4100 series are available in forging, castings, plate, pipe, and tubing to various ASTM specifications.

Cr-Mo steels are hardenable, so a welding procedure that includes preheating, and sometimes postheating, must be used. The preheat temperature depends on the amount of alloy, the carbon content, the weld material, and base metal thickness. Generally, the higher the alloy content, carbon content, or thickness, the higher the preheat temperature. *Table 6* shows minimum preheat and interpass temperatures for Cr-Mo steels including their types and ASTM steel designations.

CAUTION

Check your welding procedures for any special precautions needed when interrupting a heating cycle. Failure to follow proper welding procedures will result in defective welds.

Figure 7 Steel tubing used to construct a race car frame.
Source: Photo courtesy of The Lincoln Electric Company, Cleveland, OH, U.S.A.

TABLE 6 Minimum Preheat and Interpass Temperatures for Cr-Mo Steels

ASTM Steel	Type	Minimum Temperature		
		Up to 0.5" (13 mm)	0.5" to 1.0" (13 mm to 25 mm)	Over 1.0" (25 mm)
A335-P2	$^1/_2$ Cr - $^1/_2$ Mo	300°F (149°C)	300°F (149°C)	300°F (149°C)
A335-P12	1 Cr - $^1/_2$ Mo			
A335-P11	$1^1/_4$ Cr - $^1/_2$ Mo			
A369-FP3b	2 Cr - $^1/_2$ Mo	300°F to 350°F (149°C to 177°C)	300°F to 350°F (149°C to 177°C)	350°F (177°C)
A335-P22	$2^1/_4$ Cr - 1 Mo			
A335-P21	3 Cr - 1 Mo	300°F to 350°F (149°C to 177°C)	300°F to 350°F (149°C to 177°C)	400°F (204°C)
A335-P5	5 Cr - $^1/_2$ Mo	400°F (204°C)	400°F (204°C)	500°F (260°C)
A335-P9	9 Cr - 1 Mo			

Note:
Maximum carbon content of 0.15%. For higher content, preheat should be increased by 100°F to 200°F (38°C to 93°C).

When the welding consumables are a potential source of hydrogen, such as with SMAW, it may be necessary to raise the preheat temperature and hold the weldment at this elevated temperature for a period of time. This allows the hydrogen to escape before the base metal cools, which reduces the risk of cracking. To avoid potential problems, such as hard spots, that can occur when the heating cycle of a Cr-Mo base metal is interrupted, always plan to complete the heating cycle once it has started.

Some Cr-Mo steels containing less than 1.25% Cr and 0.05% Mo can be placed in service as-welded if a high preheat temperature was used and the section is relatively thin. Other Cr-Mo steels should be either stress relieved, annealed, or normalized and tempered.

PWHT of Cr-Mo is usually required to prevent stress cracking and brittleness. PWHT, also called **stress-relief heat treatment**, is used to reduce welding stresses and to increase the ductility and toughness of the weld metal and the **heat-affected zone (HAZ)**. The stress-relief temperature ranges from about 1,150°F to 1,400°F (621°C to 760°C) and depends on the alloy being treated. The entire weldment or the weld zone is generally held at the stress-relief temperature for one hour per inch of thickness. Cr-Mo steel is generally heated to 1,250°F to 1,300°F (675°C to 700°C) for stress relieving.

The procedures for annealing or normalizing and tempering are generally the same as those used for HTLA steels discussed earlier in this module.

Stress-relief heat treatment: A technique used to reduce welding stresses and to increase the flexibility and strength of the weld metal and the heat-affected zone (HAZ).

Heat-affected zone (HAZ): The part of the base metal that has not been melted but has been altered structurally by the weld heat.

1.3.0 | Stainless Steel

Stainless steels (corrosion-resistant steels) are iron-based alloys that normally contain at least 11% chromium. Depending on the specific type of stainless steel, other alloying elements, including nickel, carbon, manganese, and silicon, may be present in varying quantities to enhance its physical and **mechanical properties**. The main physical characteristics of all stainless steels are their resistance to corrosion and heat. Some have good low- and high-temperature mechanical properties.

When compared with mild steels, stainless steels have the following characteristics:

Mechanical properties: The characteristics or traits that indicate how flexible and strong a metal will be under stress.

- Lower **coefficient** of thermal conductivity that increases the chances of distortion
- Higher coefficient of thermal expansion that increases the chances of distortion
- Higher electrical resistance that increases the tendency to build up heat from welding current

Coefficient: A numerical measure of a physical or chemical property that is constant for a system under specified conditions, such as the coefficient of friction.

Stainless steels are classified by their grain structures. The type of grain structure is determined by the specific alloy content of the stainless steel and its heat treatment during manufacture. Based on their microcrystalline structures, the classification of stainless steels is divided into these five groups:

- Austenitic
- Ferritic
- Martensitic
- Precipitation Hardening
- Duplex

1.3.1 Austenitic Stainless Steels

Austenitic stainless steels are nonmagnetic in the annealed condition and nonhardenable by heat treatment. However, they can be hardened significantly by cold working. Austenitic stainless steels combine excellent corrosion and heat resistance with good mechanical properties over a broad temperature range. Austenitic stainless steels make up the largest of the three common grade stainless steel groups. For this reason, they are the type of stainless steel encountered most often by welders. They include the AISI 200 and 300 series stainless steels (*Table 7*) that all contain significant amounts of both chromium and nickel. Austenitic steels are sometimes further subdivided in two classifications based on their compositions: chromium-nickel (AISI 300 series) and chromium-nickel-manganese (AISI 200 series). The range of applications for austenitic stainless steels includes housewares, containers, industrial piping and vessels, and architectural facades.

TABLE 7 AISI Classification Numbers and Composition for Selected Stainless Steels

AISI No.	C (%)	Mn (%)	Si (%)	Cr (%)	Ni (%)	Other Elements (%)
Chromium-Nickel-Manganese-Austenitic (Nonhardenable)						
201	0.15	5.5–7.5	1.0	16.0–18.0	3.0–5.5	0.25 N
202	0.15	7.5–10	1.0	17.0–19.0	4.0–6.0	0.25 N
Chromium-Nickel-Austenitic (Nonhardenable)						
301	0.15	2.0	1.0	16.0–18.0	6.0–8.0	—
302	0.15	2.0	1.0	17.0–19.0	6.0–10.0	—
303	0.15	2.0	1.0	17.0–19.0	8.0–10.0	0.6 Mo
304	0.08	2.0	1.0	18.0–20.0	8.0–10.5	—
304L	0.03	2.0	1.0	18.0–20.0	8.0–12.0	—
308	0.08	2.0	1.0	19.0–21.0	10.0–12.0	—
309	0.20	2.0	1.0	22.0–24.0	12.0–15.0	—
310	0.25	2.0	1.5	24.0–26.0	19.0–22.0	—
316	0.08	2.0	1.0	16.0–18.0	10.0–14.0	2.0–3.0 Mo
316L	0.03	2.0	1.0	16.0–18.0	10.0–14.0	2.0–3.0 Mo
321	0.08	2.0	1.0	17.0–19.0	9.0–12.0	5 × %C Ti min.
347	0.08	2.0	1.0	17.0–19.0	9.0–13.0	0.80 Cb + Ta
348	0.08	2.0	1.0	17.0–19.0	9.0–12.0	0.80 Cb + Ta 0.10 Ta 0.20 Co
Chromium-Martensitic (Hardenable)						
403	0.15	1.0	0.5	11.5–13.0	—	—
410	0.15	1.0	1.0	11.5–13.5	—	—
414	0.15	1.0	1.0	11.5–13.5	1.25–2.5	—
416	0.15	1.0	1.0	12.0–14.0	—	0.15 S min.
420	0.15 min.	1.0	1.0	12.0–14.0	—	—
422	0.20–0.25	1.0	0.75	11.0–13.0	0.5–10	0.75–1.25 Mo 0.75–1.25 W 0.15–0.3 V
440A	0.60–0.75	1.0	1.0	16.0–18.0	—	0.75 Mo
440B	0.75–0.95	1.0	1.0	16.0–18.0	—	0.75 Mo
440C	0.95–1.2	1.0	1.0	16.0–18.0	—	0.75 Mo
Chromium-Ferritic (Nonhardenable)						
405	0.08	1.0	1.0	11.5–14.5	—	0.10–0.30 Al
409	0.08	1.0	1.0	10.5–11.75	—	6 × %C Ti min.
430	0.12	1.0	1.0	16.0–18.0	—	—
442	0.20	1.0	1.0	18.0–23.0	—	0.20 + 4 × (%C + %N) Ti
444	0.025	1.0	1.0	17.5–19.5	1.0	1.75–2.5 Mo 0.035 N 0.20 + 4.× (%C + %N) Ti + Nb min.
446	0.20	1.50	1.0	23.0–27.0	1.0	0.25 N
Martensitic						
501	0.11	1.0	1.0	4.0–6.0	—	0.4–0.65 Mo
502	0.10	1.0	1.0	4.0–6.0	—	0.4–0.65 Mo

Note:
Single values are maximum values unless otherwise indicated.

Austenitic stainless steels are also available in low-carbon (L-grade) and high-carbon (H-grade) types. The letter *L* after a stainless steel type, such as 304L, indicates a low-carbon content of 0.03% or under. Similarly, the letter *H* after a stainless steel type indicates a high-carbon content, ranging between 0.04% and 0.10%.

L-grades are typically used when annealing after welding is impractical, such as in the field where pipe and fittings are welded. H-grades are used when the steel will be subjected to extreme temperatures; the higher carbon content helps it retain strength.

1.3.2 Ferritic Stainless Steels

Ferritic stainless steels are always magnetic. They are hardened to some extent by cold working, not by heat treatment. Ferritic stainless steels combine corrosion resistance and heat resistance with fair mechanical properties over a narrower temperature range than austenitic steels. Ferritic stainless steels are straight chromium stainless steels containing 11.5% to 27% chromium, about 1% manganese, and little or no nickel. The carbon content is 0.20% or less. Examples of ferritic stainless steels include AISI types 405, 409, 430, 442, 444, and 446. They typically are used for decorative trim, sinks, and automotive applications, particularly exhaust systems.

1.3.3 Martensitic Stainless Steels

Martensitic stainless steels are magnetic and can be hardened by quenching and tempering. They are excellent for use in mild environments such as the atmosphere, freshwater, steam, and with weak acids. However, they are not resistant to severely corrosive solutions. Martensitic stainless steels comprise two groups: chromium martensitic and martensitic. The chromium martensitic stainless steels contain from 11.5% to 18% chromium, about 1% manganese, and in some cases 0% to 2.5% nickel. Examples of chromium martensitic stainless steels are AISI types 403, 410, 414, 420, 422, 431, and 440. Martensitic stainless steels contain only 4% to 6% chromium and no nickel. For this reason, they are not considered true stainless steels, although their corrosion resistance is much greater than that of mild-carbon steels, even at elevated temperatures. Examples of martensitic stainless steels are AISI types 501 and 502.

1.3.4 Precipitation-Hardening Stainless Steels

Precipitation-hardening (PH) stainless steels contain alloying additions, such as aluminum, that enable them to be hardened by a solution and aging heat treatment. Precipitation-hardening stainless steels were developed after World War II because the design and manufacture of jet aircraft required the use of stainless steels with a better weight-to-strength ratio. Precipitation-hardening stainless steels have designations such as 17-4PH, PH13-8MO, and AM350.

Why Stainless Steel Doesn't Rust

When steel comes in contact with oxygen and water vapor in the atmosphere, a chemical reaction occurs, and the steel begins to change to its original form of iron oxide. The scale that forms on the surfaces of metals when they are exposed to air is an oxide. The chromium added to the steel prevents it from rusting or staining, thus the name "stainless steel." The chromium reacts with the oxygen and water in the atmosphere to form a microscopic and transparent layer of chromium oxide called a *passive film*. The sizes of the chromium atoms and their oxides are about the same, so they join to form a tight, stable layer only a few atoms thick on the surface of the steel. If the stainless steel surface is cut or scratched, disrupting the passive film, more chromium oxide immediately forms and covers the exposed surface, protecting it from iron oxide (rust) again.

This is why after repeated use and sharpening, stainless steel cutlery remains bright and shiny. However, if the surface of stainless steel is contaminated by a ferrous metal, the ferrous material will rust.

1.3.5 Duplex Stainless Steels

Duplex stainless steels are supplied with approximately equal amounts of ferrite and austenite, thus the name "duplex." They also contain about 24% chromium and 5% nickel. Developed in the 1930s, with further innovation in the 1960s that increased resistance to stress corrosion cracking, duplex steels offer high yield strength. Duplex steels are valued for their superior resistance to pitting and stress cracking as compared to the 300 series stainless steels. One example of duplex stainless steel is Type 2205, which accounts for more than 80% of duplex use. Duplex stainless steels are used for heat exchangers, tubes, and pipes for applications involving the production and handling of gas and oils, and for desalinization.

Super duplex stainless steels are more highly alloyed with chromium, nickel, molybdenum, and nitrogen, resulting in higher strength and pitting corrosion resistance. These contain higher levels of chromium, molybdenum, and nitrogen. They were developed to provide outstanding resistance to acids, acid chlorides, and other caustic solutions typical of those encountered in the chemical, petrochemical, and pulp and paper industries.

1.3.6 Pickling and Passivation

Pickling and passivation are steps that can help clean and protect stainless steels from rust after the welding has been performed. Pickling and passivation involve the application of acid to remove contaminants and encourage a protective film to form on stainless steel. (Note, pickling and passivation do not remove grease or oil.)

When pickling steel, apply an acid solution to the weld and heat-affected zones. The resulting steel will look dull or etched as the surface oxide layer is removed. Passivation may occur naturally or may be done purposefully as a treatment. Passivation of steel can be done as a single process or after pickling is completed. When passivating steel, apply an oxidizing acid to dissolve carbon steel and remove contaminants from the surface. The resulting steel will have a chromium-rich passive film and resistance to corrosion.

Around the World

Recycled Steel

Steel is the world's most recycled material. According to the World Steel Association, more than 650 million tons of steel were recycled annually between 2009 and 2013.

1.4.0   Nonferrous Metals

Nonferrous metals include all metals except iron. Commonly used nonferrous metals include aluminum, magnesium, titanium, copper, nickel, zinc, tin, lead, and the precious metals. Nonferrous metals with densities lower than that of steel are considered light metals. They include, in order of increasing density, magnesium, beryllium, aluminum, and titanium. The heavier common nonferrous metals include copper, nickel, lead, tin, and zinc, as well as the precious metals platinum, gold, silver, and their alloys. Each of the nonferrous metals and alloys is classified by a different and unrelated classification system.

1.4.1 Aluminum Alloys

Pure aluminum is a soft, lightweight metal that can be used in its nearly pure form. Because aluminum is very soft and ductile with a low tensile strength, it is often alloyed with elements such as copper, manganese, silicon, magnesium, and zinc to make different alloys with specific properties for very different purposes. When aluminum is exposed to the atmosphere, a microscopic and transparent aluminum oxide layer forms immediately that protects the metal from further oxidation. This characteristic gives aluminum its high resistance to corrosion.

Aluminum is one-third the weight of steel and can be made very strong, ductile, and malleable. It is a good conductor of heat and electricity and is highly resistive to weathering and corrosion from many acids. However, aluminum can be corroded easily by alkalis because they attack the oxide layer. Aluminum is also noncombustible and nonmagnetic, so it is widely used in electrical shielding and near inflammable or explosive substances. It is decorative and easily formed, cast, and machined. Polished aluminum has the highest reflectivity of any material, including mirror glass.

Aluminum alloys are broadly classified as either casting alloys or wrought alloys for rolling, forging, or extruding. The Aluminum Association (AA) uses a three-digit plus one decimal system to designate each type of casting alloy (*Table 8*). The first digit (1 through 9) shows the main alloying element(s). The 1XX.X series is essentially pure aluminum (99.00% aluminum). Except for the 1XX.X series, the second and third digits identify the specific alloys. For the 1XX.X series only, the second and third digits give the degree of aluminum purity *above* 99.00%. For example, if the second and third digits are 60, the alloy contains a minimum of 99.60% aluminum. The decimal indicates if the alloy composition is for the final casting (0.0) or for ingot (0.1 or 0.2, depending on impurity limits). Modifications to the alloys are indicated using prefix letters such as A, B, C.

TABLE 8 AA System Code Designations for Basic Cast Aluminum Alloys

Designation	Major Alloying Element
1XX.X	Near pure aluminum
2XX.X	Copper
3XX.X	Silicon plus copper and/or magnesium
4XX.X	Silicon
5XX.X	Magnesium
6XX.X	Unused series
7XX.X	Zinc
8XX.X	Tin
9XX.X	Other element

Similarly, the AA uses a four-digit numbering system for wrought aluminum alloy types (*Table 9*) that is not related to the three-digit casting alloy designation system. In this wrought aluminum system, the first digit identifies the alloy group by the major alloying element. For the 1XXX group, the second digit designates modifications and impurity limits. A zero (0) indicates no special control on individual impurities. A second digit of 1 through 9 indicates special control over one or more impurities and is assigned consecutively. For groups 2XXX through 8XXX, the second digit indicates alloy modifications and is consecutively assigned. A second digit of 0 indicates the original alloy. The last two digits indicate the minimum percentage of aluminum.

TABLE 9 AA System Code Designations for Basic Wrought Alloy Groups

Designation	Major Alloying Element
1XXX	Nonaluminum elements are 1% or less
2XXX	Copper
3XXX	Manganese
4XXX	Silicon
5XXX	Magnesium
6XXX	Magnesium and silicon
7XXX	Zinc
8XXX	Other elements
9XXX	Unused series

The following examples explain AA wrought aluminum numbers:

- AA Number 1075:
    - 1 = Indicates an aluminum content of 99.0% or greater
    - 0 = Indicates no special control on the impurities
    - 75 = Indicates an aluminum content of 99.75%
- AA Number 1180:
    - 1 = Indicates an aluminum content of 99.0% or greater
    - 1 = Indicates impurities are limited
    - 80 = Indicates an aluminum content of 99.80%

In general, aluminum and its alloys can be welded by the following methods: oxyfuel or airfuel, electric arc, electrical resistance, and brazing.

Why Aluminum Must Be Cleaned Before Welding

The aluminum oxide coating protecting aluminum must be thoroughly cleaned from the metal before welding. The reason is that, while pure aluminum melts at about 1,200°F (650°C), the aluminum oxide coating that protects the metal does not melt until it reaches the much higher temperature of 3,700°F (nearly 2,040°C).

All cleaning materials used must be contained and disposed of in accordance with local environmental laws. Cleaning something as large as a boat presents a greater challenge when it comes to containing the runoff of cleaning materials.

Source: Photo courtesy of The Lincoln Electric Company, Cleveland, OH, U.S.A.

1.4.2 Magnesium Alloys

Magnesium is mainly used in alloy form. It is the lightest of the structural metals, having less than one-quarter the density of steel. It is mainly used in the aerospace and automotive industries. Magnesium is the eighth most common element in the world and the sixth most abundant metal, comprising about 2.5% of the earth's surface. Because seawater contains about 0.13% magnesium, some production facilities process seawater to obtain it. This is done after the precipitation of other salts from the seawater, which leaves a magnesium-rich brine. The uses of magnesium center on three properties of the metal: its ability to form intermetallic compounds with other metals, its high chemical reactivity, and its low density. Magnesium can be extruded, forged, rolled, and machined.

Magnesium alloys are broadly classified as casting alloys or wrought (rolled, forged, or extruded) alloys. Both classifications use the *ASTM Standard B951, Standard Practice for Codification of Unalloyed Magnesium and Magnesium-Alloys, Cast and Wrought*. Designations consist of one or two letters representing the alloying elements, followed by their respective percentages, and rounded off to the nearest whole number. The letters representing the alloying elements are as follows:

- *A* — Aluminum
- *B* — Bismuth
- *C* — Copper
- *D* — Cadmium
- *E* — Rare earths
- *F* — Iron
- *G* — Magnesium
- *H* — Thorium
- *K* — Zirconium
- *L* — Beryllium
- *M* — Manganese
- *N* — Nickel
- *P* — Lead
- *Q* — Silver
- *R* — Chromium
- *S* — Silicon
- *T* — Tin
- *Y* — Antimony
- *Z* — Zinc

Magnesium alloy designations also contain information indicating the temper condition. This part of the designation, separated by a hyphen from the main designation of letters and numbers, consists of a letter or a letter and number. The letters and numbers that indicate the temper are as follows:

- *F* — As fabricated
- *O* — Annealed
- *H10, H11* — Strain hardened
- *H23, H24, H26* — Strain hardened and annealed
- *T4* — Solution heat-treated
- *T5* — Artificially aged
- *T6* — Solution heat-treated and artificially aged
- *T8* — Solution heat-treated, cold worked, and artificially aged

In the alloy AZ92A-T6, the first A is aluminum, and the Z is zinc. The 9 indicates that the aluminum percentage is between 8.6 and 9.4. The 2 shows that the zinc percentage is between 1.5 and 2.5. The final A denotes that this is the first alloy to receive the AZ92 designation. The letter and number T6 indicate solution heat treatment and artificially aged tempering.

Magnesium and its alloys are best welded by the GTAW process for thin sections and by the GMAW process for thick sections. The composition of the filler metals should be compatible with the base metal. However, zinc above 1% increases **hot shortness** and can result in cracking. *Table 10* shows examples of magnesium alloys and their principal uses.

Hot shortness: The condition of metal when it proves to be very brittle and unbendable at red heat but can be bent without showing signs of brittleness when cold or at white heat. This condition is often a result of high sulfur and phosphorus content.

TABLE 10 Examples of Magnesium Alloys

ASTM Number	Nominal Alloy Composition						
	Al (%)	Mn (%)	Zn (%)	Zr (%)	Rare Earths (%)	Th (%)	Mg (%)
Sand and Permanent Mold Castings							
AZ92A-T6	9.0	0.10	2.0	—	—	—	Balance
AZ63A-T6	6.0	0.15	3.0	—	—	—	Balance
AZ81A-T4	7.6	0.13 min.	0.7	—	—	—	Balance
AZ91C-T6	8.7	0.13	0.7	—	—	—	Balance
EZ33A-T5	—	—	2.6	0.8	3.3	—	Balance
HK31A-T6	—	—	0.3	0.7	—	3.3	Balance
HZ32A-T6	—	—	2.1	0.8	—	3.3	Balance
Die Castings							
AZ91A-F	9.0	0.13	0.7	—	—	—	Balance
AM60A-F	6.0	0.13	—	—	—	—	Balance
AZ91D-F	9.0	0.13	0.7	—	—	—	Balance
Extrusions							
AZ61A-F	6.5	0.15	1.0	—	—	—	Balance
MIA-F	—	1.20	—	—	—	—	Balance
AZ80A-T5	8.5	0.12	0.5	—	—	—	Balance
AZ10A-F	1.3	0.20	0.4	—	—	—	Balance
AZ31B-F	3.0	0.20	1.0	—	—	—	Balance
ZK60A-F	—	—	5.5	0.45	—	—	Balance
Sheet and Plate							
AZ31B-H24	3.0	0.20	1.0	—	—	—	Balance
HK31A-H24	—	—	—	0.7	—	3.3	Balance

Note:
Single values are maximum values unless otherwise indicated.

Magnesium

Magnesium is widely used in specialty racing cars and commercial cars as a lightweight material to minimize weight and fuel consumption because of the 20% to 25% weight saving over aluminum. It typically is used in the construction of transmission casings, intake manifolds, and cylinder head covers. Magnesium pistons and various other engine parts are also being used in some race cars. It is difficult to visually differentiate aluminum and magnesium. Use positive metal identification (PMI) test equipment to verify these materials.

Source: Photo courtesy of The Lincoln Electric Company, Cleveland, OH, U.S.A.

1.4.3 Titanium Alloys

Titanium is mainly used in alloy form. Its alloys are high-strength, lightweight (about one-half the density of steel), corrosion-resistant structural metals that retain their strength at elevated temperatures. It is used in aircraft engines and structures, chemical processing, surgical implants, sporting goods, and marine and other applications where corrosion resistance is required.

Titanium is a reactive metal. At high temperatures, it readily combines with oxygen, hydrogen, and nitrogen to form stable compounds. Like aluminum and magnesium, titanium forms a tight, protective, microscopic oxide film on clean surfaces when in contact with oxygen at room temperature. This makes it highly resistant to corrosion at low temperatures. Based on the crystal structures that are stable at room temperature, titanium alloys are classified into three groups:

- *Alpha alloys* — Are the usual crystalline structures of pure titanium. Aluminum is added to stabilize the crystalline structure. These alloys have excellent strength and are oxidation-resistant from 600°F to 1,100°F (about 316°C to 593°C). They have good ductility and weldability and can be hardened by working, but not by heat treatment.

- *Beta alloys* — Can contain vanadium, chromium, and aluminum. Iron and chromium are added to make them heat-treatable, high-strength alloys. They have fair weldability.

- *Alpha-beta alloys* — Contain one or more of the elements including manganese, molybdenum, tin, vanadium, iron, and aluminum. They have a wide range of mechanical characteristics but have poor weldability.

Titanium is considered difficult to weld because the weld must be kept in an inert (nonreactive) gas atmosphere until it cools to below the reactive temperature. Small amounts of hydrogen or large amounts of oxygen or nitrogen can cause embrittlement in the heat-affected zone, which leads to breakability. Titanium is usually welded by the GTAW process.

About 34 different titanium alloys are currently being produced, including the ones shown in *Table 11*.

Titanium

Commercial production of titanium did not begin until the 1950s. At that time, titanium was recognized for its importance and was developed as a unique, lightweight, high-strength alloy for use in the engine and airframe components of high-performance jet aircraft. Today, titanium alloys are widely used in many applications that previously used metals such as stainless and specialty steels, copper alloys, and nickel-based alloys.

TABLE 11 Examples of Titanium Alloys

Designation	Nominal Alloy Composition				
	Al (%)	Sn (%)	Zr (%)	Mo (%)	Other (%)
Unalloyed Grades					
ASTM Grade 1	—	—	—	—	—
ASTM Grade 7	—	—	—	—	0.2 Pd
ASTM Grade 11	—	—	—	—	0.12–0.25 Pd
ASTM Grade 26	—	—	—	—	0.1 Ru
Alpha and Near-Alpha Alloys					
ASTM Grade 6	5.0	2.5	—	—	—
ASTM Grade 9	3.0	—	—	—	2.5 V
ASTM Grade 12	—	—	—	0.3	0.6–0.9 Ni
ASTM Grade 18	3.0	—	—	—	2.5 V 0.05 Pd
ASTM Grade 28	3.0	—	—	—	2.5 V 0.1 Ru
Alpha-Beta Alloys					
ASTM Grade 5	6.0	—	—	—	4.0 V
ASTM Grade 23	6.0	—	—	—	4.0 V
ASTM Grade 29	6.0	—	—	—	4.0 V 0.1 Ru
Beta Alloys					
ASTM Grade 19	3.0	—	4.0	4.0	8.0 V 6.0 Cr
ASTM Grade 20	3.0	—	4.0	4.0	8.0 V 6.0 Cr 0.05 Pd

1.4.4 Beryllium Alloys

Beryllium (Be) is a high-strength, lightweight metal. It is only about two-thirds as heavy as aluminum but is six times stiffer than steel. Beryllium's heat capacity is five times that of copper, meaning that 1 lb (0.45 kg) of it will absorb as much heat as 5 lb (2.3 kg) of copper. For this reason, it is widely used in aerospace and nuclear reactor applications.

Beryllium is a reactive metal, and it readily combines with oxygen, hydrogen, and nitrogen to form stable compounds at high temperatures. At low temperatures, it is highly resistant to corrosion. Small amounts of nitrogen and hydrogen or large amounts of oxygen cause embrittlement. *Table 12* lists beryllium alloys and some of their mechanical properties, such as tensile and yield strength.

Beryllium is considered a difficult metal to weld because it must be kept in an inert gas atmosphere while it is being welded.

> **WARNING!**

Dust, fumes, vapors, and fine machining particles of beryllium are cancer-causing agents. If inhaled, they can cause chronic beryllium disease. An independent air supply is required when working with beryllium.

TABLE 12 Examples of Beryllium Alloys and Mechanical Properties

Designation and Percentage of Be	Tensile Strength		Yield Strength		Commercial Form	Description
	(psi)	(kPa)	(psi)	(kPa)		
AlBeMet® (62%)	58,000	400,000	40,000	276,000	Extrusion, near net shapes	Hot isostatic pressed
S200F (98.5%)	47,000	324,000	35,000	241,000	Rod, bar, block	Vacuum hot pressed
S200FH (98.5%)	60,000	414,000	43,000	296,000	Rod, bar, block, near net shapes	Hot isostatic pressed
S200FC (98.5%)	38,000	262,000	25,000	172,000	Near net shapes	Cold isostatic pressed
S65 (99.2%)	42,000	290,000	30,000	207,000	Rod, bar, block	Vacuum hot pressed
S65H (99.2%)	50,000	345,000	30,000	207,000	Rod, bar, block, near net shapes	Hot isostatic pressed
I-70-H (99.0%)	50,000	345,000	30,000	207,000	Rod, bar, block, near net shapes	Hot isostatic pressed
I-220-H (98.0%)	65,000	448,000	50,000	345,000	Rod, bar, block, near net shapes	Hot isostatic pressed
C17200 (2%)	85,000	590,000	130,000	890,000	Extrusion, rod, bar, plate	Many tempers available

Beryllium

Beryllium allows the transmission of X-rays better than other metals or glass. Because of this property and its high melting point, it is widely used as the window portion of high-intensity X-ray tubes, such as those used in high-resolution X-ray machines and imaging equipment.

1.4.5 Copper and Copper Alloys

Copper, like aluminum, is often used in a nearly pure form for electrical and heat conductors. It is alloyed with zinc to make brass, and with tin to make bronze. Other bronzes are formed by alloying copper with aluminum, silicon, or beryllium. Many variations of alloys exist within each of the brass and bronze groups, made by varying the percentages of the alloying metals and by adding other metallic and nonmetallic elements. These include zinc, tin, nickel, silicon, aluminum, cadmium, and beryllium.

Copper and copper alloys are classified into the following major groups:

- *Coppers* — Coppers are metals that have a designated minimum copper content of 99.3% or higher.

- *High-copper alloys* — For the wrought-type products, these are alloys with a copper content of less than 99.3% but more than 96% that do not fall into any other copper alloy group. Cast-type high-copper alloys have a copper content in excess of 94% to which silver may be added to obtain special properties.

- *Brasses* — These alloys contain zinc as the principal alloying element, with or without other alloying elements. The wrought-type alloys are grouped into three categories of brasses: copper-zinc alloys, copper-zinc-lead alloys (leaded brasses), and copper-zinc-tin alloys (tin brasses). Cast alloys are grouped into five categories of brasses: copper-tin-zinc alloys (red, semi-red, and yellow brasses), manganese bronze alloys (high-strength yellow brasses), leaded manganese bronze alloys (leaded high-strength yellow brasses), copper-zinc-silicon alloys (silicon brasses and bronzes), and cast copper-bismuth and copper-bismuth-selenium alloys.

- *Bronzes* — Bronzes are copper alloys in which the major alloying element is not zinc or nickel. For wrought-type alloys, there are four groups of bronzes: copper-tin-phosphorus alloys (phosphor bronzes), copper-tin-lead-phosphorus alloys (leaded phosphor bronzes), copper-aluminum alloys (aluminum bronzes), and copper-silicon alloys (silicon bronzes). Cast-type alloys are divided into four groups of bronzes: copper-tin alloys (tin bronzes), copper-tin-lead alloys (leaded and high-leaded tin bronzes), copper-tin-nickel alloys (nickel-tin bronzes), and copper-aluminum alloys (aluminum bronzes). Alloys known as manganese bronzes, in which zinc is the major alloying element, are included in the brasses.

- *Copper-nickels* — Copper-nickel alloys have nickel as the principal alloying element, with or without other alloying elements.

- *Copper-nickel-zinc alloys* — These alloys contain zinc and nickel as the principal and secondary alloying elements, with or without other elements. They are commonly called *nickel silvers*.

- *Leaded coppers* — Leaded coppers comprise a series of cast alloys of copper with 20% or more lead, sometimes with a small amount of silver, but without tin or zinc.

- *Special alloys* — These are alloys whose chemical compositions do not fall into any of the above categories.

Two classification systems for copper alloys are the *ASTM Standard B224, Standard Classification of Coppers* (*Table 13*) and the Copper Development Association, Inc. designation system (*Table 14*). The Copper Development Association system has been updated to fit the UNS and includes all commercially available metals and alloys. The format for the UNS designation is CNNN00, in which NNN designates numbers. As shown in *Table 14*, the alloys are divided into wrought and cast alloy categories. For this reason, an alloy made in both a wrought and cast form can have different UNS numbers, depending on the method of manufacture.

TABLE 13 ASTM Classification System for Copper Alloys

ASTM Designation	Alloy Type	Typical Uses
Tough-Pitch Coppers		
ETP	Electrolytic, tough-pitch	Bus bars, brazing rods, wire anodes, forgings
FRHC	Fire-refined, high-conductivity, tough-pitch	Mechanical applications
FRTP	Fire-refined, tough-pitch	Sheets, strips, plate
ATP	Arsenical, tough-pitch	Roofing, radiator cores
STP	Silver-bearing, tough-pitch	Pans, printing rolls, fasteners
SATP	Silver-bearing, arsenical, tough-pitch	Pans, printing rolls, fasteners
Oxygen-Free Coppers		
OFH	Oxygen-free (without residual deoxidants)	Tubing wave guides, starting anodes, wire
OFP	Oxygen-free, phosphorus-bearing	Welding rods, forging
OFTPE	Oxygen-free, tellurium- and phosphorus-bearing	Welding rods, forging
OFS	Oxygen-free, silver-bearing	Plate, sheets, rods, forging
OFTE	Oxygen-free, tellurium-bearing	Bars, free machining
Deoxidized Coppers		
DHP	Phosphorized, high-residual phosphorus	Tubes, pipes, anodes, projectile rotating bands
DLP	Phosphorized, low-residual phosphorus	Tubes, wave guides, general use
DPS	Phosphorized, silver-bearing	Heat exchangers, steam lines, condenser tubes, tubes for general use
DPA	Phosphorized, arsenical	Heat exchangers, steam lines, condenser tubes, tubes for general use
DPTE	Phosphorized, tellurium-bearing	Free machining

TABLE 14 UNS Designations for Classifying Copper Alloys

Copper Designation	Composition
Wrought Coppers	
C11X00	99.95% oxygen-free, high-conductivity copper
C12X00	99.88% copper or more (tough-pitch copper)
C19X00	96% copper or more (high-copper alloys)
C2XX00	Copper-zinc alloys (brasses)
C3XX00	Copper-zinc-lead alloys (leaded brasses)
C4XX00	Copper-zinc-tin alloys (tin brasses)
C51X00	Copper-tin alloys (phosphor bronzes)
C54X00	Copper-tin-lead alloys (leaded phosphor bronzes)
C62X00	Copper-aluminum alloys (aluminum bronzes)
C65X00	Copper-silicon alloys (silicon bronzes)
C70X00	Copper-nickel alloys
Cast Alloy Groups	
C80X00	99% copper or more, copper alloys
C81X00	High-copper alloys (beryllium copper)
C83X00	Red brasses and leaded red brasses
C85X00	Yellow brasses and leaded yellow brasses
C86X00	Manganese and leaded manganese bronze alloys
C87X00	Silicon bronzes and brasses
C90X00	Tin bronzes

Tough-pitch copper is oxygen bearing, and most of the impurities have been oxidized during refining to increase conductivity. Continuous casting methods have allowed the oxygen content to be reduced to 0.03% or less to achieve the desired results. During fire refining, the oxygen coming from the air that was injected to cause the oxidation of impurities also combines with hydrogen to form steam. The steam creates gas pockets (porosity) in the metal during solidification. (In the past, high porosity was a characteristic of tough-pitch copper.) When tough-pitch copper is reheated by welding or annealing, the hydrogen trapped in the copper is diffused into the metal and reacts with cuprous oxide to form insoluble water vapor. The expansion associated with this reaction forces the grains apart, which causes embrittlement of the metal. For applications where copper must be resistant to gassing, the oxygen is sometimes eliminated by adding a deoxidizer such as phosphorus to the molten metal before casting. Such copper is specified as deoxidized copper.

For welding copper alloys, GMAW and GTAW are most commonly used.

Statue of Liberty

The Statue of Liberty contains 179,000 lb (81,193 kg) of copper. When copper oxidizes, it forms a blue-green self-protective coating, which is why the copper skin of the Statue of Liberty remained virtually intact after 100 years of being subjected to high sea winds, driving rains, and scorching sun. Close analysis showed that weathering and oxidation of the copper skin had caused only 0.005" (0.13 mm) of wear in a century. For this reason, the copper skin did not need to be significantly rebuilt when the statue was renovated for its centennial. However, high-alloy copper saddles and rivets were used during the restoration to fasten the copper skin to the framing underneath. The outline of the copper sheets can be seen in this image. This was done to ensure the

Source: U.S. National Park Service

structural integrity of the statue and guard against any galvanic reaction problems. Galvanic action is corrosion caused by an electrical current produced as the result of a chemical reaction between two dissimilar metals.

1.4.6 Nickel and Nickel Alloys

Nickel, a corrosion-resistant metal, is used as an alloying element as well as for plating other metals. Nickel-clad steel and nickel-based alloys containing over 50% nickel are used extensively in industry. They have good corrosion resistance and elasticity, as well as unique magnetic thermal expansion and thermal conductivity. Many high-strength, high-temperature nickel super alloys have been developed for aerospace, automotive, and hot-die applications. Alloys of nickel and chromium (Nichrome®) are used to make heating elements of high electrical resistance.

UNS and international standard designations have been assigned to many nickel alloys. However, trade names, such as Monel® 400, Inconel® 702, and Chromel® are more commonly used to identify them. *Table 15* lists some common nickel alloys and some of their uses and properties.

TABLE 15 Common Nickel Alloys and Some of Their Uses

Alloy	Ni (%)	C (%)	Mn (%)	Fe (%)	S (%)	Si (%)	Cu (%)	Cr (%)	Al (%)	Ti (%)	Other (%)	Typical Properties and Uses
Commercially Pure Nickel												
N02200	99.0	0.15	0.35	0.40	0.01	0.35	0.25	—	—	—	—	Chemical industry, electroplating
Nickel-Copper Alloys												
Monel® K-500	66.0	0.12	0.90	1.35	0.005	0.15	31.50	—	2.80	0.50	—	Corrosion resistance, toughness, high strength
Monel® 400	63.0	0.3	2.0	2.6	0.024	0.5	31.0	—	—	—	—	Corrosion resistance, high strength
Nickel-Iron Alloys												
Invar	36.0	—	—	63.0	—	—	—	—	—	—	—	Very low expansion coefficient
Platinite	46.0	0.15	—	53.85	—	—	—	—	—	—	—	Glass seal
Permalloy	80	—	—	20	—	—	—	—	—	—	—	High magnetic permeability, submarine telegraph cables
Nickel-Chromium and Nickel-Iron-Chromium Alloys												
Inconel® 702	78.0	0.02	0.05	0.30	0.007	0.15	0.05	15.85	3.0	0.60	—	Age hardenable, operates up to 2,400°F (1,316°C)
Inconel® X-750	72.9	0.04	0.70	6.8	0.007	0.30	0.05	15.0	0.80	2.5	0.9 Nb + Ta	Age hardenable, used up to 1,500°F (816°C)
Nimonic® 80A	74.5	0.05	0.55	0.55	0.007	0.20	0.05	20.0	1.30	2.50	—	Age hardenable, similar to Inconel X® X-750
Nimonic® 90	57.0	0.05	0.50	0.95	0.007	0.20	0.05	20.45	1.65	2.60	17.00 Co	Similar to Nimonic® 80A
Nichrome®	80.0	0.05	0.01	0.5	—	0.20	—	20.0	—	—	—	Heating elements
Chromel®	90.0	—	—	—	—	—	—	10.0	—	—	—	Heating elements, thermocouples
Incoloy® 901	42.5	0.04	0.45	33.75	0.007	0.30	0.5	13.50	0.25	2.5	5.90 Mo	Age hardenable, gas turbine wheels
Incoloy® 825	41.35	0.03	0.65	31.65	0.007	0.35	1.80	20.2	0.20	0.90	3.10 Mo	Resistant to certain hot acids
60Ni-12Cr	60	0.55	—	25.0	—	—	—	12.0	—	—	—	Resistance heating element

In general, nickel and nickel-based alloys can be welded by SMAW, GMAW, and GTAW processes. Welding nickel alloys is very similar to welding austenitic stainless steel, with the following exceptions:

- When welding nickel alloys, the surface oxide melts at a much higher temperature than the base metal, so surfaces should be thoroughly cleaned.
- Embrittlement can be caused by lead, sulfur, phosphorus, or other low-temperature metals. The surface should be free of all contaminants.
- Weld penetration in nickel-based alloys is shallower than in other metals and requires the use of very wide bevel and groove angles.

1.4.7 Lead, Tin, and Zinc

Lead, tin, zinc, and their alloys have very low melting points. For this reason, they are usually soldered and not welded. Zinc-plated, or *galvanized*, steels can be welded, but the fumes are toxic and must not be inhaled.

1.4.8 Platinum, Gold, Silver, and Other Precious Metals

The precious metals are osmium, iridium, rhodium, platinum, ruthenium, gold, palladium, silver, and their alloys. They are used industrially for electrical contacts, mirror coatings, catalysts, electrodes, high-temperature crucibles, and dental structures and fillings. They are usually brazed, soldered, or welded with special equipment.

1.0.0 Section Review

1. A document that a customer receives from a steel manufacturer that indicates the quality of the steel being purchased is a(n) _____.
   a. safety data sheet (SDS)
   b. ferrous metal classification bill (FMCB)
   c. certified mill test report (CMTR)
   d. evaluated properties rating index (EPRI)

2. The amount of alloy added to make HSLA steel is typically _____.
   a. less than 2%
   b. about 5%
   c. between 6% and 8%
   d. about 10%

3. Ferritic stainless steels are *always* _____.
   a. martensitic
   b. high in nickel content
   c. hardenable by tempering
   d. magnetic

4. What nonferrous metal has the *lowest* density of the structural metals and is commonly used in the aerospace industry?
   a. Magnesium
   b. Aluminum
   c. Nickel
   d. Titanium

2.0.0 Metal Characteristics

Performance Tasks

There are no Performance Tasks in this section.

Objective

Describe the physical and mechanical characteristics of metals and explain how to identify base metals.

- a. Describe the physical characteristics of different metals.
- b. Describe the mechanical properties of different metals.
- c. Explain how to identify base metals in field conditions.
- d. Describe metallurgy-related considerations for welding.

All metals have different characteristics and mechanical properties. These physical and mechanical properties can be used to identify a particular metal. In addition, a metal's properties must be taken into consideration prior to and during welding operations.

2.1.0 Physical Characteristics of Metals

Physical characteristics are distinctive for every metal. They include the following variable traits:

- Density (specific gravity)
- Electrical conductivity
- Thermal conductivity
- Thermal expansion
- Melting point
- Corrosion resistance

2.1.1 Density

Density is the mass (weight) of a specific volume of metal. Density may also be expressed as *specific gravity* or *specific density*. Specific gravity is the weight of a metal compared to the weight of an equal volume of water. The higher the density of a material, the heavier it will be.

Light metals, such as aluminum and magnesium, have low densities (low weights for a given volume) while ferrous metals, such as carbon steel and stainless steel, have much higher densities. An iron or steel item weighs nearly three times as much as an aluminum item of the same size. *Table 16* lists the densities of some common metals in pounds per cubic inch (lb/in^3) and grams per cubic centimeter (g/cm^3).

TABLE 16 Densities of Some Common Metals

Metal	Density (lb/in^3)	Density (g/cm^3)
Aluminum	0.098	2.71
Copper	0.324	8.96
Gold	0.698	19.32
Iron or steel	0.284	7.87
Magnesium	0.063	1.74
Manganese	0.267	7.43
Molybdenum	0.369	10.20
Nickel	0.309	8.90
Tin	0.264	7.30
Titanium	0.163	4.54
Tungsten	0.697	19.30
Zinc	0.258	7.13

2.1.2 Electrical Conductivity

Electrical conductivity is a metal's ability to conduct electricity. All metals can conduct electricity to some degree. Silver is one of the best conductors, but it is too expensive to use for commercial power transmission. Copper is not quite as good a conductor as silver, but it is a better conductor than aluminum for the same diameter conductor or wire. However, for conductors weighing the same per unit of length, aluminum will carry more current than copper because it has a larger cross section.

The electrical conductivity of a metal is calculated from its *resistivity*, measured resistance to current flow. Resistivity is the inverse (opposite) of conductivity. The lower the resistivity value, the higher the conductivity value and the more current a metal can carry for its size. The resistivity of any metal increases as its temperature rises while its conductivity decreases. This means that metals with higher electrical resistances build up heat from welding current faster. They also do not transfer heat away as efficiently, resulting in higher distortion and stresses in the weld zone. *Table 17* lists the electrical resistivity of some common metals in microhms per cubic centimeter.

2.1.3 Thermal Conductivity

Thermal conductivity is a metal's ability to conduct heat. Metals that are good conductors of electricity are also good conductors of heat. As temperature increases, a metal's resistance to the flow of heat increases. Because a metal with poor heat conductivity does not carry the weld heat away from the weld zone efficiently, the temperature rises and distortion is greater. Silver and copper are excellent heat conductors, iron and nickel are fair conductors, and titanium and manganese are poor conductors.

2.1.4 Thermal Expansion

Thermal expansion is the change in size that occurs in a material as its temperature changes. Solids expand in all dimensions when heated and contract from all dimensions when cooled. The increase in unit length when a solid is heated one degree is called its *coefficient of linear expansion*. The larger the coefficient of expansion, the greater the dimensional increase of the solid for each degree of temperature increase. Large dimensional changes cause distortion and stresses in weldments. *Table 18* lists the coefficients of linear expansion for a variety of metals. *Figure 8* shows the expansion of common metal pipes in inches per 100' for a range of temperatures in degrees Fahrenheit.

TABLE 17 Electrical Resistivity of Some Common Metals

Metal	Electrical Resistivity (microhm cm)
Aluminum	2.92
Copper	1.67
Gold	2.19
Iron or steel	9.71
Magnesium	4.46
Manganese	185.00
Molybdenum	5.17
Nickel	6.84
Titanium	40.00
Zinc	5.91

TABLE 18 Coefficients of Linear Expansion of Some Common Metals

Metal	Coefficient α (10^{-6}/°F)	Coefficient α (10^{-6}/°C)
Aluminum	12.8	23.1
Copper	9.2	16.5
Iron	6.6	11.8
Magnesium	13.8	24.8
Manganese	12.1	21.7
Molybdenum	2.7	4.8
Nickel	7.4	13.4
Titanium	4.8	8.6
Zinc	16.8	30.2

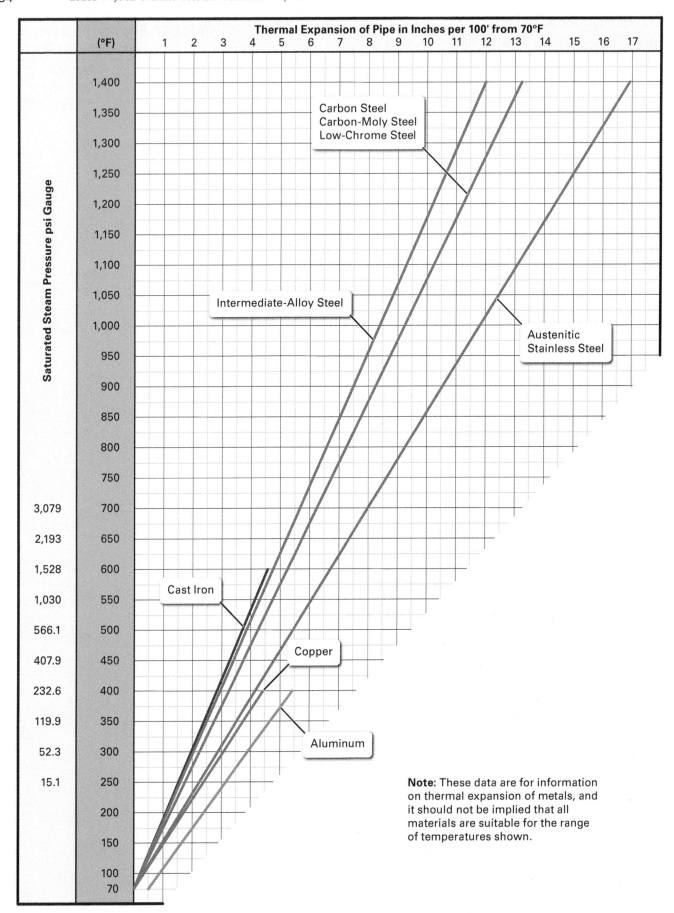

Figure 8 Expansion of different types of pipe.

2.1.5 Melting Point

The melting point of a solid is the temperature at which the solid converts into a liquid. *Table 19* lists a number of metals and their melting points.

TABLE 19 Melting Points of Some Common Metals

Metal	Melting Point
Aluminum	1,215°F (657°C)
Copper	1,981°F (1,083°C)
Iron	2,800°F (1,538°C)
Magnesium	1,202°F (650°C)
Manganese	2,273°F (1,245°C)
Molybdenum	4,760°F (2,627°C)
Nickel	2,651°F (1,455°C)
Titanium	3,300°F (1,816°C)
Zinc	787°F (419°C)

2.1.6 Corrosion Resistance

Metals and alloys are corroded by different agents at varying rates. Corrosion is important to consider because it can severely reduce the tensile strength of a metal. The tensile strength, which is pounds per square inch (psi), is a fixed material property. What will change is that the corrosion reduces the thickness and thereby lowers the cross-sectional area (square inch). As a result, less force is needed to bring the piece to failure. The tensile strength remains the same, but the wall thicknesses is less, thereby less force for failure. Certain metals may not react to specific chemicals. For example, iron is often used to handle strong caustics. Other metals produce a protective oxide coating that retards further corrosion. Both aluminum and lead form weather-resistant oxide coatings in the atmosphere. The stainless steels and nickel-based alloys are designed to withstand corrosion from such things as the atmosphere, elevated temperatures, and strong chemicals.

2.2.0 Mechanical Properties of Metals

The mechanical properties of metals determine how they will react to or be affected by the external forces of tension, compression, torsion, shear, impacts, and cold shaping (working). The ability of metals to permanently change or deform is called *plasticity*. Plasticity is different from *elasticity*, which refers to the ability of a material to change temporarily and then revert to its original form.

2.2.1 Stress-Strain Relationship

Stress is the force or load applied to a specimen that causes it to deform. Strain is the magnitude of deformation caused by stress. Within certain limits, strain increases directly with stress. The four types of stress are as follows:

- *Tension* — Opposed, in-line forces pulling away from each other at opposite sides of a specimen

- *Compression* — Opposed, in-line forces pushing toward each other against opposite sides of a specimen

- *Torsion* — Opposed, separated, twisting forces acting on the same axis of a specimen

- *Shear* — Opposed, nonintersecting forces acting opposite each other and perpendicular to a specimen

Figure 9 shows the force directions that result from different types of stress. Often, several different stress types act on a specimen at the same time.

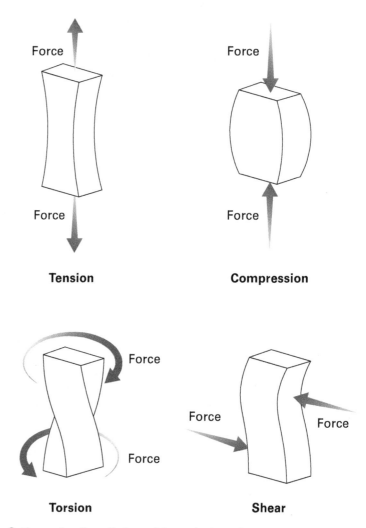

Figure 9 Force directions that result in each stress type.

2.2.2 Elasticity and Elastic Limit

Elasticity is the ability of a material to be deformed temporarily. If a material remains within its elastic range, the material will return to its original size and shape when the deforming force (stress) is removed. While within the elastic range, the strain will be directly proportional to the stress that produces it.

Every material has its elastic limit. When the elastic limit is exceeded, the strain is no longer proportional to the stress, and the material will not return to its original size or shape. The material has been distorted. Elastic limits are commonly measured by an extensometer.

2.2.3 Modulus of Elasticity

Every material strains (deforms) in direct proportion to the applied stress while it remains within its elastic range, but the amount each material strains for a given stress is not the same. The ratio of stress to strain is called the *modulus of elasticity*. It remains the same for any material under the same proportional conditions. For example, a material that will stretch 1" (25.4 mm) under a 10 lb (4.5 kg) tension will stretch 2" (50.8 mm) under a 20 lb (9 kg) tension, if the elastic limit is not exceeded. Similarly, a material that will stretch $1/4$" (6.4 mm) under a 10 lb (4.5 kg) tension will stretch $1/2$" (12.7 mm) under a 20 lb (9 kg) tension as long as the elastic limit is not exceeded. The modulus of elasticity of a metal can be used to predict how much the metal will deform under a specified load, provided the elastic limit is not exceeded.

2.2.4 Yield Strength and Tensile Strength

Yield strength is the maximum tensile stress a material can withstand before irreversibly deforming. Once stressed beyond its yield strength, the material will not return to its original shape when the stress is removed. Tensile strength, on the other hand, is the maximum tensile stress a material can withstand without breaking. The maximum load on a tensile material when it fractures is called the *ultimate load*. Brittle materials tend to break at values near their yield strength, while more elastic materials may deform significantly before breaking. Both yield and tensile strength are expressed in force per cross-sectional unit area. Engineers are interested in both values during the design process.

2.2.5 Ductility

Ductility is the plasticity exhibited by a material under tension loading. Ductility is measured by dividing the change in length of a material by the original length at the point of failure. Ductility is the characteristic that allows metals to be drawn into thinner sections without breaking. Copper and aluminum are examples of common ductile metals.

2.2.6 Hardness

The hardness of a material can be defined as its resistance to indentation. It is related to its elastic and plastic properties. The greater the hardness, the greater the resistance to indentation, penetration, and wear. Hardness is measured by pressing a steel ball or diamond point against a surface with a calibrated force. The diameter or depth of the impression is then measured to determine the hardness. Standard hardness testers include the Brinell, Rockwell, and Vickers instruments. *Figure 10* shows a portable hardness tester suitable for testing on the jobsite.

The Brinell hardness test is performed by pressing a hardened 10 mm steel ball into the metal being tested. Force is applied by standard weights: 500 kg for aluminum, copper, and other soft materials; and 3,000 kg for iron, steel, and other hard metals. The diameter of the indentation is measured under a microscope and then converted to the Brinell hardness number.

The Rockwell hardness tester uses a $\frac{1}{16}$" (1.6 mm) hardened steel ball with a 100 kg weight on soft metals, and a diamond cone and 150 kg weight on hard metals. Depth of penetration is read directly from hardness scales on the tester.

The Vickers hardness tester uses a pyramid-shaped diamond penetrator with a variable 1 kg to 120 kg load. The hardness is read from standard tables by comparing the width of the dent with the pressing load.

Figure 10 Hardness tester based on Rockwell testing criteria.
Source: The L.S. Starrett Company

2.3.0   Field Identification of Base Metals

Many of the common base metals and alloys can be generally identified by simple tests and observation of their characteristic metallic features. This is especially true of the ferrous metals, which can be identified by magnetic attraction, surface appearance, structural form or shape, and grinding sparks.

2.3.1 Metal Labeling

Because it is difficult to identify base metal types, most sites have a system to label or color-code the various types of metals being used. Labeling is performed by printing the ASTM number on the material (shape or plate) with an indelible marker or paint stick. Color coding involves assigning a color to a base metal type and then spraying that color on the material. When color coding is used, a master chart is often posted near the storage area. The master chart identifies the color assigned to each base metal.

For stainless steel and nickel alloys, special care should be taken to mark only with low-chloride markers. Using markers other than low-chloride markers will leave a residue on the metal that can contaminate the weld. Specialty markers, which usually contain white or yellow ink, are available for marking metals such as stainless steel (*Figure 11*).

Figure 11 A typical low-chloride marker for specialty metals.
Source: ITW Pro Brands

NOTE

Check your site quality standards for the type of material labeling that is used at your location. Follow those standards to ensure material traceability. When cutting a piece of material that is labeled or color-coded, it is very important to make sure that the label or color code is visible on the material that is being left behind. If possible, make the cut in such a way as to leave the label or color code still visible. If this is not possible, be sure to mark the material being left behind to ensure that others can correctly identify the material. It is also a good idea to mark the material being removed so that everyone involved can identify it as the material specified for the job. Some codes require that a witness be present when marking the two pieces.

2.3.2 Identification by Magnet

All carbon steels, cast steels, and cast irons are attracted to a magnet. Of the stainless steels, ferritic, martensitic, duplex, and precipitation-hardening stainless steels are magnetic. Austenitic stainless steels can be magnetized. Cold working or welding austenitic stainless steels can change the crystal structure, causing austenite to be converted to the ferromagnetic martensite or ferrite forms of iron, thus making it magnetic.

2.3.3 Identification by Appearance

Both hot-rolled and cold-finished carbon steel are milled in standard structural shapes, such as rods, bars, beams, channels, and angles. Unweathered, hot-rolled carbon steel has a blue-gray, smooth-to-scaly surface, but it oxidizes to a rough, red-brown rust when exposed to the weather. The corners of hot-rolled shapes are slightly rounded or chamfered.

Unweathered, cold-rolled carbon steel has a smooth, gray finish that oxidizes to a fine red-brown rust when weathered. Except for round shapes, the sides and edges are very flat and smooth, and the corners are well formed, not rounded or chamfered.

Cast steel and cast iron look similar on the surface. They both exhibit the shape characteristics typical of castings: curved surfaces and rounded edges, thin and thick regions, reinforcing ribs or raised areas, spokes, and irregular shapes. The cast surfaces are usually slightly textured and can contain mold seams. When broken, cast iron has a grainy appearance, while cast steel has a smooth texture. Cast iron is used to make everything from water meter covers to engine crankshafts and fan blades (*Figure 12*).

CAUTION

Magnetic testing is sometimes unreliable because of alloys or changes that can occur during welding. Do not rely solely on magnetic testing.

Figure 12 A compressor's cooling fan blade made of cast iron.

There are three types of cast iron: gray, white, and malleable. The exterior surface of all the cast irons has a slightly irregular texture that comes from the use of sand molds. Malleable iron is sometimes forged. If it has been forged, it will have a smoother appearance and die marks.

Gray cast iron contains a high percentage of carbon and silicon. Gray cast iron is only formed by slow cooling, which forms large crystals and a soft, easily machined metal. During the slow cooling process, the high silicon content forces the carbon to separate out in the form of graphite flakes (free carbon). This gives a fresh fracture its characteristic grainy and gray appearance. If you rub a clean finger or white cloth against a freshly fractured gray cast iron surface, the graphite will leave a dark smear. Gray cast iron can be welded by following special procedures.

White cast iron is chilled rapidly, which forms small crystals and a very hard and brittle metal. The carbon is united with the iron and is not free carbon. Freshly fractured cast iron has an irregular, fine, silvery-white, crystalline surface. Because the weldability of white cast iron is poor, it is generally not welded.

Malleable cast iron is white cast iron that has been annealed. The annealing process results in a casting with an outer shell of white cast iron that grades into a core of gray cast iron. The resulting casting is tougher and stronger than gray cast iron and less brittle than white cast iron. Malleable cast iron can be welded, but special procedures must be used or it will revert back to white cast iron.

Identification of Nonferrous Metals

The electrical conductivity measuring instrument shown can be used to determine the composition of nonferrous metals, including titanium alloys and silver. By using eddy currents to determine the electrical conductivity of a nonferrous metal under test, conclusions can be made about the composition, structure, and condition of the material.

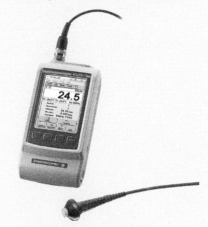

Source: Fischer Technology, Inc.

2.3.4 Identification by X-Ray Fluorescence Spectrometry

Portable metal analyzers that use X-ray fluorescence (XRF) spectrometry provide a fast and accurate analysis for metal and metal alloy identification. An X-ray of sufficient energy is emitted from a miniaturized X-ray tube or from a small sealed radioactive isotope. A fluorescent X-ray is created when the energy strikes an atom in the sample, dislodging an electron from one of the atom's inner orbital shells. The atom regains stability, filling the vacancy left in the inner orbital shell with an electron from one of the atom's higher energy orbital shells. The electron drops to the lower energy state by releasing a fluorescent X-ray, and the energy of this X-ray is equal to the specific difference in energy between two quantum states of the electron.

When a sample is measured using X-ray fluorescence, each element present in the sample emits its own unique fluorescent X-ray energy spectrum. By simultaneously measuring the fluorescent X-rays emitted by the different elements in the sample, handheld XRF analyzers can rapidly determine those elements that are present in the sample, along with their relative concentrations—in other words, the elemental chemistry of the sample. For samples with known ranges of chemical composition, such as common grades of metal alloys, the XRF analyzer also identifies most sample types by grade name through comparing the analyzed concentrations with known values contained in an alloy library programmed within the analyzer. This entire process is typically completed in less than five seconds and is completely nondestructive.

Thirty or more elements may be analyzed simultaneously by measuring the characteristic fluorescence X-rays emitted by a sample. *Figure 13* shows a handheld analyzer, its screen, and the analyzer in use.

NOTE

Analyzing samples with known ranges of chemical composition is NOT full chemical analysis. The process is not an absolute chemical test but rather a type of positive material identification confirming what you should already know. There are known limitations to quantifying lighter elements with this approach. For example, the handheld XRF analyzer cannot accurately measure any element lighter than magnesium, including lithium, beryllium, and carbon.

2.3.5 Identification by Laser Induced Breakdown Spectroscopy

The Laser Induced Breakdown Spectroscopy (LIBS) analyzer is another relevant technology. While the XRF analyzer cannot be used to detect carbon, the LIBS analyzer has this capability. The LIBS analyzer can be used to confirm alloy type and grade of stainless steels and low-alloy steels including carbon analysis down to 70 ppm. The LIBS analyzer performs instant carbon equivalent (CE) calculations for welding as well as material verification pre- and post-fabrication and validating mill test reports.

The LIBS analyzer operates by using a pulsed, focused laser that is fired at a sample with sufficient pulse energy as to create a plasma around the area struck. Bound atomic electrons are striped from the atoms comprising the material. As the plasma cools, atoms recombine with electrons and emit light in the ultraviolet (UV), optical and infrared (IR) regimes in the process. The LIBS analyzer has been used for years as a laboratory technique, capable of analyzing any element in the periodic table. Recently, the technique has been miniaturized into a handheld device capable of analyzing any element, depending on the spectrometer range chosen for the device.

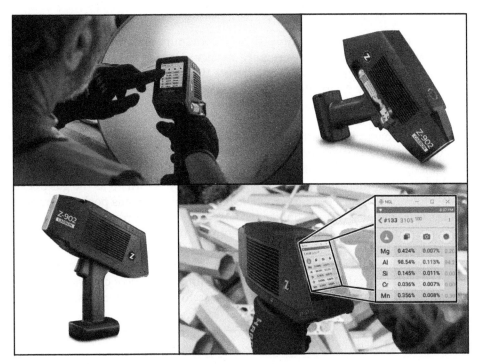

Figure 13 SciAps Z-902 handheld analyzer.
Source: SciAps

2.4.0 Metallurgical Considerations for Welding

Many metallurgical variables must be considered when determining the effects of welding on both the base metal and the weld. Base metal composition and thickness, filler metal, and joint configuration all affect the weld quality. There are also several processes and design considerations used to change the metal characteristics to produce a more desirable weld. The following sections explain the metallurgical factors that must be considered to make acceptable welds:

- Base metal preparation
- Joint design considerations
- Filler metal selection
- Filler metal and electrode selection considerations
- Preheating and interpass temperature control
- Postweld heat treatment

Determining the Weight and Density of Materials

At 39°F (4°C), water is at its maximum density of 62.42 lb/ft^3 ($\approx$1,000 kg/m^3). Since water is the standard to which other materials are compared, it is given a specific gravity value of 1.

If the specific gravity of a material is known, its weight can be calculated by multiplying its specific gravity by the weight of an equal volume of water. For example, the specific gravity of aluminum is 2.71. Its weight in pounds per cubic foot (lb/ft^3) can be found using the formula 2.71 × 62.42 lb/ft^3 = 169.16 lb/ft^3. In other words, one cubic foot of aluminum has a density (weight) of 169.16 lb. In metric units, the calculation is 2.71 × 1,000 kg/m^3 = 2,710 kg/m^3.

Similarly, if the weight of a material is known, its specific gravity can be calculated by dividing its weight by the weight of an equal volume of water. For example, one cubic foot of aluminum weighs 169.16 lb. The specific gravity of aluminum can then be determined using the formula 169.16 lb/ft^3 ÷ 62.42 lb/ft^3 = 2.71. In other words, aluminum has a specific gravity of 2.71. In metric units, the calculation is 2,710 kg/m^3 ÷ 1,000 kg/m^3 = 2.71.

2.4.1 Base Metal Preparation

Because of the defects caused by surface contaminants and the possibility of toxic fumes, all welding codes require surface cleaning prior to any cutting or welding activity. In general, the codes state that the surface to be thermally cut or welded must be clean and free from paint, oil, rust, scale, and other material that would be detrimental to either the weld or the base metal when heat is applied.

The most common defect caused by surface contamination is porosity, which occurs when gas pockets or voids appear in the weld metal. When the porosity has a length greater than its width and is approximately perpendicular to the weld face, it is called *piping porosity*. Piping porosity is formed as the gas pocket floats toward the surface of the weld, leaving a void. The gas pocket is trapped as the weld metal solidifies, but it continues to float up as the next layer of weld is deposited. Porosity and piping porosity may not be visible on the surface of the weld.

Removing surface contaminants from base metal is done using mechanical and/or chemical means. Mechanical cleaning is the most common method. It can involve the use of hand tools, power tools, and special sandblasting equipment.

> **WARNING!**
>
> When performing mechanical cleaning, be sure to wear safety glasses and a face shield for protection from the flying particles produced during the cleaning operation. In addition, special clothing is required for sandblasting. Respiratory protection may also be needed.

Several different types of chemicals can be used to clean metal before cutting or welding. Their primary purpose is to clean oil and grease from metal surfaces that require high-quality welds. The cleaning chemical to be used should be selected specifically for the type of base metal and the application. Using the wrong chemical can cause a reaction with the base metal that results in burning, pitting, or discoloration of the surface. All chemicals used for cleaning metal are very strong, which enables them to do their job quickly. Because they are so strong, they can be very hazardous if mishandled.

> **WARNING!**
>
> Never use a chemical without specific and recent training. The mishandling of cleaning chemicals can cause severe burns to the skin, lungs, and eyes. Some chemicals can also cause death or blindness. Always use approved solvents. Procedures and methods for using, storing, and disposing of most chemicals are given in the SDS/MSDS for the specific chemical being used. These are available at the shop or jobsite.

2.4.2 Joint Design Considerations

Welded joints are selected primarily for the safety and strength required for the service conditions. When selecting the joint, groove, and weld to use, there are many variables to take into account. The following sections explain those considerations.

A welding code is a detailed listing of the rules and principles that apply to specific welded products. Codes ensure that safe and reliable welded products will be produced and that the weldments will be reasonably safe. Clients specify which codes to follow when placing orders or letting contracts. These orders or contracts impose severe penalties for not conforming to the code(s) specified. In addition, when codes are specified, the use of these codes is mandated with the force of law by one or more government jurisdictions. Always check the contract, order, or project specification for the code(s) specified.

Codes require that a WPS be written for all critical welds. A WPS is a written set of instructions for producing sound welds. It includes the type of joint to use, as well as the groove preparation, if required. Each WPS is written and tested in accordance with a particular welding code or specification. All welding requires that acceptable industry standards are followed, but not all welds require a WPS. If a weld does require a WPS, it must be followed. The requirement to use a WPS is often listed on project drawings as a note or in the tail of the welding symbol. If you are unsure whether the welding being performed requires a WPS, do not proceed until you check with your supervisor.

One of the easiest joint preparations is the square groove joint because it only requires butting the edges of the plate together. Butt joints are welded with a square groove weld that can have a partial or complete joint penetration weld. A partial joint penetration weld has much less strength than a complete penetration weld. The *AWS D1.1 Structural Welding Code — Steel* imposes restrictions for complete penetration welds. With SMAW, the maximum base metal thickness is $\frac{1}{4}$" (6 mm plate), and welding from both sides is required. In addition, a root opening of half the thickness of the base metal is required, and the root of the first weld must be gouged before the second weld is made. For GMAW and FCAW, the maximum base metal thickness is $\frac{3}{8}$" (10 mm plate) with the same requirements for welding from both sides and for back gouging.

Fillet welds also require very little joint preparation. When using a fillet weld on outside corner welds, two types of fit-up can be used: corner to corner or half lap. The corner-to-corner joint is difficult to assemble because neither plate can support the other, and care must be taken not to burn through the corner when welding. The half-lap joint is easier to assemble, requires less weld metal, and has less danger of corner burn-through. The half-lap requires a second weld on the inside of the corner. If a half-lap fit-up is used, allowances in the plate dimensions for the lap must be made. Several methods of joint preparation are explained in the following items:

- *Single- and double-groove welds (Figure 14)* — When possible, use a double-groove joint in place of a single groove. The double groove requires half the weld metal of the single groove. Also, welding from both sides reduces distortion because the forces of distortion work against each other.

- *U- and J-groove preparation* — On thick joints, U- and J-grooves often require less weld metal than V or bevel joints. The major disadvantage of U- and J-grooves is that they require more preparation time. For this reason, V and bevel joints are used much more frequently than J- or U-grooves.

- *Groove angles and root openings (Figure 15)* — The purpose of the groove angle is to allow electrode access to the root of the weld. If the groove angle is larger than necessary, it requires additional weld metal to fill. This increases the time and cost to complete the weld and increases the chance of distortion. If the groove angle is too small, it can result in weld discontinuities. The root preparation is sized to control melt-through. Increasing or decreasing the root preparation (root opening and root face) results in excess melt-through or insufficient root penetration. Generally, as the groove angle decreases, the root opening increases to compensate. Root faces are used with open-root joints, but not when metal backing strips are used (except for aluminum).

- *Groove welds with root openings* — For groove welds on plate, the groove angle is 60°, and the maximum size of the root opening and root face is $\frac{1}{8}$" (3.2 mm). For open-root welds on pipe, the groove angle is 60° or 75°, depending on the code or specifications used. The maximum size of the root opening and root faces is referred to in the WPS. *Figure 16* shows open-root joint preparation, including the groove angles.

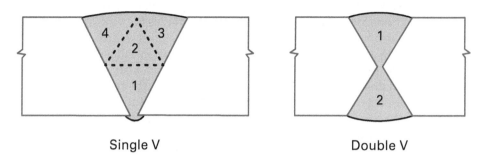

Single V   Double V

Figure 14 Single- and double-groove welds.

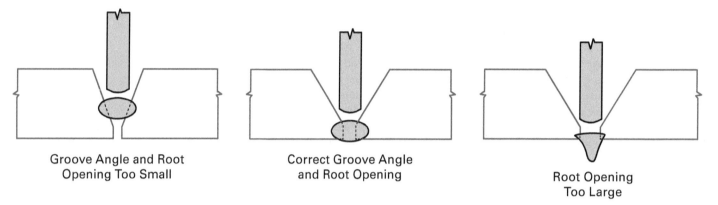

Groove Angle and Root
Opening Too Small

Correct Groove Angle
and Root Opening

Root Opening
Too Large

Figure 15 Proper and improper groove angles and root openings.

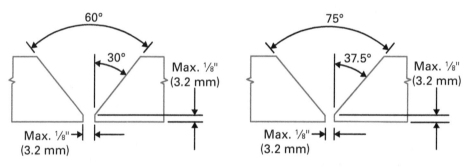

Figure 16 Open-root joint preparation.

2.0.0 Section Review

1. The mass, or weight, of a specific volume of metal is commonly expressed as _____.
   a. purity
   b. hardness
   c. density
   d. resistivity

2. A type of stress in which opposed, in-line forces pull away from each other at opposite sides of a specimen is _____.
   a. torsion
   b. compression
   c. shear
   d. tension

3. A handheld analyzer that can be used to accurately identify most types of metal relies on the use of _____.
   a. X-ray fluorescence spectrometry
   b. magnetic resonance imagery
   c. the modulus of elasticity
   d. tension loading ductility

4. A(n) _____ is a written set of instructions for producing sound welds.
   a. code
   b. WPS
   c. mill report
   d. LIBS

3.0.0 Common Structural Steel Shapes

Objective

Identify the common structural steel shapes of metal.

   a. Identify the most common structural steel shapes.
   b. Identify different structural beam shapes.
   c. Identify pipe and tubing types.
   d. Identify other common metal forms, including rebar.

Performance Tasks

There are no Performance Tasks in this section.

Metals are commercially available in many shapes and sizes. They can be shaped by casting, rolling, drawing, forging, cutting, machining, extruding, and spinning. Structural steel and other common structural shapes are formed by rolling hot or cold metal between a succession of rollers. The rollers are specially configured for the shapes they will produce.

3.1.0 Common Structural Shapes

Structural steel includes a wide range of steel types, classes, and shapes. To ensure uniform standards, the ASTM specifies the properties of strength, weight, corrosion resistance, and weldability for various steel classifications. *Table 20* lists some typical structural steel types and classifications. Note that the minimum yield stress is expressed in the table as ksi. This is one of the standard notations used for structural steel. The term *ksi* is defined as kips per square inch, where kips means kilopounds (1,000 lb) of dead load. Hence, a yield stress of 36 ksi means 36,000 lb per square inch of dead load.

TABLE 20 Common Types and Classifications of Structural Steel

Steel Type	ASTM Class	Minimum Yield Strength (ksi)	Form	Description
Carbon	A36	36	Plate, shapes, bars, sheets, strips, rivets, bolts, nuts	For buildings and general structures; available in high-toughness grades
	A516	30	Plate, sheet, bars	High notch toughness for pressure vessels
	A529	50	Plate, shapes, bars	For buildings and similar construction
HSLA	A441	40 to 50	Plates, shapes, bars	Primarily for lightweight welded buildings and bridges
	A572	42 to 65	Plate, shapes, bars	Lightweight high-toughness for buildings, bridges, and similar structures
Corrosion-resistant HSLA	A242	42 to 50	Plate, shapes, bars	Lightweight and added durability weathering grades available
	A588	42 to 50	Plate, shapes, bars	Lightweight, durable in high thickness; weathering grades available
Quenched and tempered alloy	A514	90 to 100	Plate, shapes, bars	Strength varies with thickness and type

NOTE

It is important to remember that dimensions for various structural steel shapes that are given in US customary units (such as inches) do not directly correspond to metric dimensions for those same shapes. In other words, a steel plate that measures $1/4$" × 4' × 10' will not have an exact metric equivalent. An actual metric steel plate size that is close to the same size might measure 6 mm × 1,250 mm × 3,000 mm.

The shape or type of structural steel is identified on drawings by a symbol or abbreviation. Sizes and dimensions are always given in a specified order. The specification format for designating shape and size is shown with the relevant shape in the following sections.

3.1.1 Plate

Plate is rolled metal with a uniform thickness equal to or greater than $3/16$" (4.8 mm), and its identifying symbol is PL. Its thickness is given in inches or millimeters, and its length and width are typically listed in feet or meters. An example of a specification would be: PL $1/4$" × 4' × 10'.

3.1.2 Sheet Metal

Sheet metal is rolled metal with a uniform thickness less than $3/16$" (4.8 mm). Thickness is given in inches or millimeters or indicated by an American Wire Gauge (AWG) number. Length and width are given in feet or a metric equivalent. Sheet metal is often rolled into coils. An example of a specification would be: Sheet No. 12 AWG × 4' × 8'.

3.1.3 Bars

Bars are rolled to a variety of cross-sectional shapes and sizes (*Figure 17*). Standard bar shapes include the following:

- *Round* — Round bar is specified by BAR, followed by the bar diameter, often with a diameter symbol (a circle with a slash through it), and then the length.
- *Square* — Square bar is specified by BAR, followed by the face width, often with a face width symbol (a square with a slash through it), and then the length.
- *Bar (flat)* — Flat bar is specified by BAR, followed by the width dimension, the thickness dimension, and then the length.
- *Z-bar* — Z-bar is specified by Z, followed by the flange width, the web depth, the flange and web thickness, and then the length. The flange and the web are always the same thickness.
- *Hexagonal* — Hexagonal bar is specified by HEX, followed by the bar thickness (measured across the flats), and then the length.
- *Octagonal* — Octagonal bar is specified by OCT, followed by the bar thickness (measured across the flats), and then the length.

Shape		Specification Format
Round	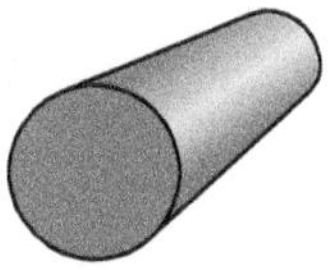	BAR (Diameter Ø) × (Length) BAR ¾" Ø × 8'3" or BAR 19 mm Ø × 3 m
Square	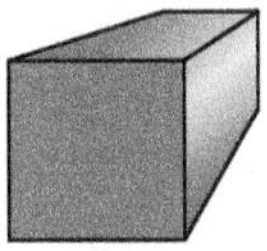	BAR (Face Width ⊠) × (Length) BAR ¾" ⊠ × 8'3" or BAR 19 mm ⊠ × 3 m
Bar	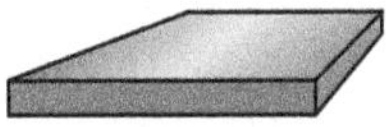	BAR (Width) × (Thickness) × (Length) BAR 2" × ¼" × 8'3" or BAR 50 mm × 6 mm × 3 m
Z-bar		Z (Flange Width) × (Web Depth) × (Flange and Web Thickness) × (Length) Z 2" × 2" × ½" × 8'3" or Z 50 mm × 50 mm × 13 mm × 3 m
Hexagonal	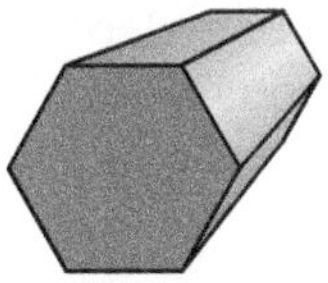	HEX (Bar Thickness) × (Length) HEX ⅞" × 10' or HEX 22 mm × 3 m
Octagonal	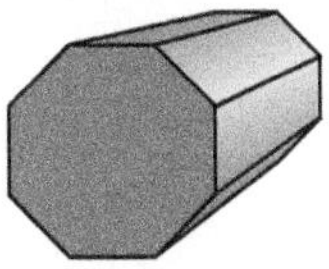	OCT (Bar Thickness) × (Length) OCT 1" × 6' or OCT 25 mm × 2 m

Figure 17 Bar shapes and specification formats.

3.1.4 Angles

Angles are L-shaped bars that have either equal-sized or unequal-sized legs that are always at 90° to each other. *Figure 18* shows both types of angle bar stock and the specification formats for each type. The dimensions for the angle bar stock examples are given in US customary units and metric units. Remember, though, these two units of measurement do not directly correspond. The specification format represents the size and thickness of the angle bar stock legs. For example, a specification of L 2" × 2" × $\frac{3}{8}$" would indicate a 2" × 2" angle of $\frac{3}{8}$" thickness in US customary units. For unequal-leg stock, the longest leg is listed first, as shown in *Figure 18*.

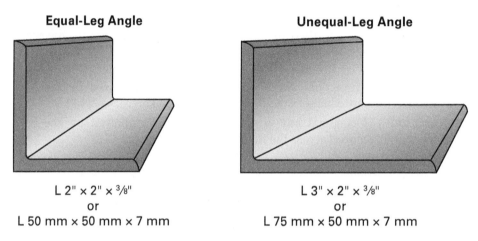

Equal-Leg Angle

L 2" × 2" × $\frac{3}{8}$"
or
L 50 mm × 50 mm × 7 mm

Unequal-Leg Angle

L 3" × 2" × $\frac{3}{8}$"
or
L 75 mm × 50 mm × 7 mm

L = Angle Bar Stock

Figure 18 Angle bar stock and specification formats.

3.1.5 Channels

Channels are U-shaped forms made of two flanges connected by a common web. The flanges extend from the same side of the web. The flanges can be of uniform thickness or tapered toward the outer edges.

There are several channel variations, each having a unique specification format designation. *Figure 19* shows two different channels and their related American Standard Channels specification format designations. It also shows similar channels with metric dimensions. The channel C 8 × 11.5 has a nominal depth of 8" and a weight of 11.5 lb/ft. Similarly, the channel C 8 × 18.75 has a nominal depth of 8" and a weight of 18.75 lb/ft. The metric channels both have a nominal depth of 200 mm. The thinner channel has a weight of 22 kg/m, while the thicker channel has a weight of 25.3 kg/m. As shown in the figure, the channel C 8 × 11.5 has a flange width of $2\frac{1}{4}$" and a web thickness of $\frac{1}{4}$", while the C 8 × 18.75 channel has a flange width of $2\frac{1}{4}$" and a web thickness of $\frac{1}{2}$". The metric channels both have a flange width of 75 mm. The thinner channel has a web thickness of 10.5 mm, while the thicker channel has a web thickness of 11.5 mm. The specific dimensions for the beam depth, flange width, and web thickness for all common channel designations can be found in related tables of dimensions and properties in most structural engineering and architectural standards reference books.

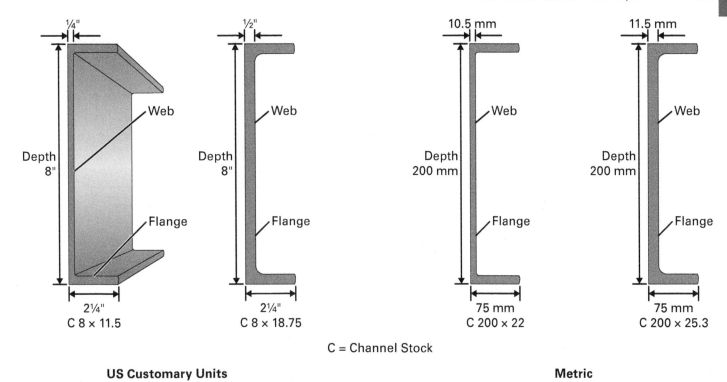

C = Channel Stock

US Customary Units   **Metric**

Figure 19 Channels and specification formats.

3.2.0 Beams and Shapes from Beams

Beams (*Figure 20*) are made in I-shaped, H-shaped, and T-shaped cross sections, and they are made with flat or tapered flanges. In addition to different flange widths and web depths, the thickness of the flanges and webs varies with beam size. Also, beam weight (per linear foot or meter) for a given dimension can be increased by adding thickness to the web and flanges with very little change in beam width or depth. The specification designator describes the beam type, its nominal depth, and its weight per foot of length or kilogram per meter. An example specification designator for a beam is S 24 × 120, where S 24 represents the beam type (S) and the nominal depth in inches (24). The 120 value is the weight in pounds per foot of length. A comparable metric beam might have a 300 mm nominal depth and have a weight of 42.2 kg/m. The specific dimensions for the depth, web thickness, and flange thickness for the various designations of beams are given in related tables of dimensions and properties in most structural engineering and architectural standards books.

Beams are identified with a standard set of symbols. The I-beam and H-beam letter symbols include the following:

- *S (American Standard)* — I-shaped beams with tapered flanges
- *W (Wide flange)* — Wider flanges than S-beams with thinner webs and nontapered flanges
- *M* — Similar to W-beams but with short flanges
- *HP* — H-shaped beams with nontapered flanges

T-beams include structural tees and tee shapes. Structural tees are made by cutting S-, W-, and M-beams down the center, usually by shearing. The conventional symbol for a structural tee is the original beam letter followed by a T. For example, a tee cut from an S 8 × 18.4 beam would be identified as ST 4 × 9.2 in US customary units.

T-shapes are rolled into their final tee shape. The symbol for tee shapes is a T without any other letter. The specification gives the nominal depth, flange width, thickness, and length. An example specification is T 3 × 7.8. T-shapes are short pieces, commonly used as connections or supports.

I-Beams

The older American Standard I-beams have been replaced by the American Institute of Steel Construction (AISC) wide flange (W) beams. These wide flange shapes are now used for most beams and columns. The wider flange provides a stronger beam than the older I-beam with the same web width.

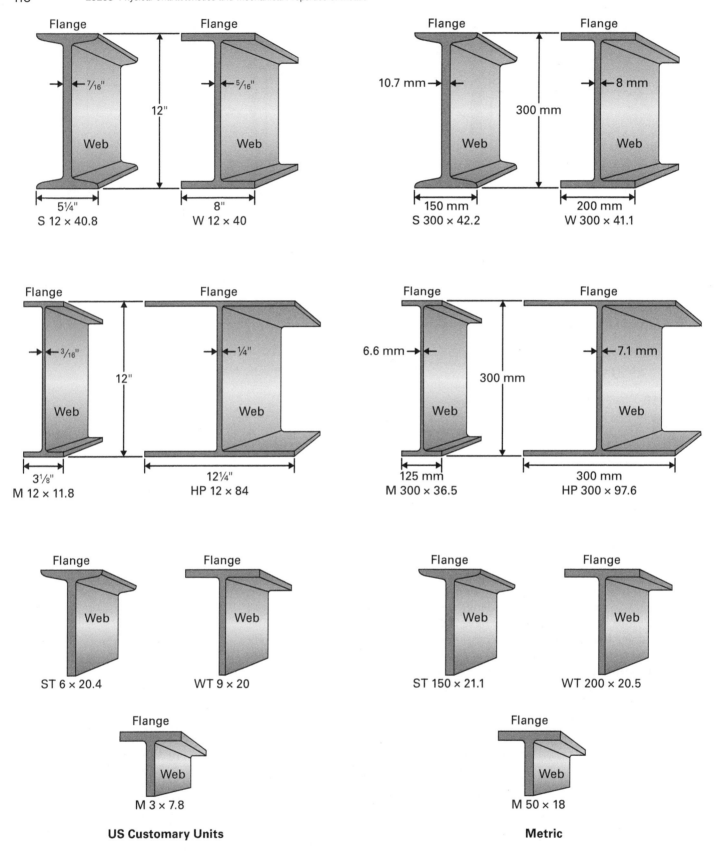

US Customary Units

Metric

Figure 20 Structural beam and T-beam shapes.

3.3.0   Pipe and Tubing

Pipe and tubing share some basic similarities, but they are identified using different measuring methods and standards. This section focuses on various sizes of pipe, some common tubing shapes, and different methods for producing pipe and tubing.

3.3.1 Pipe

Steel pipe size is specified as a nominal inside diameter (ID) up through 12" (or 300 mm). The word *nominal* is used to identify the design size, but the actual ID will not be exactly that given size. Pipe 14" (350 mm) and larger is specified by the actual outside diameter (OD). Pipe weight, which is determined by the wall thickness, is specified by a schedule number or strength. Common schedule numbers are 5, 10, 20, 30, 40, 60, 80, 100, 120, 140, and 160. Strengths are specified as standard (STD), extra strong (XS), and extra-extra strong (XXS). Standard falls between Schedules 30 and 40. Extra strong falls between Schedules 60 and 80. Extra-extra strong is Schedule 160 or higher.

Because the OD of a given pipe size is always the same, the ID becomes smaller as the pipe schedule increases. The additional wall thickness is taken from the ID while the OD remains fixed. *Table 21* shows common pipe sizes and schedules with nominal wall thicknesses in US customary units and similarly sized metric pipe measurements. As with other steel shapes, nominal size dimensions for pipe that are given in US customary units do not directly correspond to metric pipe. For instance, a 12" pipe is not identical to a 300 mm pipe.

Why Doesn't the Pipe Size Match Pipe ID or OD?

When pipe was first made, it was made from lead, which required a thick wall to hold the specific pressure for a given ID. When the switch was made from lead pipe to steel pipe, a decision had to be made whether to maintain the OD so the steel pipe would fit existing fittings, or to change the OD and keep the ID size. The decision was made to maintain the OD to provide direct replacement with the existing fittings. At that point, neither nominal ID nor OD corresponded to size. Note, once pipe hits the 14" size, the OD will match.

3.3.2 Tubing

Tubing is manufactured in square, rectangular, and round cross sections. Round tubing can be distinguished from pipe by its dimensions. Unlike pipe, standard round tubing is measured by its actual OD.

The cross sections of square and rectangular tubing have even outside dimensions and slightly rounded corners. *Figure 21* shows three standard tubing shapes and their specification formats in inches, as well as three similarly sized metric tubing shapes.

3.3.3 Seamed and Seamless Tubing and Pipe

Both standard and nonstandard shapes can also be produced by other processes. For example, tubing can be produced by extrusion or by rolling and welding. Extrusions can be produced in an almost limitless variety of shapes. When tubing is produced by rolling and welding, the most common welding process used is electronic resistance welding (ERW).

TABLE 21 Standard Pipe Sizes and Schedules with Nominal Wall Thicknesses

Nominal Wall Thickness (in)

Nom. Pipe Size	OD.	Sch. 10	Sch. 20	Sch. 30	STD	Sch. 40	Sch. 60	XS	Sch. 80	Sch. 100	Sch. 120	Sch. 140	Sch. 160	XXS
1/8	0.405	—	—	—	0.068	0.068	—	0.095	0.095	—	—	—	—	—
1/4	0.540	—	—	—	0.088	0.088	—	0.119	0.119	—	—	—	—	—
3/8	0.675	—	—	—	0.091	0.091	—	0.126	0.126	—	—	—	—	—
1/2	0.840	—	—	—	0.109	0.109	—	0.147	0.147	—	—	—	0.188	0.294
3/4	1.050	—	—	—	0.113	0.113	—	0.154	0.154	—	—	—	0.219	0.308
1	1.315	—	—	—	0.133	0.133	—	0.179	0.179	—	—	—	0.250	0.358
1 1/4	1.660	—	—	—	0.140	0.140	—	0.191	0.191	—	—	—	0.250	0.382
1 1/2	1.900	—	—	—	0.145	0.145	—	0.200	0.200	—	—	—	0.281	0.400
2	2.375	—	—	—	0.154	0.154	—	0.218	0.218	—	—	—	0.344	0.436
2 1/2	2.875	—	—	—	0.203	0.203	—	0.276	0.276	—	—	—	0.375	0.552
3	3.500	—	—	—	0.216	0.216	—	0.300	0.300	—	—	—	0.438	0.600
3 1/2	4.000	—	—	—	0.226	0.226	—	0.318	0.318	—	—	—	—	—
4	4.500	—	—	—	0.237	0.237	—	0.337	0.337	—	0.438	—	0.531	0.674
5	5.563	—	—	—	0.258	0.258	—	0.375	0.375	—	0.500	—	0.625	0.750
6	6.625	—	—	—	0.280	0.280	—	0.432	0.432	—	0.562	—	0.719	0.864
8	8.625	—	0.250	0.277	0.322	0.322	0.406	0.500	0.500	0.594	0.719	0.812	0.906	0.875
10	10.750	—	0.250	0.307	0.365	0.365	0.500	0.500	0.594	0.719	0.844	1.000	1.125	1.000
12	12.750	—	0.250	0.330	0.375	0.406	0.562	0.500	0.688	0.844	1.000	1.125	1.312	1.000
14 OD	14.000	0.250	0.312	0.375	0.375	0.438	0.594	0.500	0.750	0.938	1.094	1.250	1.406	—
16 OD	16.000	0.250	0.312	0.375	0.375	0.500	0.656	0.500	0.844	1.031	1.219	1.438	1.594	—
18 OD	18.000	0.250	0.312	0.438	0.375	0.562	0.750	0.500	0.938	1.156	1.375	1.562	1.781	—
20 OD	20.000	0.250	0.375	0.500	0.375	0.594	0.812	0.500	1.031	1.281	1.500	1.750	1.969	—
22 OD	22.000	0.250	0.375	0.500	0.375	—	0.875	0.500	1.125	1.375	1.625	1.875	2.125	—
24 OD	24.000	0.250	0.375	0.562	0.375	0.688	0.969	0.500	1.218	1.531	1.812	2.062	2.344	—
26 OD	26.000	0.312	0.500	—	0.375	—	—	0.500	—	—	—	—	—	—
28 OD	28.000	0.312	0.500	0.625	0.375	—	—	0.500	—	—	—	—	—	—
30 OD	30.000	0.312	0.500	0.625	0.375	—	—	0.500	—	—	—	—	—	—
32 OD	32.000	0.312	0.500	0.625	0.375	0.688	—	0.500	—	—	—	—	—	—
34 OD	34.000	0.312	0.500	0.625	0.375	0.688	—	0.500	—	—	—	—	—	—
36 OD	36.000	0.312	0.500	0.625	0.375	0.750	—	0.500	—	—	—	—	—	—
42 OD	42.000	—	—	—	0.375	—	—	0.500	—	—	—	—	—	—

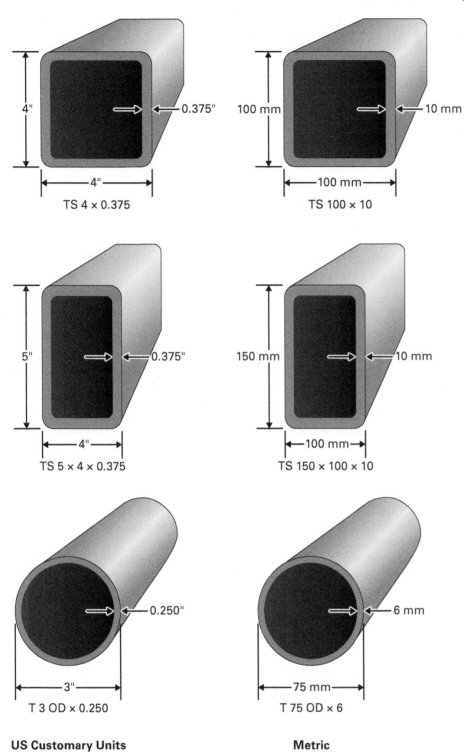

TS 4 × 0.375

TS 100 × 10

TS 5 × 4 × 0.375

TS 150 × 100 × 10

T 3 OD × 0.250

T 75 OD × 6

US Customary Units   **Metric**

Figure 21 Standard tubing shapes and specification formats.

3.4.0 | Other Common Forms

In addition to common structural steel shapes, beams, pipe, and tubing, there are other forms of steel that are commonly used in industry. These other forms include reinforcing bars, forged objects, cast objects, and powdered metals.

3.4.1 Reinforcing Bars

Reinforcing bars (rebars), sometimes called *rerods*, are used for concrete reinforcement. They are available in several grades. These grades vary in yield strength, ultimate strength, percentage of elongation, bend test requirements, and chemical composition. Reinforcing bars can be coated with different compounds, such as epoxy, for use in concrete where corrosion can be a problem.

The ASTM International has established standard specifications for reinforcing bars. These grades appear on bar-bundle tags, in color coding, in rolled-on markings on the bars, and/or on bills of materials. The ASTM specifications are as follows:

- *A615, Standard Specification for Deformed and Plain Carbon-Steel Bars for Concrete Reinforcement*
- *A706, Standard Specification for Deformed and Plain Low-Alloy Steel Bars for Concrete Reinforcement*
- *A996, Standard Specification for Rail-Steel and Axle-Steel Deformed Bars for Concrete Reinforcement*

The standard configuration for reinforcing bars is the deformed bar. Different patterns may be impressed upon the bars depending on the manufacturer, but all are rolled to conform to ASTM specifications. The purpose of the deformation is to improve the bond between the concrete and the bar and to prevent the bar from moving within the concrete.

Plain bars are smooth and round without deformations on them. Plain bars are used for special purposes: dowels at expansion joints, where the bars must slide in a sleeve; for expansion and contraction joints in highway pavement; and for column spirals.

Deformed bars are designated by a number in 11 standard sizes (metric or inch-pound) as shown in *Table 22*. The number indicates the approximate diameter of the bar in eighths of an inch or millimeters. For example, a #16 metric bar has an approximate diameter of 16 mm. The nominal dimension of a deformed bar (the nominal diameter does not include the deformation) is equivalent to that of a plain bar having the same weight per foot.

TABLE 22 ASTM Standard Metric and Inch-Pound Reinforcing Bars

Bar Size		Nominal Characteristics					
		Diameter		Cross-Sectional Area		Weight	
Metric	Imperial	(mm)	(in)	(mm²)	(in²)	(kg/m)	(lb/ft)
#10	#3	9.5	0.375	71	0.11	0.560	0.376
#13	#4	12.7	0.500	129	0.20	0.944	0.668
#16	#5	15.9	0.625	199	0.31	1.552	1.043
#19	#6	19.1	0.750	284	0.44	2.235	1.502
#22	#7	22.2	0.875	387	0.60	3.042	2.044
#25	#8	25.4	1.000	510	0.79	3.973	2.670
#29	#9	28.7	1.128	645	1.00	5.060	3.400
#32	#10	32.3	1.270	819	1.27	6.404	4.303
#36	#11	35.8	1.410	1006	1.56	7.907	5.313
#43	#14	43.0	1.693	1452	2.25	11.38	7.65
#57	#18	57.3	2.257	2581	4.00	20.24	13.60

As shown in *Figure 22*, bar identification is made by ASTM specifications. These require that each bar manufacturer roll the following information onto the bar:

- A letter or symbol to indicate the manufacturer's mill
- A number corresponding to the size number of the bar (*Table 22*)
- A symbol or marking to indicate the type of steel (*Table 23*)
- A marking to designate the grade (*Table 24*)

TABLE 23 Reinforcement Bar Steel Types

Symbol/ Marking	Type of Steel
A	Axle (ASTM A996)
S	Billet (ASTM A615)
R	Rail (ASTM A996)
W	Low alloy (ASTM A706)

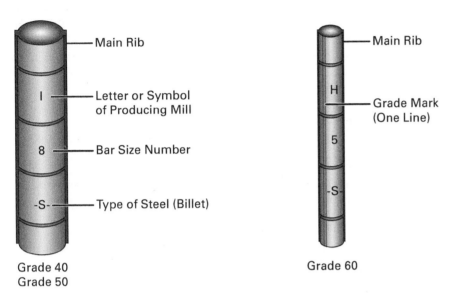
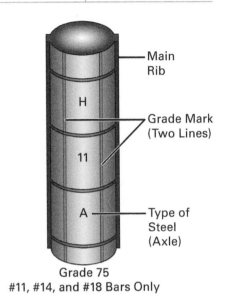

Line System Grade Marks

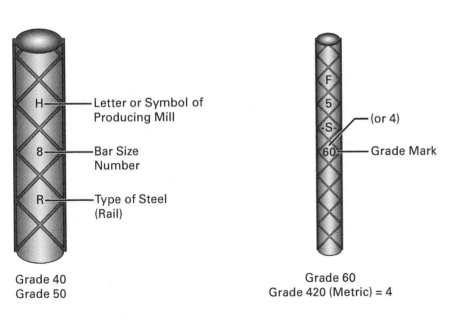

Number System Grade Marks

Figure 22 Reinforcement bar identification.

TABLE 24 Reinforcement Bar Grades

Grade	Identification	Minimum Yield Strength
40 and 50	None	40,000 psi to 50,000 psi (40 ksi to 50 ksi)
60	One line or the number 60	60,000 psi (60 ksi)
75	Two lines or the number 75	75,000 psi (75 ksi)
420	The number 4	420 MPa (60,000 psi or 60 ksi)
520	The number 5	520 MPa (75,000 psi or 75 ksi)

The grade represents the minimum yield or tension strength, measured in kips per square inch (ksi) or megapascals (MPa), that the type of steel will withstand before it permanently stretches (elongates). Today, Grade 420 is the most used rebar. Rebars are normally supplied from the mill bundled in 60' lengths (12 m lengths are common for metric rebar) that are cut in the field to the required length.

3.4.2 Forged Shapes

Forging is used to produce specialized shapes where high strength is required. An example is forged high-pressure pipe fittings such as flanges and elbows.

3.4.3 Cast Shapes

Casting is another method of producing shapes. Cast iron and steel are used for machine parts, automobile components, and large castings.

3.4.4 Powdered Metals

Sintered: A process of making parts from a powdered metal in which the metal is molded and then heated in a furnace without melting to fuse it into a solid metal.

Commercially prepared powdered metals are used for welding fillers, surface coatings, and mold castings. When cast to make parts, the molded powder is usually **sintered** (heated in a furnace without melting) to fuse it into a solid metal. This is a common technique for manufacturing complex-shaped or porous parts. It is also used with high-temperature metals that cannot be liquid-cast.

3.0.0 Section Review

1. Steel plate is rolled metal with a uniform thickness equal to or greater than _____.
    a. $\frac{1}{8}$" (3.2 mm)
    b. $\frac{3}{16}$" (4.8 mm)
    c. $\frac{1}{4}$" (6.4 mm)
    d. $\frac{5}{16}$" (7.9 mm)

2. The three steel beam cross section shapes are _____.
    a. A, H, and I
    b. I, L, and T
    c. H, T, and V
    d. I, H, and T

3. Pipe that is 14" (350 mm) and larger is specified by its _____.
    a. actual ID
    b. nominal ID
    c. actual OD
    d. nominal OD

4. The steel form that is *most* likely to be used for expansion and contraction joints in highway pavement is _____.
    a. forged high-pressure pipe
    b. plain reinforcing bars
    c. sintered powdered metal
    d. deformed rebar rods

1. Common welding base materials can be made of one metallic element or a(n) _____ of metallic and nonmetallic components.
   a. quench
   b. notch toughness
   c. iron
   d. alloy

2. The *largest* group of the ferrous-based metals includes the _____.
   a. cast irons
   b. alloy steels
   c. heavy metals
   d. carbon steels

3. Quench-and-tempered steels and chromium-molybdenum steels are examples of _____.
   a. common-grade stainless steels
   b. low-alloy steels
   c. high-alloy steels
   d. specialty-grade stainless steels

4. A lower coefficient of thermal conductivity along with a higher coefficient of thermal expansion and a higher electrical resistance are all characteristics of _____.
   a. nickel alloy
   b. carbon steel
   c. stainless steel
   d. aluminum alloy

5. A type of stainless steel known for its superior resistance to pitting and stress cracking is _____.
   a. superferritic
   b. duplex
   c. superaustenitic
   d. precipitation-hardening

6. Pickling and _____ are steps that can help protect stainless steels from corrosion after the welding has been performed.
   a. passivation
   b. fabricating
   c. austenitizing
   d. electrode selection

7. The *principal* alloying element in brasses is _____.
   a. aluminum
   b. zinc
   c. beryllium
   d. bronze

8. Corrosion resistance is an important characteristic of a metal because corrosion can severely reduce a metal's _____.
   a. tensile strength
   b. interpass ductility
   c. thermal expansion
   d. residual stress

9. The ability of a material to be strained (deformed) without permanent deformation is known as _____.
   a. ductility
   b. modulus
   c. elasticity
   d. tensile strength

10. The resistance to indentation of a material is called _____.
    a. ductility
    b. hardness
    c. tensile strength
    d. modulus of elasticity

11. The three types of cast iron are malleable, white, and _____.
    a. ductile
    b. green
    c. blue
    d. gray

12. Malleable cast iron is white cast iron that has been _____.
    a. annealed
    b. cooled slowly
    c. forged
    d. chilled rapidly

13. The *most* common welding defect caused by surface contamination of a metal is _____.
    a. rust
    b. porosity
    c. corrosion
    d. toxic fumes

14. A detailed listing of the rules and principles that apply to specific welded products is called a _____.
    a. WPS
    b. welding code
    c. mill report
    d. bill of materials

15. The ASTM specifies the properties of strength, weight, corrosion resistance, and weldability for various steel classification to _____.
    a. ensure uniform standards
    b. provide guidance on removal of surface contaminants from base metals
    c. calculate porosity
    d. provide certified mill test reports

16. A channel that has a designation of C 12 × 30 is a(n) _____.
    a. American Standard Channel with a depth of 12"
    b. Miscellaneous Channel with a depth of 30"
    c. Miscellaneous Channel with a weight of 12 lb/ft
    d. American Standard Channel with a weight of 12 lb/ft

17. Shearing an S-, W-, or M-beam down the center creates a(n) _____.
    a. wide flange beam
    b. H-beam
    c. structural tee
    d. seamed I

18. Tubing can be produced by rolling and welding or by _____.
    a. casting
    b. compression
    c. annealing
    d. extrusion

19. The standard configuration for reinforcing bars is the _____.
    a. dowel bar
    b. deformed bar
    c. column bar
    d. plain bar

20. Molded powder used to cast to make parts should be _____ to fuse it into a solid metal.
    a. sintered
    b. sanded
    c. melted
    d. webbed

Answers to odd-numbered questions are found in the Review Question Answer Keys at the back of this book.

Answers to Section Review Questions

Answer	Section Reference	Objective
Section 1.0.0		
1. c	1.1.2	1a
2. a	1.2.1	1b
3. d	1.3.2	1c
4. a	1.4.2	1d
Section 2.0.0		
1. c	2.1.1	2a
2. d	2.2.1	2b
3. a	2.3.4	2c
4. b	2.4.2	2d
Section 3.0.0		
1. b	3.1.1	3a
2. d	3.2.0	3b
3. c	3.3.1	3c
4. b	3.4.1	3d

User Update

Did you find an error? Submit a correction by visiting **https://www.nccer.org/olf** or by scanning the QR code using your mobile device.

SCAN ME

Preheating and Postheating of Metals

Source: © Miller Electric Mfg. LLC

Objectives

Successful completion of this module prepares you to do the following:

1. Describe the relationship between heat and metal and identify preheating methods.
   a. Describe the relationship between heat and metal.
   b. Identify and describe methods used to preheat metal prior to welding.
   c. Identify and describe devices and products used to measure temperatures.
2. Describe interpass temperature control and postheating processes.
   a. Describe interpass temperature control.
   b. Describe various postheating processes.

Performance Task

Under supervision, you should be able to do the following:

1. Preheat base metal to 350°F (177°C) and verify preheat using a temperature-indicating device.

Overview

Welders need to understand how heat affects metals and how to control heating during each stage of a welding operation to make strong, enduring welds. This module explains the relationship between heat and metal and describes various devices and methods welders use to heat metals and measure temperatures.

NCCER Industry-Recognized Credentials

If you are training through an NCCER-accredited sponsor, you may be eligible for credentials from NCCER. The ID number for this module is 29204. Note that this module may have been used in other NCCER curricula and may apply to other level completions. Contact NCCER at 1.888.622.3720 or go to **www.nccer.org** for more information.

You can also show off your industry-recognized credentials online with NCCER's digital credentials. Transform your knowledge, skills, and achievements into credentials that you can share across social media platforms, send to your network, and add to your resume. For more information, visit **www.nccer.org**.

Digital Resources for Welding

Scan this code using the camera on your phone or mobile device to view the digital resources related to this craft.

1.0.0 Metal and Preheating

Performance Task

1. Preheat base metal to 350°F (177°C) and verify preheat using a temperature-indicating device.

Objective

Describe the relationship between heat and metal and identify preheating methods.

a. Describe the relationship between heat and metal.
b. Identify and describe methods used to preheat metal prior to welding.
c. Identify and describe devices and products used to measure temperatures.

Preheat: The process of heating the base metal before making a weld.

Heat-affected zone (HAZ): The portion of the base metal workpiece that is not melted but has had its properties affected by welding or other heat-intensive operation such as oxyfuel cutting.

Preheat is the controlled heating of the base metal immediately before welding begins. The main purpose of preheating is to keep the weld and the base metal along the weld line in the **heat-affected zone (HAZ)** from cooling too fast. In some cases, the whole structure is preheated. When it is not possible or practical to heat the whole structure, only the section near the weld is preheated.

Heat Treatments

Different forms of heat treatments are used to alter the properties of metals. Soft metals can be made hard, brittle, or strong using the correct heat treatment. These metals can be returned to their original soft form with heat treatment. Welding specifications often require that welded joints be heat-treated before welding or after fabrication. To avoid mistakes, heat treatments made to metals of any kind must be done in a planned and systematic way. This careful focus is important! Using temperatures that are too low or too high can do more harm than good.

1.1.0   Temperature and Metal Relationship

Crystalline: Like a crystal; having a uniform atomic structure throughout the entire material.

Constituents: The elements and compounds, such as metal oxides, that make up a mixture or alloy.

At room temperature, most metals are solid and have a **crystalline** structure (*Figure 1*). They become liquids if heated to a high enough temperature. Temperature is a measure of molecular activity. When the temperature is high enough, the forces that hold the molecules in their crystalline structures break down, and the metal becomes a liquid. At even higher temperatures, metals become gases. The temperature at which a metal becomes liquid is its melting point. The melting point is different for every metal. In the case of alloys (metal mixtures), the various **constituents** melt at different temperatures. Some, like carbon, may not melt at all.

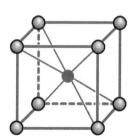

Body-Centered Cubic (BCC)

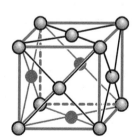

Face-Centered Cubic (FCC)

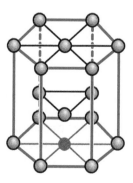

Hexagonal Close Packed (HCP)

Figure 1 Crystal structures of iron.

Below the melting point, there is a temperature zone at which crystalline changes take place. Depending on the metal or alloy, these changes can seriously affect the metal's physical structure and mechanical characteristics. Quick cooling (quenching) of metal in this temperature zone can freeze some of the modified crystalline structure. As a result, the mechanical properties and physical traits of the cooled metal will be very different from those of the same metal cooled more slowly. Quenching does not have much of an effect on mild steels because they do not contain much carbon. However, quenching medium- and high-carbon steels produces a hard and brittle crystalline structure, which reduces the metal's toughness.

Table 1 shows the minimum preheat temperatures for various base materials, as well as the welding processes and material thicknesses. The material thickness figures are for general reference only. Different codes have dramatically different material thicknesses, so always refer to the codes that relate to your work.

NOTE

Tables in this module should not be relied upon as a resource for work outside of the learning environment. Always consult job specifications and other named project resources for current and accurate guidance.

TABLE 1 Minimum Preheat Temperatures for Various Base Metals

Base Materials	Welding Processes	Material Thicknesses	Minimum Preheat Temperatures
ASTM A36 ASTM A53 GR. B ASTM A106 GR. B API 5L GR. B API 5L GR. X42	Shielded Metal Arc Welding with Other than Low Hydrogen Electrodes	≤³⁄₄" (19 mm)	None (See Note)
		Between ³⁄₄" and 1¹⁄₂" (19 mm and 38 mm)	150°F (66°C)
		Between 1¹⁄₂" and 2¹⁄₂" (38 mm and 64 mm)	225°F (107°C)
		>2¹⁄₂" (64 mm)	300°F (149°C)
ASTM A36 ASTM A53 GR. B ASTM A106 GR. B ASTM A572 GR. 42 ASTM A572 GR. 50 ASTM A586 API 5L GR. B API 5L GR. X42	Shielded Metal Arc Welding with Low Hydrogen Electrodes, Submerged Arc Welding, Gas Metal Arc Welding, and Flux-Cored Arc Welding	≤³⁄₄" (19 mm)	None (See Note)
		Between ³⁄₄" and 1¹⁄₂" (19 mm and 38 mm)	50°F (10°C)
		Between 1¹⁄₂" and 2¹⁄₂" (38 mm and 64 mm)	150°F (66°C)
		>2¹⁄₂" (64 mm)	225°F (107°C)
ASTM A572 GR. 60 ASTM A572 GR. 65 ASTM 5L GR. X52	Shielded Metal Arc Welding with Low Hydrogen Electrodes, Submerged Arc Welding, Gas Metal Arc Welding, and Flux-Cored Arc Welding	≤³⁄₄" (19 mm)	50°F (10°C)
		Between ³⁄₄" and 1¹⁄₂" (19 mm and 38 mm)	150°F (66°C)
		Between 1¹⁄₂" and 2¹⁄₂" (38 mm and 64 mm)	225°F (107°C)
		>2¹⁄₂" (64 mm)	300°F (149°C)
ASTM A514 ASTM A517	Shielded Metal Arc Welding with Low Hydrogen Electrodes, Submerged Arc Welding, Gas Metal Arc Welding, and Flux-Cored Arc Welding	≤³⁄₄" (19 mm)	50°F (10°C)
		Between ³⁄₄" and 1¹⁄₂" (19 mm and 38 mm)	125°F (52°C)
		Between 1¹⁄₂" and 2¹⁄₂" (38 mm and 64 mm)	175°F (79°C)
		>2¹⁄₂" (64 mm)	275°F (135°C)

Note: If below 50°F (10°C), preheat to 70°F (21°C).

Critical temperature: The temperature at which iron crystals in a ferrous-based metal transform from being face-centered to body-centered. This dramatically changes the strength, hardness, and ductility of the metal.

When medium- and high-carbon steel are welded, the temperature in the weld zone always reaches the melting point (above the **critical temperature** or transformation temperature), and the crystalline structure in the weld zone changes. If the metal has not been preheated enough, the large mass of the cool base metal will quench the weld zone and cause localized hardness and brittleness. If the metal has been adequately preheated, the weld zone will cool much more slowly, allowing the crystalline structure to return to the crystalline form of the surrounding base metal.

Preheating may be required to do the following:

- Reduce localized shrinkage in the weld and the adjacent base metal.
- Prevent excessive hardening and reduced ductility of the weld filler and base metals in the HAZ.
- Allow hydrogen gas the time to escape in the weld zone.
- Ensure compliance with required welding procedures.

The critical temperature of a metal is somewhere below its melting point. The temperature range between a metal's critical temperature and its melting point is called the *transformation temperature range*. This range, generally considered to lie between 932°F and 1,040°F (500°C and 560°C), is the lowest temperature point at which the molecular structure of the metal will begin to change during preheating, welding, and postheating. The melting point, strength, and hardness of steel are determined by the amount of carbon in the steel.

Table 2 shows the classifications of steel based on carbon content.

TABLE 2 Classification of Carbon Steel

Common Name	Carbon Content
Low-carbon steel	0.15% maximum
Mild steel	0.15% to 0.30%
Medium-carbon steel	0.30% to 0.50%
High-carbon steel	0.50% to 1.00%

When preheating metal, apply heat to the region of the intended weld long enough to sufficiently heat the area. Insufficient heating will cause a relatively large weld to cool too quickly. To ensure proper preheating, allow heat to be absorbed through the thickness of the part. Sometimes this may take 15 to 20 minutes for very thick parts.

1.1.1 Reducing Shrinkage

Residual stresses: Strains that occur in a welded joint after the welding has been completed, as a result of mechanical action, thermal action, or both.

Because restrained assemblies are held firmly in position to maintain their shape, they can build up **residual stresses** following welding. The welded section expands due to the heat. When it cools, it shrinks, and stresses may occur if the surrounding area does not shrink along with it. With steel weldments, this shrinkage can cause distortion, warping, and even cracking. With cast iron weldments, shrinkage usually causes the base metal to crack or the weld to fail. For example, if a broken cast iron pulley spoke is being repaired, the rim and other spokes must be adequately preheated, or they will be placed under heavy strain by the thermal expansion of the welded spoke. When the weld cools, the residual strains could cause the rim to crack or the welded spoke to crack again. If the rim and spokes near the weld are adequately preheated, they will shrink along with the welded spoke, reducing or eliminating most of the stress on the rim. *Figure 2* shows the areas of heavy strain resulting from inadequate preheating of a cast iron pulley.

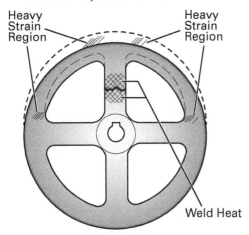

Figure 2 Strain caused by inadequate preheating of a cast iron pulley.

1.1.2 Preventing Excessive Hardening and Reduced Ductility

All welding processes use or generate temperatures higher than the melting point of the base metal. If the welded metal is high-carbon alloy steel or cast iron that has been quenched, a hard, brittle metal called **martensite** will form in the weld zone. This causes the entire HAZ parallel to the weld fusion line to become hard and brittle (*Figure 3*). The large temperature difference also causes differential thermal expansion and creates stresses in the weld region. Heavy plates and castings have a high heat absorption capacity that will cause the weld metal and adjacent base metal to quench unless they are adequately preheated.

1.1.3 Reducing Hydrogen Gas in the Weld Zone

The welding arc can cause moisture in the weld zone to separate into hydrogen and oxygen. The hydrogen dissolves in the molten base metal, and the oxygen combines with deoxidizers in the flux, becoming part of the slag. As the metal begins to cool, the hydrogen forms gas bubbles in the base metal along the weld boundary. These bubbles create pressure that can result in porosity, assuming the metal is ductile enough to stretch. Otherwise, this hydrogen may move to the HAZ, straining the metal and causing **underbead cracking**, also called *hydrogen cracking*, along the weld boundary and at the toe. Preheating helps to remove moisture from the weld zone. This results in slower cooling, allowing the hydrogen more time to diffuse up through the filler metal and down into the adjacent base metal. The more time the hydrogen has to diffuse, the less pressure it causes and the less chance there is for underbead cracking. *Figure 4* is a simplified diagram showing underbead cracking caused by hydrogen bubbles.

Underbead Cracking

Always consider preheating and postheating a weld when there is the danger of underbead cracking. The risk of underbead cracking increases when welding carbon and low-alloy steels. Underbead cracking can occur when there is a combined presence of the following conditions:

- Diffusible (atomic) hydrogen in the steel
- A steel with a partly or wholly martensitic microstructure
- Tensile stress at the weldment and HAZ
- A temperature below 300°F (149°C)

Proper preheat, interpass temperature maintenance, and postweld heat treatment, together with the proper electrode/shielding gas combination, can reduce the risk of underbead cracking.

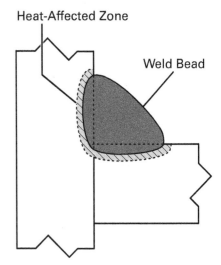

Figure 3 Heat-affected zone (HAZ).

Martensite: A solid solution of carbon in alpha-iron that is formed when steel is cooled so rapidly that the change from austenite to pearlite is suppressed; responsible for the hardness of quenched steel.

Underbead cracking: Subsurface cracking in the base metal under or near the weld.

NOTE

Some metals absorb hydrogen from their environment during normal use. Later when they are welded, this hydrogen can cause cracking. To address this problem, the metal is preheated for several hours—a process known as *hydrogen bakeout*. This action removes much of the absorbed hydrogen, so normal welds are less likely to crack.

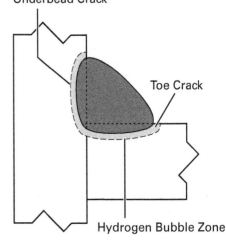

Figure 4 Underbead cracking caused by hydrogen bubbles.

1.1.4 Complying with Required Welding Procedures

If a site has a Welding Procedure Specification (WPS) for particular welding procedures, it is mandatory that the specifications be followed exactly. Check with your supervisor if you are not sure of the specifications that apply to your job.

1.1.5 Metals That Require Preheating

Determining whether metal assemblies and conditions require preheating can be very difficult. Preheating and **interpass temperature control** depend on the base metal's composition, thickness, degree of restraint, and the code the work is being performed under. Mild-carbon steel, for example, requires preheating if its carbon content exceeds 0.30%. If the carbon content is lower than 0.30%, it must be heated to a temperature above 70°F (21°C). Some base metals require preheating only when they exceed a certain thickness or are too cold. For example, metals exposed to freezing winter temperatures usually require some preheating. As a general rule, the temperature of the metal to be welded should be above the dew point temperature at the time it will be welded. Welding assembles and castings with complex shapes usually require preheating to avoid severe stresses or warping. Some alloys, such as high-carbon or alloy steels, require preheating to avoid brittleness and cracking. Not all metals require preheating. Low-carbon steels rarely need preheating. Always consult the WPS for specifics.

Table 3 shows the metals that usually require preheating and their conditions and forms.

Interpass temperature control: The process of maintaining the temperature of the weld zone within a given range during welding.

TABLE 3 Metals That Usually Require Preheating

Metal or Alloy	Conditions or Forms
Aluminum	Large or thick section castings
Copper	All (prevents too rapid heat loss)
Bronze, copper-based	All (prevents too rapid heat loss)
Mild-carbon steels	Restricted joints, complex shapes, freezing temperatures, and carbon content over 0.30%
Cast irons	All types and all shapes
Cast steels	Higher carbon content or complex shapes
Low-alloy steels	Thicker sections or restricted joints Heating temperature dependent on carbon or alloy content
Manganese steels	Heating temperature dependent on carbon or alloy content
Martensitic stainless steels	Carbon content over 0.10%
Ferritic stainless steels	Thick plates

Preheating requirements are affected by alloy composition and thickness of the base metal, the welding process to be used, and the ambient temperature. The hardenability of steel is directly related to its carbon content and alloying elements. Because different types of steel have different amounts of carbon and alloying elements, they have different preheat requirements. Generally, the higher the carbon content, the higher the preheating temperature required. *Table 4* lists preheat requirements for steels, including the steel type, shapes or conditions, and preheat temperature.

NOTE

The combination of elements in certain steel work together to affect the required preheat temperature. This effect is calculated using a carbon efficiency (CE) formula. Welding-related organizations promote the use of different CE formulas. The following are two commonly used examples.

International Institute of Welding (IIW):

$CE-IIW = C + Mn/6 + (Cr + Mo + V)/5 + (Cu + Ni)/15$

American Welding Society (AWS):

$CE-AWS = C + (Mn + Si)/6 + (Cr + Mo + V)/5 + (Cu + Ni)/15$

TABLE 4 Preheat Requirements for Steels

Steel Type	Shapes or Conditions	Preheat Temperature
Low-carbon steel (0.10% to 0.30% carbon)	Freezing temperatures	Above 70°F (21°C)
Low-carbon steel (0.10% to 0.30% carbon)	Complex large shapes	100°F to 300°F (38°C to 149°C)
Medium-carbon steel (0.30% to 0.45% carbon)	Complex large shapes	300°F to 500°F (149°C to 260°C)
High-carbon steel (0.45% to 0.80% carbon)	All sizes and shapes	500°F to 800°F (260°C to 427°C)
Low-alloy steel	Lower carbon or alloy percent	100°F to 500°F (38°C to 260°C)
Low-alloy steel	Higher carbon or alloy percent	500°F to 800°F (260°C to 427°C)
Manganese steel	Lower carbon or alloy percent	100°F to 200°F (38°C to 93°C)
Manganese steel	Higher carbon or alloy percent	200°F to 500°F (93°C to 260°C)
Cast iron	All shapes	200°F to 400°F (93°C to 204°C)
Martensitic stainless steel	0.10% to 0.20% carbon	500°F (260°C)
Martensitic stainless steel	0.20% to 0.50% carbon	500°F (260°C) and postheat
Ferritic stainless steel	$1/4$" (6.4 mm) and thicker	200°F to 400°F (93°C to 204°C) and postheat

1.2.0 Preheating Methods

Preheating is most effective when the entire welding assembly is preheated (general heating) in an oven or preheating furnace because the heating is even. However, because of site conditions or the size or shape of the welded assembly, preheating by general heating is not always possible. Therefore, a variety of heating devices can be used for preheating. Localized heating of specific regions can be done with gas torches, blowtorches, electric resistance heaters, radio frequency induction heaters, or partial ovens built only around the region to be heated. Parts of the oven or enclosure can be removed so the welder can have access and still maintain the preheating temperature.

Most welding shops have one or more of the following types of preheating devices:

- Oxyfuel torches
- Portable preheating torches
- Open-top preheaters
- Electric resistance heaters
- Induction heaters
- Enclosed furnaces

Preheating Steel in Low-Temperature or High-Humidity Conditions

Steel should not be welded if the steel temperature is below 32°F (0°C). If the structural steel temperature is below 32°F (0°C), steel should be heated to at least 70°F (21°C) prior to welding. Piping codes require a minimum temperature. Additionally, under humid conditions, steel should be heated to a higher preheat temperature to dry any moisture from its surface.

NOTE

Did you know it is a myth that moisture comes out of the pores of metal? Metal is not porous, meaning it does not absorb or release moisture. This is important to remember at the beginning of the heating operation when moisture may condense and bead on the base of metal. While the moisture may appear to have been released by the metal itself, it's a trick of the eye. Trust your knowledge of metal properties.

1.2.1 Oxyfuel Torches

If properly monitored, oxyfuel torches equipped with specialized heating tips can be used for localized preheating. *Figure 5* shows a torch handle to which heating tips can be attached.

The specialized heating tips are designed to produce blowtorch flame patterns. Heating tip attachments with multi-flame heating heads are often referred to as *rosebuds*. They are designed to mount directly to torch handles like welding tips. Heating tips designed to replace the cutting tips on straight or combination cutting torches are also available. *Figure 6* shows several types of heating tips.

WARNING!

Always use the correct type of tip for the fuel gas being used. Using a tip with the wrong gas can cause the tip to explode. Observe the pressures and flow rates recommended by the manufacturer.

Combination Torch Handle

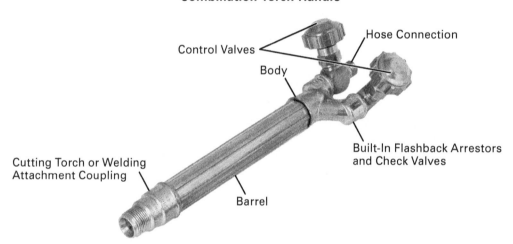

Figure 5 Combination torch handle.
Source: Courtesy of Uniweld Products

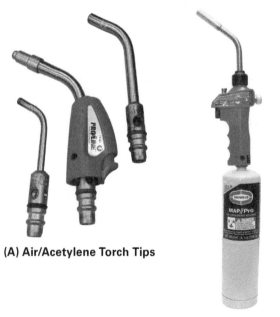

(A) Air/Acetylene Torch Tips

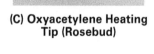

(B) Air/Fuel Torch (Suitable for Propane or Mappro Gases)

(C) Oxyacetylene Heating Tip (Rosebud)

(D) Oxypropylene Heating Tips

Figure 6 Types of heating tips.
Sources: Courtesy of Uniweld Products (6B); Photo courtesy of The Lincoln Electric Company, Cleveland, OH, U.S.A. (6C–6D)

1.2.2 Portable Preheating Torches

Portable preheating torches are designed to burn either oil or gas combined with compressed air. Some use only gas and have an air blower to increase the intensity of the flame. *Figure 7* shows an air torch suitable for preheating large weldments. The combustion chamber head can easily be removed with an extension added to increase the length. This torch operates on various fuel gases combined with compressed air.

Figure 7 Gas preheating torch.
Source: Belchfire Corp.

Air Tube Burner

The gas fuel / compressed air tube burner can be used for the heat treatment of large weldments. The tube section has evenly spaced multiple heating tips along its length. A burner of this type provides a more even heat distribution to a large weldment.

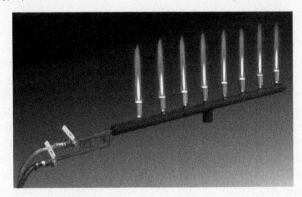

Source: Belchfire Corp.

1.2.3 Open-Top Preheaters

Open-top preheaters come in several basic designs. The simplest is a large, horizontal gas burner with an adjacent or overhead support structure. The object to be heated is mounted on the support structure directly over the gas burners. A variation is the open flat-top preheater, which consists of a horizontal grate or series of bars with gas burners underneath (*Figure 8*). The item to be preheated is placed on the grating. This type of preheater works similarly to a gas-powered kitchen stove.

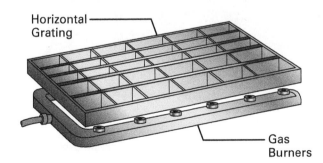

Horizontal Grating

Gas Burners

Figure 8 Open flat-top preheater.

1.2.4 Electric Resistance Heaters

Electric resistance heaters have tubular, or tile-shaped, resistance elements formed in several shapes or styles to fit various applications (*Figure 9*). These elements are temporarily fastened against the surface of the metal to be heated and covered with insulation. Electricity passing through the elements creates heat, which is transferred to the weldment.

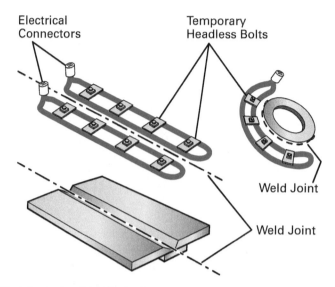

Figure 9 Resistance heating elements.

1.2.5 Induction Heaters

In induction heating, an alternating magnetic field is generated within the metal body being heated. A magnetic coil is placed around the metal object, and a strong alternating current is passed through the coil. The resulting alternating magnetic field generated within the base metal creates circulating electrical currents known as *eddy currents*. The interaction of these currents with the electrical resistance of the base metal produces an electrical heating effect, raising the temperature of the base metal. *Figure 10* shows an air-cooled induction heating system consisting of a power supply, rolling inductor, and the associated power and control cables. Air-cooled systems are typically used for preheating applications up to 400°F (204°C). This type of system typically uses a controller to monitor and automatically control the temperature.

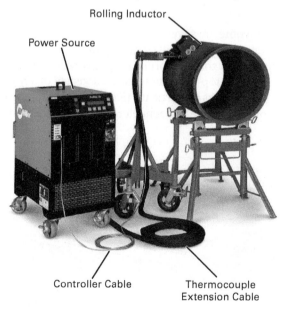

Figure 10 Induction heating system.
Source: © Miller Electric Mfg. LLC

Induction Heaters

Because water cools more efficiently than air, water-cooled induction heating systems can be used for high-temperature preheating and stress-relieving applications. They have the ability to get to 1,450°F (788°C). This type of system is normally equipped with a temperature controller and temperature recorder, which are important components in stress-relieving applications.

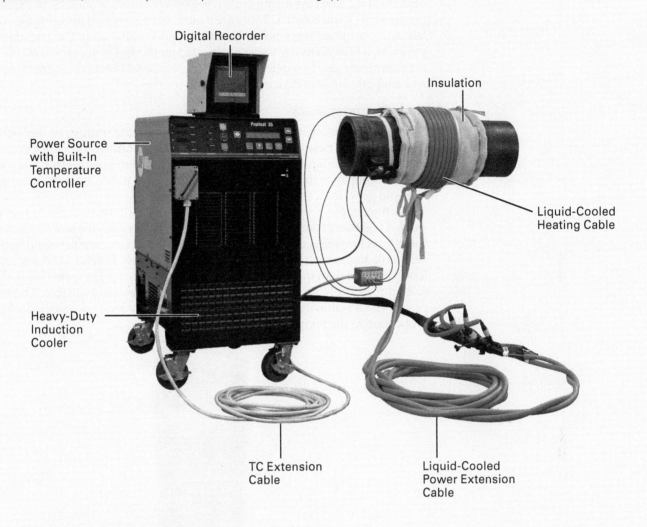

Source: © Miller Electric Mfg. LLC

1.2.6 Enclosed Furnaces

Enclosed furnaces or ovens include permanent and temporary structures or partial enclosures that must be made of firebrick and clay, sheet metal, or other noncombustible material. The item to be heated may be totally or partially enclosed. With very large items, it may not be practical or possible to enclose them entirely. Furnaces and ovens are heated by gas or oil burners, torches, or electric resistance heaters. *Figure 11* shows an example of an electric-powered, inert-atmosphere dual-chamber furnace.

Figure 11 Inert-atmosphere, dual-chamber furnace.
Source: Lucifer Furnaces, Inc.

1.3.0   Measuring Temperature

Control of preheating, interpass, and postheating temperatures depends on accurately determining the temperature of the weld zone or assembly. Both underheating and overheating can negatively affect a metal's physical and mechanical characteristics. There are a number of ways to determine the surface temperature of an item being heated. Some involve complex manufactured products, while others use less sophisticated methods. Devices and materials that are commonly used to determine surface temperatures include thermometers and temperature-sensitive indicators.

1.3.1   Thermometers

Electronic thermometers (*Figure 12*) now dominate industrial temperature measurement. These thermometers display temperatures on either an analog meter or a digital liquid crystal display (LCD) readout. They generally use either a thermocouple or thermistor type of temperature probe, or both, to sense the heat and generate a temperature indication.

Thermocouple probes use a sensing device made of two dissimilar wires welded together at one end called a *junction*. When heated, the junction generates a low-level DC voltage that produces a proportional temperature reading on the electronic thermometer indicator. Thermocouple probes tend to be rugged and less expensive than thermistor probes. The probe shown in *Figure 12* is designed to clamp securely to small pipe; many probe styles are available. Thermistor probes use a semiconductor electronic element in which resistance changes in proportion to the temperature change.

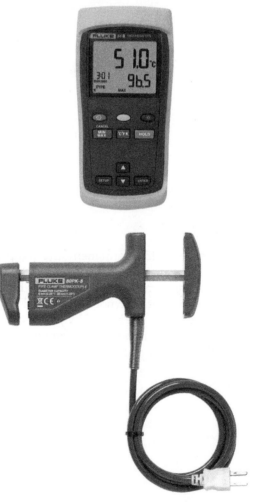

Figure 12 Electronic thermometer.
Source: Fluke Corporation, reproduced with permission

Often, several different probes are used with the same electronic thermometer to allow temperature measurement in different applications. Many electronic thermometers can also accommodate two or more probes at the same time, so simultaneous measurements can be made at several locations. Most thermometers of this type can calculate and display the difference in temperature between the locations being measured.

Electronic thermometers are precise measuring instruments. Read and follow the manufacturer's instructions for operating electronic thermometers. Also, be sure to follow the manufacturer's instructions for calibration of the instrument.

Dial thermometers are available in various forms. They are popular because they are rugged, small, and typically inexpensive. They come in many stem lengths and dial sizes. Because pocket-style dial thermometers have smaller calibrated scales, it is sometimes difficult to get accurate readings. An electronic version of the pocket thermometer, such as the one shown in *Figure 13*, works best when accuracy is important. Remember that most thermometers, regardless of the type, are generally more accurate in the midsection of their range. They are often somewhat less accurate when reading temperatures close to the minimum and maximum values in their range.

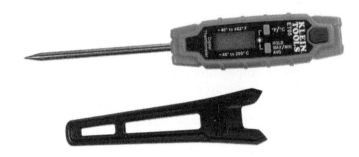

Figure 13 Electronic pocket thermometer.
Source: Klein Tools

Handheld infrared thermometers (*Figure 14*) are also very popular. Infrared thermometers allow the user to measure temperature without contacting the surface of an object. For this reason, they are also called *noncontact thermometers*. They work by detecting infrared energy emitted by the object at which they are pointing. Laser sighting devices allow the user to aim the thermometer like a pistol at the surface to be measured. The temperature is displayed instantly. This type of thermometer is useful for taking temperature readings in hard-to-reach areas. However, it may not be effective on highly reflective surfaces. The unit may have an adjustable setting that allows the measurement of reflective surfaces. A thermometer with no such adjustment can still be used to measure the temperature of a shiny surface by applying matte paint or nonreflective tape to the surface.

The distance-to-spot (D:S) ratio of infrared thermometers is important to understand. The D:S is the ratio of the distance to the object being measured compared to the diameter of the temperature measurement area. For example, if the D:S ratio is 12:1, measuring from 12" (30.5 cm) away will give the average temperature of a spot 1" (25 mm) in diameter (*Figure 15*). The greater the distance, the larger the spot becomes. When using an infrared unit with a 12:1 D:S from 12' away, the spot becomes 1' in diameter. As a result, the temperature measurement may be inaccurate unless the surface being measured is at least as large as the spot. Otherwise, the device will average the temperature of two or more unrelated surfaces.

Figure 14 An infrared noncontact thermometer.

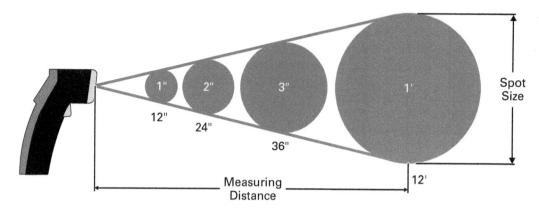

Figure 15 The D:S ratio illustrated.

Furnace Thermocouples

The rate of temperature rise, holding time at temperature, and rate of cooling are critical to proper heat treatment. For this reason, thermocouples used in furnaces must measure the temperature of the heat-treated metal, not the ambient temperature inside the furnaces. To do this, the thermocouples must, of course, be in contact with the metal.

1.3.2 Temperature-Sensitive Indicators

Temperature-sensitive indicators are devices that change form or color at specific temperatures or temperature ranges. These include commercially made materials and devices such as crayon sticks (*Figure 16* [A]), liquids (*Figure 16* [B]), chalks, powders, pellets, and labels (*Figure 16* [C]). These indicators are designed to change color or form at precise temperatures, often within a 1% tolerance.

(A) Temperature Sticks

(B) Temperature Liquid

(C) Temperature Labels

Figure 16 Temperature-sensitive indicators.

CAUTION

Do not mark with the crayon in the weld zone. The recommended practice is to mark from the toe of the weld out 1" to 3" (25.4 mm to 76.2 mm), depending on the code.

Temperature sticks are among the most widely used temperature-sensitive indicators. The sticks consist of a series of graduated temperature-indicating crayons that cover the temperature range from 100°F to 1,150°F (38°C to 621°C). These sticks are used to mark the base metal. At a specific temperature, the mark melts or changes color. *Table 5* is an example of a preheating chart that could be used when selecting temperature-indicating crayons for various metals.

TABLE 5 Preheating Chart for Selecting Temperature-Indicating Crayons

Metal Group	Metal Description	Approximate Composition Percentage								Recommended Preheat Temperature
		C	Mn	Bi	Cr	Ni	Mo	Cu	Other	
Plain-Carbon Steels	Plain-Carbon Steel	Below 0.20%	—	—	—	—	—	—	—	Up to 150°F (Up to 65°C)
		0.20% to 0.30%	—	—	—	—	—	—	—	150°F to 300°F (65°C to 150°C)
		0.30% to 0.45%	—	—	—	—	—	—	—	250°F to 450°F (120°C to 230°C)
		0.45% to 0.80%	—	—	—	—	—	—	—	450°F to 750°F (230°C to 400°C)
Plain-Moly Steels	Carbon Moly Steel	0.10% to 0.20%	—	—	—	—	0.50%	—	—	150°F to 250°F (65°C to 120°C)
		0.20% to 0.30%	—	—	—	—	0.50%	—	—	200°F to 400°F (95°C to 205°C)

Determining Metal Temperatures by Color

The approximate temperature of a metal heated in a furnace can be determined by the different colors exhibited by the metal. Reference charts are available for welders to interpret metal temperatures by observing the color of the metal.

The chart shown in the image, based on carbon steel, is an example only. The information should not be relied upon for actual temperature measurement. Also, note that the colors shown in the image will appear differently in various environments. Sunlight, for example, will cause the colors to appear differently than they would in a darkened room.

Color		Approximate Temperature	
		°F	°C
Faint Red		930	500
Blood Red		1075	580
Dark Cherry		1175	635
Medium Cherry		1275	690
Cherry		1375	745
Bright Cherry		1450	790
Salmon		1550	845
Dark Orange		1630	890
Orange		1725	940
Lemon		1830	1000
Light Yellow		1975	1080
White		2200	1205

1.0.0 Section Review

1. The temperature range that exists between a metal's critical temperature and its melting point is called its _____.
   a. transformation temperature range
   b. preheat span
   c. HAZ
   d. discontinuity arc

2. Open-top preheaters work similarly to a(n) _____.
   a. convection oven
   b. oxyfuel torch
   c. box furnace
   d. gas kitchen stove

3. Handheld infrared thermometers are also called _____.
   a. temperature sticks
   b. finger elements
   c. noncontact thermometers
   d. optical thermocouples

2.0.0 Temperature Maintenance and Postheating Processes

Performance Tasks

There are no Performance Tasks in this section.

Objective

Describe interpass temperature control and postheating processes.
   a. Describe interpass temperature control.   b. Describe various postheating processes.

In addition to proper preheating, it is equally important to maintain the proper temperature of a weldment or assembly during welding operations and after the welding has been completed. This section introduces several postheating processes that help prevent weld failures by relieving stresses, improving corrosion resistance, restoring ductility, and reducing brittleness.

2.1.0 Interpass Temperature Control

Given the impact of the interpass temperature on weld strength, interpass temperature control is an important process for welders to understand and master. Welders use interpass temperature control to maintain the temperature of the weld zone during welding, particularly when the cooling rate is too high or too low to maintain the correct temperature of the weldment between weld passes. Variables affecting interpass temperatures include the time between weld passes, the thickness of the base metal, and heat-related characteristics of the metal. Interpass temperature control is practiced in multiple different welding situations.

If welding is ever interrupted, the temperature of the weldment must be brought back to the interpass temperature before welding is continued. Minimum interpass temperature control is an important strategy to prevent cracking. The minimum interpass temperature is usually the same as the preheat temperature. Usually, there is also a maximum interpass temperature, which is especially important in preventing overheating from heat buildup in smaller weldments.

While preheating is important, interpass temperature for multipass welds is usually also important. Interpass temperature control is often simply a continuation of preheating with the same heating equipment and techniques.

With large, high-mass weldments and infrequent weld passes, as well as low-temperature working conditions, the base metal can cool below the required preheat temperature if additional heat is not supplied. When the base metal cools below the specified minimum preheat temperature, it must be reheated to the preheat temperature before welding continues.

Importance of Maximum Interpass Temperatures

Austenitic stainless steel typically has a 350°F (177°C) maximum interpass temperature because it is subject to intergranular migration of alloying elements (sensitization) between 800°F and 1,600°F (427°C and 871°C). Migration decreases the corrosion resistance of stainless steel. Adhering to the maximum interpass temperature is especially critical when welding on components or piping systems that will be exposed to cryogenic applications such as liquid oxygen.

2.2.0 Postheating

Postweld heat treatment (PWHT) is the heating of a weldment or assembly after welding has been completed. It can be performed on most types of base metals. Postheat equipment and techniques are the same as those used for preheating.

Postweld heat treatments include the following:

- Stress relieving
- **Annealing**
- Normalizing
- Tempering

2.2.1 Stress Relieving

Stress relieving is the most commonly used postheat treatment. Stress relieving is done below the lower transformation temperature, usually in the range of 1,050°F to 1,200°F (566°C to 649°C).

Stress relief treatment is done for the following reasons:

- To reduce residual (shrinkage) stresses in weldments and castings, which is especially important with highly restrained joints
- To improve resistance to corrosion and stress corrosion cracking
- To improve dimensional stability during machining operations
- To improve resistance to impact loading and low-temperature failure

Annealing: A postweld heat treatment for stress-relieving weldments that is done at temperatures approximately 100°F (38°C) above the critical temperature with a prolonged holding period.

Stress Relief

The most commonly used stress relief method of weldments is postweld heat treatment. Stress relief can also be accomplished mechanically. For instance, *peening* is a popular technique for mechanical stress relief in which a peening hammer hits a small metal ball on the surface of metal to increase the surface's density. Another emerging method involves attaching a mechanical vibrating device to the weldment during or immediately after welding. When activated, the device vibrates the weldment at a specific resonant frequency. The sustained force of the vibrations deforms but does not fracture the shape of the metal's grains. This is called *plastic deformation*. The plastic deformation of the metal's grains evens the stress distribution within the weldment. The weight of the weldment generally determines the length of time the weldment will be subjected to the vibrations.

NOTE

Research demonstrates the applicability of the vibration technology on small parts. However, the research findings are mixed for the applicability of the vibration technology regarding geometrically complex structures between many kilograms and tons. Given the current research findings, caution should be used in discussing this technology as uniformly useful for all shapes and sizes of structures.

Some codes cover stress relieving. These codes specify the heating rate, holding time, and cooling rate. The heating rate is usually 300°F to 400°F (149°C to 204°C) per hour. The holding time is typically one hour for each inch (25.4 mm) of thickness. The cooling rate is usually 300°F to 500°F (149°C to 260°C) per hour. Stress relieving requires the use of temperature indicators and temperature-control equipment.

2.2.2 Annealing

Annealing relieves much more stress than a stress-relieving postheat treatment can, but it results in steel of lower strength and higher ductility. Annealing is done at temperatures approximately 100°F (38°C) above the critical temperature, with a prolonged holding period. The heating is followed by slow cooling in the furnace or covering, wrapping, or burying the item. Austenitic stainless steel is an exception because it requires rapid cooling. Annealing is used to relieve the residual stresses associated with welds in carbon-molybdenum pipe and in welded castings containing casting strains. Generally, annealing leaves the metal in its softest condition with good ductility.

2.2.3 Normalizing

Normalizing is a process in which heat is applied to remove strains and reduce grain size. Normalizing is done at temperatures and holding periods comparable to annealing. However, the cooling is usually done in still air outside the furnace, and the cooling rate is slightly faster than the slow cooling used in annealing. Normalizing usually results in higher strength and less ductility than annealing. The process can be used on mild steel weldments to form a uniform austenite solid solution, soften the steel, and make it more ductile. Normalized metal is not as soft and free of stresses as fully annealed metal.

2.2.4 Tempering

Tempering, also called *drawing*, increases the toughness of quenched steel and helps avoid breakage and failure of heat-treated steel. The process reduces both the hardness and brittleness of hardened steel. Tempering is done at temperatures below the critical temperature, much lower than those used for annealing, normalizing, or stress relieving. Tempering is commonly used on tool steel.

2.2.5 Hardening

The hardness of a material is its resistance to indentation. In metals, hardness is related to their elastic and plastic properties. These properties can be altered when mechanical forces or temperature changes exert stresses or strains on the metal. The deformation, or interruption, of the crystal structure inside the metal is called **discontinuity**. Discontinuity is also called *dislocation*.

Discontinuity: The deformation, or interruption, of the crystal structure inside a metal.

When a piece of metal is hit with a hammer, deformation occurs with each stroke. Hammering metal can form it into a new shape, but the amount of metal will still be the same. The reshaping of metal at ambient temperature is called *cold forming* or *cold working*. During cold forming, the metal is strained beyond its yield point where plastic deformation occurs, and it will not return to its original form. When a metal is strained or stressed beyond its yield point at ambient temperature, it is said to have been work hardened. **Strain hardening**, also called *work hardening*, tends to make the component stronger (harder) and less prone to failure. Parts that have been hardened using the cold forming process, called *cold hardening*, may look brighter in appearance.

Strain hardening: A method, also called *work hardening*, used to increase the strength of a metal through permanent deformation.

Plastic deformation is the permanent distortion of a metal after its elastic limits have been exceeded because of stress. Metals that have been cold worked have undergone plastic deformation. The ability of metals to flow in a plastic manner without breaking apart is the basis for all metal-forming processes. Plastic flow is the process in which metals are stretched or compressed permanently without breaking or cracking.

When metal is heated to a point just below its melting point, the crystals in it are forced to deform to adapt to the strain caused by the heat. If the crystals in a metal are deformed during the welding process, the weldment will gain strength and become harder. In some situations, metals are intentionally heated again after they have been welded to gain more strength.

2.2.6 Heat Treatment Devices

Metals can be heated by a variety of methods and devices before or after welding. Properly controlling heat is a major consideration. Water-cooled induction heaters are used in some applications, but they are not always available or may not be the process required for the application. *Figure 17* shows a portable induction heater on top of its power supply / control box. A rear view of the power supply is also shown.

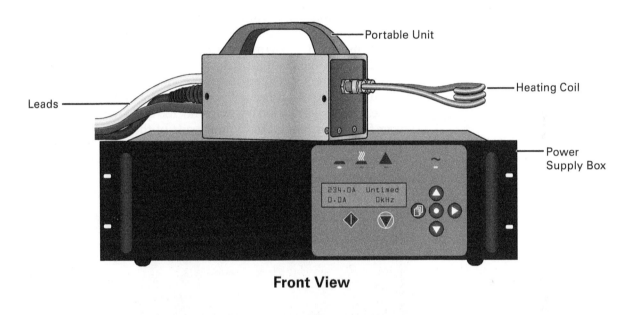

Front View

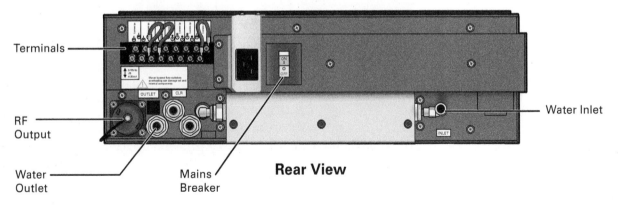

Rear View

Figure 17 Portable induction heater and its power supply / control box.

Depending on the length of the leads between the power supply and the handheld unit (work head), the operator can move the induction coils to the object rather than move the object to the coils. These portable induction units can be used to heat or cool the object. They can also be programmed to operate at different power levels and times. This variety of options allows better control of the preheat or postheat treatment.

Ceramic heating pads are electrical devices often used to heat and relieve stress in metals. In a ceramic heating pad (*Figure 18*), the individual ceramic components are threaded together as needed to cover the metal being heated. The heating pads can be placed on plate steel or laid over pipe. Electrical wires are threaded through the individual ceramic pieces, generating heat when power is applied. Each individual piece is rated at a specific amperage level. The number of pieces used is dependent upon the desired heating output and the physical size of the pad needed.

The wires from the heating pads must be connected to an electrical device that will control the amount of electrical current flowing through the wires. Temperature-sensing thermocouples attached to the metal let the controlling device detect the temperature of the object being heated. The controller compares the thermocouple's signal against a preprogrammed temperature setting and determines if the metal needs to be warmer or cooler. If the controller senses that the metal is getting too hot or too hot too quickly, it can reduce the electrical power flowing through the ceramic pads of the heating device. On the other hand, if the controller determines that the metal is not heating fast enough, the controller can increase the current flow through the pads. This causes the pads to become warmer and the metal to heat more quickly. Ceramic heating pads are electrical, so they may be plugged into the AC outlet of a welding power supply.

Figure 18 Ceramic heating pads.
Source: © Miller Electric Mfg. LLC

2.2.7 Time-at-Temperature Considerations

In most heat treatment procedures, temperature and time-at-temperature are both specified by code. Time is very important because metallurgical changes are sluggish below critical temperatures. The maximum temperature is determined by the metal alloy. The holding time at the maximum temperature is based on the metal's thickness, usually one hour for each inch (25.4 mm) of thickness. The cooling rate is determined by the specific treatment or by any applicable code. Temperature control must also be used for postheating. Postheating temperatures are determined in the same manner as preheating temperatures.

2.0.0 Section Review

1. An additional interpass temperature for small weldments and frequent weld passes is often given, specified as a _____.
    a. modal value
    b. maximum temperature
    c. calculated distribution
    d. critical temperature

2. For stress relief on metals being welded, the holding time is typically how many hour(s) per inch (25.4 mm) of thickness?
    a. 14
    b. 12
    c. 2
    d. 1

Module 29204   Review Questions

1. The controlled heating of the base metal immediately before welding begins is called _____.
    a. progressive base quenching
    b. thermal stress conditioning
    c. interpass temperature control
    d. preheat

2. The temperature at which a metal turns liquid is its _____.
    a. melting point
    b. preheating point
    c. welding temperature
    d. postheating temperature

3. Hydrogen gas in the weld zone will cause _____.
    a. slow melting of the base metal
    b. underbead cracking
    c. hardening
    d. slag

4. The hardenability of steel is directly related to its alloying elements and _____.
    a. thickness
    b. carbon content
    c. preheat temperature
    d. welding temperature

5. Preheating is *most* effective when the welding assembly is heated _____.
    a. rapidly
    b. with a torch
    c. in an oven
    d. with induction heaters

6. Air-cooled induction heating systems are typically used for preheating applications *up to* _____.
    a. 300°F (149°C)
    b. 400°F (204°C)
    c. 600°F (316°C)
    d. 1,000°F (538°C)

7. Temperature-indicating crayons can be used to indicate temperatures ranging from 100°F (38°C) to _____.
   a. 400°F (204°C)
   b. 980°F (527°C)
   c. 1,150°F (621°C)
   d. 1,350°F (732°C)

8. What temperature is usually the same as the *minimum* interpass temperature?
   a. Cutting
   b. Annealing
   c. Welding
   d. Preheat

9. For stress relief on metals being welded, the heating rate is usually how many degrees per hour?
   a. 100°F to 200°F (38°C to 93°C)
   b. 200°F to 300°F (93°C to 149°C)
   c. 300°F to 400°F (149°C to 204°C)
   d. 400°F to 500°F (204°C to 260°C)

10. Annealing is done at temperatures that are approximately how many degrees above the critical temperature?
    a. 100°F (38°C)
    b. 200°F (93°C)
    c. 300°F (149°C)
    d. 400°F (204°C)

Answers to odd-numbered questions are found in the Review Question Answer Keys at the back of this book.

Answers to Section Review Questions

Answer	Section Reference	Objective
Section 1.0.0		
1. a	1.1.0	1a
2. d	1.2.3	1b
3. c	1.3.1	1c
Section 2.0.0		
1. b	2.1.0	2a
2. d	2.2.7	2b

GMAW and FCAW – Equipment and Filler Metals

Source: Photo courtesy of The Lincoln Electric Company, Cleveland, OH, U.S.A.

Objectives

Successful completion of this module prepares you to do the following:

1. Describe basic GMAW/FCAW processes and related safety practices.
   a. Identify GMAW/FCAW-related safety practices.
   b. Identify GMAW/FCAW processes.
   c. Describe the various GMAW metal transfer modes.
   d. Describe the FCAW metal transfer process.
2. Describe GMAW and FCAW equipment and explain how to prepare for welding.
   a. Identify common GMAW/FCAW welding equipment.
   b. Describe power source control considerations.
   c. Identify and describe welding cables and terminations.
   d. Identify and describe external wire feeders and their controls.
   e. Identify and describe GMAW and FCAW guns, contact tips, and nozzles.
   f. Identify various shielding gases and their related equipment.
   g. Explain how to set up welding equipment for GMAW and FCAW welding.
3. Identify GMAW and FCAW filler metals.
   a. Identify various GMAW filler metals.
   b. Identify various FCAW filler metals.

Performance Task

Under supervision, you should be able to do the following:

1. Set up GMAW and FCAW equipment with appropriate shielding gases and filler metals.

Overview

Gas metal arc welding (GMAW) and flux-cored arc welding (FCAW) are two common welding processes. This module describes GMAW and FCAW processes and related basic safety practices. An overview of GMAW and FCAW equipment and setup is also provided. Various types of filler metals used in GMAW and FCAW are introduced.

Digital Resources for Welding

Scan this code using the camera on your phone or mobile device to view the digital resources related to this craft.

1.0.0 GMAW and FCAW Welding

Performance Task

1. Set up GMAW and FCAW equipment with appropriate shielding gases and filler metals.

Objective

Describe basic GMAW/FCAW processes and related safety practices.

a. Identify GMAW/FCAW-related safety practices.
b. Identify GMAW/FCAW processes.
c. Describe the various GMAW metal transfer modes.
d. Describe the FCAW metal transfer process.

To perform gas metal arc welding (GMAW) and flux-cored arc welding (FCAW) safely and effectively, a good understanding of how both processes work is required. It is also critical to understand the safety practices necessary for all welding operations.

1.1.0 Safety Summary

The following sections summarize the safety procedures and practices that welders must observe when cutting or welding. Complete safety coverage is provided in NCCER Module 29101, *Welding Safety*, which should be completed before continuing with this module. The safety procedures and practices are in place to prevent potentially life-threatening injuries from welding accidents. Above all, be sure to wear appropriate protective clothing and equipment when welding or cutting.

1.1.1 Protective Clothing and Equipment

Welding activities can cause injuries unless you wear all the protective clothing and equipment designed specifically for the welding industry. The following safety guidelines about protective clothing and equipment should be followed to prevent injuries should be followed to prevent injuries:

- Wear a face shield over snug-fitting cutting goggles or safety glasses for gas welding or cutting. The welding hood should be equipped with an approved shade for the application. A welding hood equipped with a properly tinted lens is best for all forms of welding.

- Wear proper protective leather and/or flame-retardant clothing along with welding gloves that protect the welder from flying sparks, molten metal, and heat (*Figure 1*).

- Wear high-top safety shoes or boots. Make sure the tongue and lace area of the footwear will be covered by a pant leg. If the tongue and lace area is exposed or the footwear must be protected from burn marks, wear leather spats under the pants or chaps over the front and top of the footwear.

- Wear a 100% cotton cap with no mesh material included in its construction. The bill of the cap points to the rear. If a hard hat is required for the environment, use one that allows the attachment of rear deflector material and a face shield. A hard hat with a rear deflector is generally preferred when working overhead and may be required by some employers and jobsites.

- Wear earmuffs, or at least earplugs, to protect your ear canal from sparks.

Figure 1 A fully protected welder wearing leather, gloves, and a hood.
Source: © Miller Electric Mfg. LLC

WARNING!

Using proper personal protective equipment (PPE) for the hands and eyes is particularly important. The most common injuries that welders experience during GMAW and FCAW operations are injuries to the fingers and eyes.

1.1.2 Fire/Explosion Prevention

Welding activities usually involve the use of fire or extreme heat to melt metal. Whenever fire is used in a weld, the fire must be controlled and contained. Welding or cutting activities are often performed on vessels that may have once contained flammable or explosive materials. Residues from those materials can catch fire or explode when a welder begins work on such a vessel. The following fire and explosion prevention guidelines associated with welding contribute to a safe work zone:

- Never carry matches or gas-filled lighters in your pockets. Sparks can cause the matches to ignite or the lighter to explode, causing serious injury.

- Never use oxygen to blow dust or dirt from clothing. The oxygen can remain trapped in the fabric for a time. If a spark hits clothing during this time, the clothing can burn rapidly and violently out of control.

- Make sure any flammable material in the work area is moved or shielded by a fire-resistant covering. Approved fire extinguishers must be available before attempting any heating, welding, or cutting operations. If a hot work permit and a fire watch are required, be sure those items are in place before beginning and that all site requirements have been observed.

- Never release a large amount of oxygen or use oxygen as compressed air. The presence of oxygen around flammable materials or sparks can cause rapid and uncontrolled combustion. Keep oxygen away from oil, grease, and other petroleum products.

- Never release a large amount of fuel gas, especially acetylene. Propane tends to concentrate in and along low areas and can ignite at a considerable distance from the release point. Acetylene is lighter than air but is even more dangerous than propane. When mixed with air or oxygen, acetylene will explode at much lower concentrations than any other fuel.

- To prevent fires, maintain a neat and clean work area, and make sure that any metal scrap or slag is cold before disposing of it.

- Before cutting containers such as tanks or barrels, check to see if they have contained any explosive, hazardous, or flammable materials, including petroleum products, citrus products, or chemicals that decompose into toxic fumes when heated. As a standard practice, always clean and then fill any tanks or barrels with water or purge them with a flow of inert gas such as nitrogen to displace any oxygen. Containers must be cleaned by steam cleaning, flushing with water, or washing with detergent until all traces of the material have been removed.

WARNING!

Welding or cutting must never be performed on drums, barrels, tanks, vessels, or other containers until they have been emptied and cleaned thoroughly, eliminating all flammable materials and all substances (such as detergents, solvents, greases, tars, or acids) that might produce flammable, toxic, or explosive vapors when heated. Clean containers only in well-ventilated areas, as vapors can accumulate during cleaning, causing explosions or injury.

Proper procedures for cutting or welding hazardous containers are described in the *American Welding Society (AWS) F4.1, Safe Practices for the Preparation of Containers and Piping for Welding, Cutting, and Allied Processes.*

1.1.3 Work Area Ventilation

Vapors and fumes tend to rise in the air from their sources. Welders must often work above the welding area where the fumes are being created. Welding fumes can cause personal injuries including long-term respiratory harm. Good work area ventilation helps to remove the vapors and protect the welder. The following is a list of work area ventilation guidelines to consider before and during welding activities:

- Follow confined space procedures before conducting any welding or cutting in a confined space.

- Never use oxygen for ventilation in confined spaces.

- Always perform cutting or welding operations in a well-ventilated area. Cutting or welding operations involving zinc or cadmium materials or coatings result in toxic fumes. For long-term cutting or welding of such materials, always wear an approved full-face, supplied-air respirator that uses breathing air supplied from outside of the work area. For occasional, very short-term exposure, a high-efficiency particulate arresting-rated (also called HEPA-rated) or metal fume filter may be used on a standard respirator.

- Make sure confined spaces are properly ventilated for cutting or welding purposes. Use powered extraction systems when available. *Figure 2* shows a welder working with a portable extraction (vacuum) system designed to remove and filter welding fumes and vapors.

Figure 2 Welder working with a portable extraction system.
Source: Photo courtesy of The Lincoln Electric Company, Cleveland, OH, U.S.A.

WARNING!

Backup flux may contain silica, fluorides, or other toxic materials. Alcohol and acetone used for the mixture are flammable. Weld only in well-ventilated spaces. Failure to provide proper ventilation could result in personal injury or death.

1.2.0 GMAW and FCAW Processes

GMAW (*Figure 3*) uses an electric arc to melt a consumable wire electrode and fuse it with the base metal. The arc generates an intense heat of 6,000°F to 10,000°F (3,316°C to 5,538°C). The arc is stabilized, and the molten filler and base metals are protected from oxidation by a shielding gas dispensed from a GMAW gun nozzle. The wire is automatically fed into the weld puddle as it melts. The process usually operates with direct current electrode positive (DCEP).

FCAW uses an electric arc to melt a flux-cored wire electrode and fuse it with the base metal to form a weld. The wire is tubular, allowing a flux material to be inserted inside the electrode. A wire feeder automatically feeds the flux-cored wire as it is being consumed in the arc. The equipment used for FCAW is basically the same as that used for GMAW. The main difference between the two processes is that GMAW uses solid or **composite metal-cored wire electrodes** with a shielding gas, while FCAW uses a tubular flux-cored wire electrode with or without a shielding gas.

Composite metal-cored wire electrodes: Filler wire electrodes with hollow cores containing powdered materials, primarily metals.

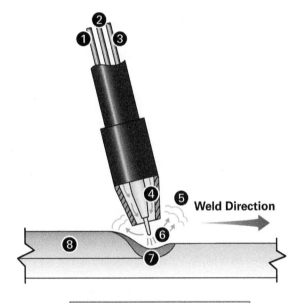

Weld Direction

1. Current Conductor
2. Wire Electrode
3. Shielding Gas
4. Wire Guide and Contact
5. Shielding Gas
6. Arc
7. Molten Weld Metal
8. Solidified Weld Material

Note: Other gun configurations and styles are available.

Figure 3 Basic GMAW process.

In the two basic FCAW processes, one uses an external shielding gas while the other is self-shielding. Whether a shielding gas is used is determined by the type of flux-cored wire used. When the flux-cored wire does not require an external shielding gas, the process is called self-shielding FCAW or FCAW-S. When an external shielding gas is required, the process is called gas-shielded FCAW or FCAW-G. Equipment specifically designed for self-shielding FCAW does not have provisions for shielding gas (gas connections, solenoids, and gas preflow and postflow timers), and the guns do not have gas nozzles. Equipment used for gas-shielded FCAW welding is the same as that used for GMAW, except that the wire feed drive assemblies have to be able to handle the cored wire, which is softer and generally larger in diameter than solid wire.

> **WARNING!**
>
> Do not apply a shielding gas to a self-shielding, flux-cored electrode. The resulting weld is likely to be brittle and risk cracking. Through exposure to air, FCAW electrodes oxidize and form the intended weld deposit and mechanical properties. The use of a shield gas stops the oxidizing process, leaves higher levels of alloys, and destroys the mechanical properties of the weld. Even though the weld may run smoother when using a shielding gas, the resulting weld will not be a good weld.

FCAW is presently limited to the welding of ferrous metals and nickel-based alloys. These include low- and medium-carbon steels, some low-alloy steels, cast irons, nearly every stainless steel alloy, and many hard-facing and wear-resistant alloys. Additionally, many custom alloys exist due to the ease of designing specific chemistry in small batches. When critical welds are required, the gas-shielded FCAW-G process is often used because the combination of flux and separate shielding gas produces welds of very high quality.

Synergic GMAW Welding

Innovation in GMAW technologies is being used in the welding industry to increase the quality of welds as well as address labor shortages. The technologies of synergic welding machines automate more of the welding process, making welding work more able to be performed by operators with beginning levels of welding experience.

GMAW increasingly uses synergic operations, given the ease of use of semiautomatic welding machines with single control output for welding parameters. Synergic welding machines allow the welder to preset the variables of material, diameter of the wire, and filler metal. The machine responds with the corresponding volts and speed of wire. Potential benefits include increased productivity. Potential challenges include the loss of control to customize weld or job needs.

In the United States, two of the leading controlled short-circuiting pulse GMAW processes include Miller Electric's Regulated Metal Deposition™ (RMD™) and Lincoln Electric's Surface Tension Transfer® (STT®). The goal of RMD™ and STT® is uniform and code-meeting welds created more quickly by a wider range of welders, including apprentice operators. The processes control the short circuit and reduce the welding current to produce a consistent metal transfer with uniform droplets.

1.3.0   GMAW Metal Transfer Modes

GMAW can use several methods of metal transfer to conduct the filler metal to the molten weld pool. The principal modes include the following:

- **Spray transfer**
- **Globular transfer**
- **Short-circuiting transfer**
- **Pulsed transfer**

Spray transfer: A GMAW welding process in which the electrode wire does not touch the base metal, and the filler metal is transferred axially (in a straight line) in the arc as fine droplets.

Globular transfer: A GMAW process in which the electrode wire does not touch the base metal, and the filler metal is transferred randomly in the arc as large, irregular globules.

Short-circuiting transfer: A GMAW welding process in which the electrode wire shorts against the base metal. The welding current repeatedly and rapidly melts the wire and establishes an arc, which is extinguished when the wire again touches the base metal.

Pulsed transfer: A GMAW welding process in which the welding power is cycled from a low level (background level) to a higher level (welding level), resulting in lower average voltage and current; this is a variation of the spray-transfer process.

1.3.1 Spray Transfer

In the spray-transfer mode (*Figure 4*), the welding voltage is set in the higher ranges, between 26 volts (V) to 32 V, to keep the end of the wire electrode away from the work and maintain a continuous arc. The gas mixture must have a minimum 80% threshold level of argon to enter spray-transfer mode. The high heat of the arc melts the end of the wire electrode, and fine droplets are carried by the arc **axially** to the molten weld puddle on the base metal. The spray transfer deposits filler metal at a high rate and produces very little spatter. However, the weld puddle is relatively large, and the mode works best in the flat or horizontal fillet positions. Also, only certain shielding gases or gas combinations can be used with spray transfer. Shielding gases are covered in detail later in this module.

Axially: In a straight line along an axis.

1.3.2 Globular Transfer

In the globular-transfer mode (*Figure 5*), the welding voltage, generally 22 V to 25 V, is set lower than for spray transfer. The end of the wire electrode remains a short distance from the work to maintain a continuous arc. Because of its weight, the electrode wire tip periodically melts off in a large glob, which the arc then carries to the workpiece, where it mixes with the weld puddle. Globular transfer can be used to deposit filler metal at high rates and produce high-quality welds in all positions. However, bead appearance can be rough, and spatter may be heavy in some applications.

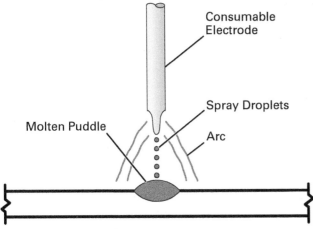

Figure 4 Spray transfer.

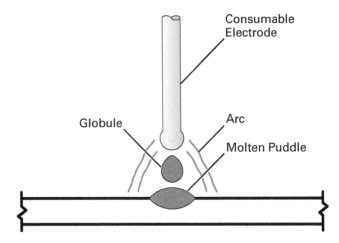

Figure 5 Globular transfer.

1.3.3 Short-Circuiting Transfer

Short-circuiting transfer (GMAW-S), at 16 V to 22 V, uses the lowest voltage settings of the three modes (*Figure 6*). It is the best mode for welding light-gauge sheet metal. The wire electrode is fed against the workpiece. When contact is made, current flow heats the wire electrode until it melts enough for a droplet to separate and mix with the weld puddle. When the droplet separates, a gap is formed, and an arc is initiated. The continuously feeding wire closes the gap and contacts the workpiece, shorting out the arc. The current again heats the wire until another droplet separates, and the arc is again established. This process repeats many times each second. The short-circuiting process can be used to weld thin materials and is suitable for all positions. However, it does not penetrate deeply and may produce incomplete fusion on thicker materials because of the low heat input.

Several manufacturers have introduced proprietary versions of short-circuiting transfer that rely on intelligent inverter power sources. These adjust the voltage and current moment-by-moment throughout each stage of the short-circuiting transfer process so the droplets form and then transfer to the weld puddle optimally. The result is minimal spatter, excellent fusion, and higher-quality welds. These new technologies are especially popular in settings involving stainless steel pipe welds.

Adaptive Control Welding

Some newer inverter power sources can use adaptive control welding during pulse transfer. These units sense and alter the frequency of the pulse, its width, the peak current, and wire speed automatically to maintain optimum welding characteristics during welding.

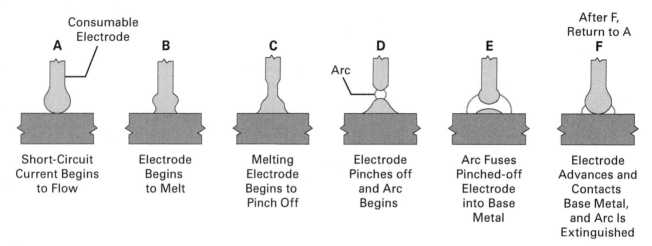

Figure 6 Short-circuiting transfer.

1.3.4 Pulsed Spray Transfer

Pulsed-transfer (GMAW-P) welding is a variation of spray-transfer welding (*Figure 7*). The process transfers molten metal in droplets while maintaining an arc with the base metal. The GMAW-P process also allows the transfer action to occur while lower current levels are used. The advantages of the lower current levels are less spatter and better penetration without melting through the base metal. Pulsed transfer is especially good for aluminum welding because it results in a smaller heat-affected zone (HAZ).

In pulsed-transfer welding, the welding power is electronically controlled by a special power source that pulses the output power to the consumable electrode. Although different GMAW machines may have different types of power sources, the results are the same for each. The welding current is generated in pulses of high current at specified time intervals. The frequency of these pulses varies from one power source to the next, but some power sources can produce pulses over a wide time range. During the period between each pulse, the welding current is reduced but is still maintained at a high enough level to keep the wire heated and an arc established. This lowered current level is called the background current.

After an arc has been established and maintained between the consumable electrode and the base metal, the welding current to the electrode is momentarily increased to form a pulse, and then it returns to its lower background level.

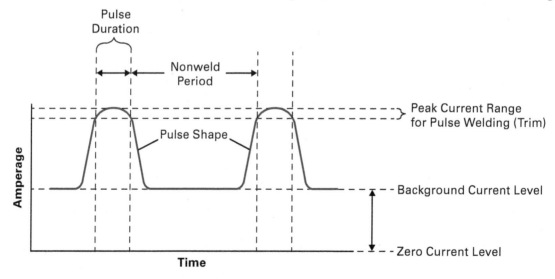

Note: Different waveforms and durations are used for various applications.

Figure 7 Output pulses of high welding current occurring over time.

Each pulse of welding current must peak strongly enough and have enough duration to make a single drop of molten metal separate from the wire that is maintaining an arc with the base metal.

Figure 8 shows the GMAW-P process relative to one of the pulse shapes. Points A through E on the diagram correspond to the labels in the electrode arc illustrations above the diagram.

When the single drop of molten metal separates from the electrode, it is transferred across the arc to form the molten puddle on the base metal. This transfer process creates less spatter and reduces the chance of melt-through. When these pulses are repeated many times per second, the result is a high-quality weld with minimal heat affecting the welding zone.

The following is a summary of the advantages of GMAW-P over welding processes using continuous currents:

- Rivals gas tungsten arc welding (GTAW) for weld quality
- Permits all-position welding
- Allows smooth welds to be made on thin materials without melt-through
- Controls thick plate weld pool more easily
- Reduces cost because larger-diameter electrode wire can be used
- Can be used on base metals such as aluminum, stainless steel, copper alloy, nickel steels, and carbon steels
- Reduces weld contamination from surface oxides because less surface area is exposed during the weld
- Creates a smaller HAZ, which favors its use with aluminum

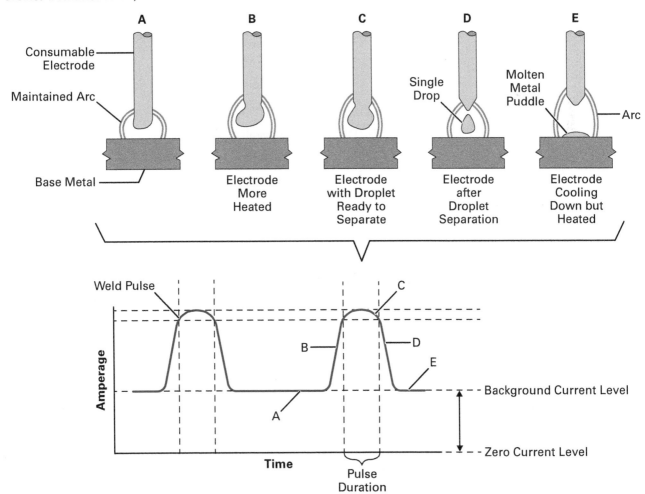

Figure 8 Typical pulsed welding process associated with one pulse shape.

1.4.0 FCAW Metal Transfer Process

The FCAW process (*Figure 9*) uses an electric arc to melt the filler and base metals and transfer fine droplets of filler metal to the weld puddle. Globular transfer is commonly used in FCAW processes, but axial spray can be achieved if the correct electrode and shielding gas are used.

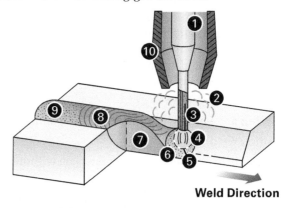

Weld Direction

❶ Contact Tip

❷ Optional Shielding Gas Envelope

❸ Flux-Cored Electrode

❹ Arc

❺ Flux-Generated Gas Envelope

❻ Molten Metal

❼ Weld Metal

❽ Molten Slag

❾ Solidified Slag

❿ Optional Nozzle

Note: Other gun configurations and styles are available.

Figure 9 FCAW process.

The electromagnetic force that propels the metal spray is strong, but the process produces very little spatter. The flux inside the core contains various ingredients that perform the following functions:

- Ionizers to stabilize the arc
- Deoxidizers to purge the weld of gases
- Additional metals and elements to enhance the quality of the weld metal
- Ingredients to generate shielding gas
- Ingredients to form a protective slag over the weld bead

The gas shield generated by flux in the wire core, as well as the external gas shield, protects the arc and molten weld metal. The slag cover produced by the flux helps to support the weld metal and allows the process to be used for making high-quality welds in all positions. With larger diameter flux-cored wires, deposition rates of about 25 pounds per hour (~11 kg per hour) can be achieved. Note the same wire should not be used across the board for all applications or treated in the same manner. For example, the dual-shield method and the self-shielding wire method require different types of FCAW wire, and FCAW-S wire should not be used with gas shielding as a reaction to the atmosphere is needed for the shield to properly form.

1.4.1 FCAW Weld Penetration

FCAW welds made with carbon dioxide (CO_2) shielding gas have a much deeper penetration than either shielded metal arc welding (SMAW) or FCAW without a shielding gas. For this reason, edge preparation is often not needed for double-welded butt joints on plates up to $^3/_8$" (~10 mm) thick. When making FCAW fillet welds using CO_2 shielding, a smaller size fillet weld will have as much strength as a larger fillet made with SMAW because of the deep penetration achieved with FCAW and the CO_2 shielding.

1.4.2 FCAW Joint Design

FCAW uses the same basic joint designs (*Figure 10*) as SMAW. However, when making groove welds with FCAW using a CO_2 shielding gas, the square groove (square butt) joint can be used for plate thicknesses up to $^3/_8$" (~10 mm). Above this thickness, beveled joints should be used.

Because FCAW electrode wire is much thinner than comparable SMAW electrodes and FCAW penetration with CO_2 is superior to SMAW, the included angles of V-joints can be reduced to half the angle required for SMAW. For example, an included angle of 60° can be reduced to 30° when using FCAW with CO_2 shielding. This can save approximately 50% on filler metal and a considerable amount of welding time.

CAUTION

Always refer to the Welding Procedure Specification (WPS) or site quality standards for specific information on joint requirements. Information in this module is provided as a general guideline only.

NOTE

Open-root passes are not typically done with FCAW. Short-circuit GMAW, GTAW, or SMAW processes are typically used for root passes on open V-groove welds. If using FCAW for a root pass, use a backing bar.

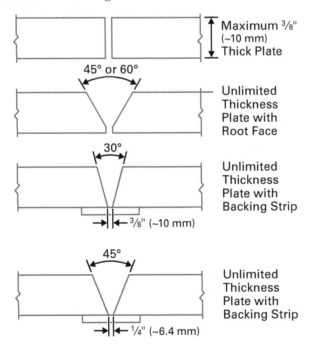

Figure 10 Joint designs for FCAW with CO_2 shielding.

1.0.0 Section Review

1. One good way to minimize fire hazards and vent a confined space before a cutting or welding procedure is to _____.
   a. spray the entire space with water
   b. use a powered extraction system
   c. blow compressed oxygen into the area
   d. purge the area with methane

2. FCAW uses a tubular flux-cored wire electrode with or without a(n) _____.
   a. oxidizer tip
   b. flux insulator
   c. shielding gas
   d. pulse generator

3. The GMAW metal transfer method that uses the *lowest* voltage settings is the _____.
   a. globular-transfer method
   b. pulsed-transfer method
   c. spray-transfer method
   d. short-circuiting transfer method

4. In an FCAW metal transfer process, the flux inside the core of the electrode contains ionizers that are used to _____.
   a. purge the weld of gases
   b. stabilize the electric arc
   c. generate a shielding gas
   d. form a protective slag

2.0.0 GMAW and FCAW Equipment

Performance Task

1. Set up GMAW and FCAW equipment with appropriate shielding gases and filler metals.

Objective

Describe GMAW and FCAW equipment and explain how to prepare for welding.

a. Identify common GMAW/FCAW welding equipment.
b. Describe power source control considerations.
c. Identify and describe welding cables and terminations.
d. Identify and describe external wire feeders and their controls.
e. Identify and describe GMAW and FCAW guns, contact tips, and nozzles.
f. Identify various shielding gases and their related equipment.
g. Explain how to set up welding equipment for GMAW and FCAW welding.

Typical GMAW and FCAW systems (*Figure 11*) are made of four interconnected pieces of equipment, including the following:

- Welding machine
- Wire feeder
- Welding gun
- Shielding gas supply (if required)

In some systems, equipment functions may be integrated or combined. For example, the wire feeder can be built into the same cabinet as the power source, or it may be incorporated into the welding gun. Welding guns are shown later in *Section 2.5.0*.

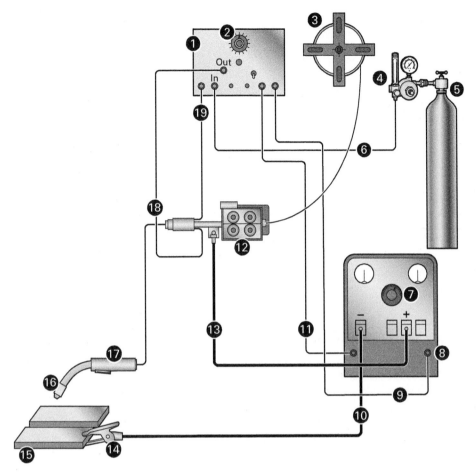

1	Control Unit
2	Wire Feed Speed Control
3	Electrode Wire Reel
4	Regulator/Flowmeter
5	Optional Shielding Gas Source
6	Gas-In Hose
7	Voltage Control
8	Power Source
9	Power to Control Unit
10	Workpiece Lead
11	Contactor Control Cable
12	Wire Feed Drive Assembly
13	Electrode Lead
14	Workpiece Clamp
15	Workpiece
16	Optional Gas Nozzle
17	GMAW or FCAW Gun
18	Gas-Out Hose
19	Gun Control Cable

Note: The polarity of the gun and workpiece leads is determined by the type of filler metal and application.

Figure 11 Simplified GMAW/FCAW system diagram.

Most systems have a separate wire feeder (push system) that pushes the wire from the feeder through the gun cable and gun. Other systems may incorporate the wire feeder into the welding gun itself (pull system) or do both (push-pull system). There are even systems that have synchronized wire feeders located at intervals along very long feeder conduits. Various types of wire feeders are covered later in *Section 2.4.0*. Heavy-duty, water-cooled guns may also be equipped with a closed-loop water-cooling system. Systems used for FCAW-S do not require gas cylinders since the shielding gas comes from the flux-core electrode.

2.1.0   Welding Power Source Types

Direct current (DC) welding machines are usually designed to produce either constant-current (CC) or constant-voltage (CV) DC power. A CC machine produces a CC output over a wide voltage range; this is typical of a SMAW welding machine. A CV machine maintains a CV as the output current varies; this is the typical welding machine used in GMAW or FCAW welding machines. The output voltages and output currents of a welding machine can be plotted on a graph to form a curve. These curves show how the output voltage relates to the output current as either change (*Figure 12*).

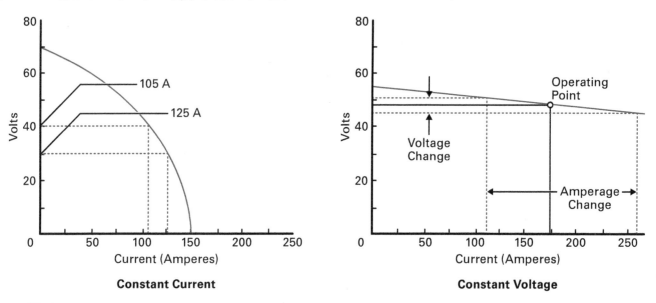

Figure 12 CC and CV output curves.

Different types of welding machines (power sources) are available for GMAW, including the following:

- Transformer-rectifier welding machines
- Engine-driven generator welding machines
- Inverter welding machines

2.1.1 Characteristics of Welding Current

The current produced by a welding machine has different characteristics from the current flowing through utility power lines. Welding current has low voltage and high amperage, while the power line current has high voltage and low amperage.

Voltage is the measure of the electromotive force or pressure that causes current to flow in a circuit. The specific unit of measure is the volt (V). There are two types of voltage associated with welding current: open-circuit voltage and operating voltage. Open-circuit voltage is the voltage present when the machine is on, but no welding is being done. There are usually ranges of open-circuit voltages that can be selected up to about 40 V. Operating voltage, or arc voltage, is the voltage after the arc has been struck.

Arc voltage is generally slightly lower than the open-circuit voltage. The arc voltage is typically 2 V to 3 V lower than the open-circuit voltage for every 100 amperes (A) of current, but it depends on the range selected.

Amperage is the measure of electric current flow in a circuit. The specific unit of measure for amperage is the ampere, commonly abbreviated to amp.

In welding, the current flows in a closed loop through two welding cables; the workpiece lead connects the power source to the base metal, and the other cable connects the power source to the wire electrode. During welding, an arc is established between the end of the wire electrode and the workpiece. The arc generates intense heat at 6,000°F to 10,000°F (3,316°C to 5,538°C), melting the base metal and the wire electrode to form the weld. The amount of amperage produced by the welding machine determines the intensity of the arc and the amount of heat available to melt the workpiece and the electrode.

2.1.2 Transformer-Rectifier Welding Machines

The transformer-rectifier welding machine uses a transformer to convert the primary current to welding current and a rectifier to change the current from alternating current (AC) to DC. Transformer-rectifier welding machines can be designed to produce either AC and DC or DC only. Transformer-rectifiers that produce both AC and DC are usually lighter duty than those that produce DC only. Transformer-rectifier welding machines that produce DC only are sometimes called rectifier power sources or just rectifiers. Depending on the size, transformer-rectifier welding machines may require 120/240 volts of alternating current (VAC) single-phase input power, 208/230/240 VAC three-phase input power, or 480 VAC three-phase input power.

> **WARNING!**
>
> Contact with the primary voltage of a welding machine can cause major shock or electrocution. Make sure that all welding machines are properly grounded to prevent injury.

Transformer-rectifiers used for GMAW and FCAW have an On/Off switch and a voltage control. If the machine has a CV/CC or mode switch, it also has an amperage control for SMAW or GTAW. When the switch is set to CV for either GMAW or FCAW, the amperage control is disabled or becomes the voltage control.

The welding cables (electrode and workpiece) are connected to terminals marked Electrode and Ground, or Positive (+) and Negative (−). Welding machines often have selector switches to select direct current electrode negative (DCEN) or direct current electrode positive (DCEP). If there is no selector switch, the cables must be manually changed on the machine terminals to select the type of current desired.

Figure 13 shows a typical industrial transformer-rectifier CV/CC power source with a wire feeder. These units usually have a 100% duty cycle at their rated output and a 60% duty cycle at their maximum output. They require a wire feeder and a GMAW/FCAW gun or spool gun.

Figure 13 Industrial transformer-rectifier CV/CC power source with wire feeder.
Source: Photo courtesy of The Lincoln Electric Company, Cleveland, OH, U.S.A.

2.1.3 Engine-Driven Generator Welding Machines

Welding machines can also be powered by gasoline or diesel engines. The engine is connected to a generator that produces the power needed for the welding machine. The size and type of engine used depends on the requirements of the welding machine.

To produce sufficient electrical power, the engine generator must turn at a required number of revolutions per minute (rpm). The engines of power generators have governors to control the engine speed. Most governors have a welding speed switch. The switch can be set to idle the engine when no welding is taking place. When the electrode is touched to the base metal, the governor automatically increases the engine's speed to the required welding speed. If no welding takes place for about 15 seconds, the engine will automatically return to idle. The switch can also be set for the engine to run continuously at the welding speed.

Figure 14 shows a diesel engine-driven welding machine. This unit can produce CC/AC or DC and CV/DC outputs up to 400 A at a 100% duty cycle for SMAW, GTAW, GMAW, and FCAW. For GMAW and FCAW, it is available with a portable wire feeder that provides a constant wire-feed speed for use with CV power sources. When used with CV power sources, the wire feeder provides wire as needed when it senses that an arc is struck. The wire feeder is shown in *Figure 14* with an FCAW-S gun, but other guns can also be used.

Engine-driven units often have an auxiliary power unit that produces 120 VAC for lighting, power tools, and other electrical equipment. When 120 VAC is required, the engine-driven generator or alternator must run continuously at its normal welding speed.

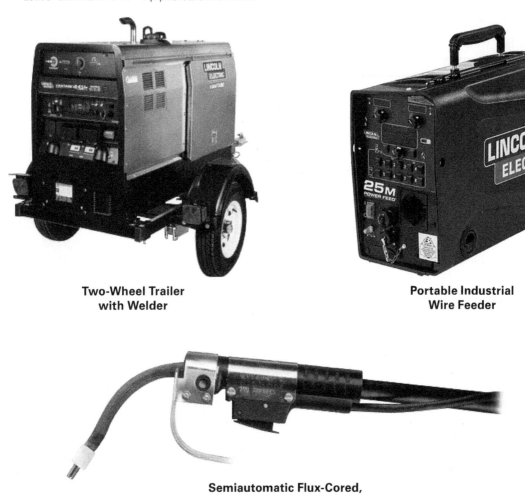

**Two-Wheel Trailer
with Welder**

**Portable Industrial
Wire Feeder**

**Semiautomatic Flux-Cored,
Self-Shielding (FCAW-S) Gun**

Figure 14 Diesel engine-driven welding machine with wire feeder and FCAW-S gun.
Source: Photo courtesy of The Lincoln Electric Company, Cleveland, OH, U.S.A.

Engine-driven units have both engine controls and welding machine controls. The engine controls vary with the type and size of the unit but normally include the following:

- Starter
- Voltage meter
- Temperature gauge
- Fuel gauge
- Hour meter
- Oil pressure indicator

Engine-driven welding machines have a voltage/amperage control. They may have a polarity or mode selector switch, or the welding cables at the welding current terminals might have to be changed to change the polarity. If the machine has a CV/CC or mode selector switch, it also has an amperage control. When the switch is set to CV or GMAW/FCAW mode, the amperage control is disabled and becomes a voltage control.

The advantage of engine-driven generators is that they are portable and can be used in the field where electricity is unavailable for other types of welding machines. The disadvantage is that engine-driven generators are more costly to purchase, operate, and maintain.

GMAW/FCAW Feeders for Engine-Driven Welding Machines

If possible, select wire feeders or spool guns that operate from 24 volts direct current (VDC) battery power and arc voltage. If AC-powered feeder or spool gun controls are used, the power source will have to operate in the continuous run mode to supply AC power to operate the feeder or spool gun.

2.1.4 Inverter Welding Machines

Inverter welding machines (*Figure 15*) increase the frequency of the incoming primary power. This provides a smaller, lighter power source with a faster response time and more control for pulse welding. Inverter welding machines are used where space is limited, and portability is important. The controls on these welding machines vary according to size and application. Smaller inverter machines are often mounted on lightweight carts for greater portability. The compact size, maneuverability, cooling abilities, and efficiency of inverters make them increasingly popular choices for on-site power sources.

Typical industrial advanced inverter welding machines designed specifically for GMAW or FCAW operations usually have controls like those shown in *Figure 16*. For specific control functions for the power supply you are using, refer to the operator's manual.

A typical program sequence for one of the internal welding programs available in an advanced inverter welding machine is shown in *Figure 17*.

Figure 15 Inverter welding machine.
Source: © Miller Electric Mfg. LLC

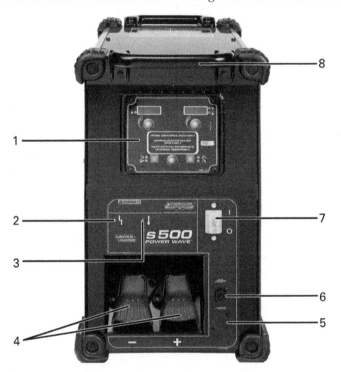

Front
1. Optional S-Series User Interface Kit For Stick, Tig and CV Mig with Voltage Sensing Feeder
2. Status Light
3. Thermal Fault Indicator Light
4. Output Studs
5. 12-Pin Output Control Receptacle Knockout Plate
6. Work Sense Lead Receptacle
7. Main Power Switch
8. Reversible Handles

Figure 16 Typical industrial inverter unit controls.
Source: Photo courtesy of The Lincoln Electric Company, Cleveland, OH, U.S.A.

2.1.5 Welding Machine Ratings

The rating of a welding machine is determined by the amperage output of the machine at a given duty cycle. The duty cycle of a welding machine is based on the percentage of a 10-minute period during which the machine can continuously produce its rated amperage without overheating. For example, a machine with a rated output of 300 A at a 60% duty cycle can deliver 300 A of welding current for 6 minutes out of every 10 minutes (60%) without overheating.

The duty cycle of a welding machine will be 10%, 20%, 30%, 40%, 60%, or 100%. A welding machine with a duty cycle of 10% to 40% is considered a light- to medium-duty machine. Most industrial, heavy-duty machines for manual welding have 60% to 100% duty-cycle ratings. Machines designed for automatic welding operations have 100% duty-cycle ratings.

Except for 100% duty-cycle machines, the maximum amperage that a welding machine can produce is always higher than its rated capacity. A welding machine rated at 300 A at a 60% duty cycle will generally put out a maximum of 375 A to 400 A. But, since the duty cycle is a function of its rated capacity, the duty cycle will decrease as the amperage is raised over 300 A.

Multiprocess Welding Machines

Portable, light industrial GMAW/FCAW welding machines are also available. The MIG unit shown in the image is equipped with an internal wire feeder and has dual voltage inputs of 120 VAC or 240 VAC. With a 120 VAC input, it is rated at 100 A with a 40% duty cycle. With a 240 V input, it is rated at 200 A with a 25% duty cycle. This machine can also be used to perform TIG welding.

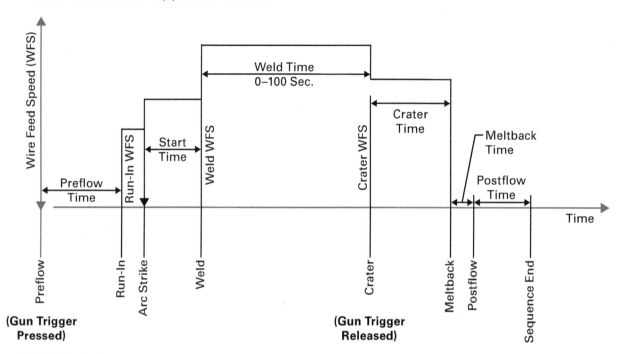

Figure 17 Typical welding program sequence.

Welding at 375 A with a welding machine rated 300 A at a 60% duty cycle will lower the duty cycle to about 30%. If welding continues for more than 3 minutes out of 10 minutes, the machine will overheat. Most welding machines have a heat-activated circuit breaker that shuts off the machine automatically when it overheats. Leave the machine plugged in and turned on to allow the cooling fans to continue running when this occurs. The machine cannot be turned back on until it has cooled below a preset temperature.

On the other hand, if the amperage is set below the rated amperage, the duty cycle increases. Setting the amperage at 200 A for a welding machine rated at 300 A at a 60% duty cycle will increase the duty cycle to 100%. *Figure 18* shows the relationship between the amperage and duty cycle.

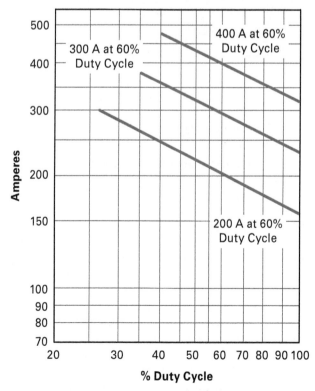

Figure 18 Relationship between amperage and duty cycle.

2.2.0 Power Source Control

As described previously, GMAW and FCAW are best performed with a CV or CV/CC DC welding machine. In CV welding machines, the open-circuit and welding voltage are nearly the same. CC variable-voltage DC welding machines, the type used for SMAW in which the output voltage varies between high open-circuit voltage and lower arc voltage, are sometimes used. However, they require specially designed voltage-sensing wire feeders. The use of CC machines for GMAW and FCAW is prohibited by some codes. CV/CC welding machines are required for GMAW-P operations.

Welding machines usually contain controls for adjusting arc voltage, slope, and **inductance**, and they have meters to monitor amperage and voltage. The simplest machines are power sources with built-in wire feeders. A simple combination welder and wire feeder may have only a wire-feed speed control and voltage control.

Inductance: An electrical circuit property, adjustable on some FCAW power supplies, that influences the pinch effect.

2.2.1 Slope

The CV power sources used for GMAW and FCAW are not precisely constant in their voltage output because the voltage always drops a little as the current increases. This voltage-to-amperage relationship forms a slight curve when plotted as a graph. The general angle of the volt-ampere curve is known as the slope. This slope is adjustable on some units. A flatter (more horizontal) slope is better for spray-transfer welding, while a steeper slope is better for short-circuiting welding (*Figure 19*).

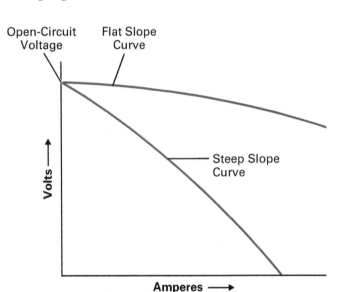

Figure 19 Volt-ampere curve slope profiles for GMAW.

2.2.2 Pinch Effect and Inductance

When using short-circuiting welding in GMAW, the current passing through the wire electrode heats and melts the wire. This is sometimes called the **pinch effect** because the molten wire appears to be pinched off, forming a droplet of molten metal (*Figure 20*).

Pinch effect: The tendency of the end of the wire electrode to melt into a droplet during short-circuiting GMAW welding.

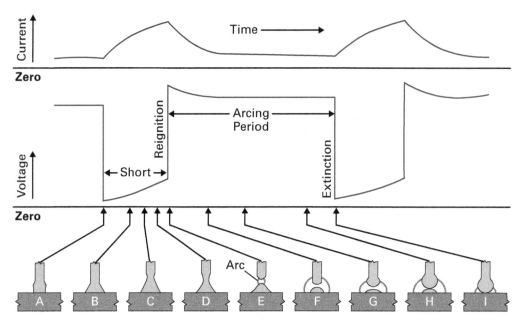

Figure 20 Pinch effect.
Source: AWS C5.6:1989, Figure 6. Reproduced with permission from the American Welding Society (AWS), Miami, FL, USA

An electrical circuit property called inductance influences the pinch effect. Low inductance produces high pinch effect. Conversely, high inductance produces low pinch effect. With low circuit inductance, the current rise is very fast, and the high pinch effect can cause the wire droplet to explode or spatter. When the circuit inductance is high, the current rises more slowly, decreasing the number of short circuits per second and extending the arc duration. The pinch effect forms the droplet more gently, which results in a more fluid puddle, smoother weld, and less spatter.

CV power sources are self-regulating and automatically produce current as it is required. Inductance controls the rate at which the current output rises. Some inductance is desirable in FCAW because inductance prevents explosive arc starts by slowing down the rate of current rise. Many FCAW power sources have a control to adjust the inductance. In general, lower thermal conductivity materials use a higher inductance, and higher thermal conductivity materials use a lower inductance.

2.2.3 Arc Blow

When operating in GMAW or FCAW spray-transfer mode, arc blow can sometimes be a problem. Arc blow is the deflection of the arc from its normal course because of the arc magnetic field's attraction to or repulsion from the weld current's magnetic field in the base metal. The amount and direction of arc deflection depend upon the relative position, direction, and density of the base metal's magnetic field. Arc blow can result in excess spatter and weld defects. It can be minimized by relocating the workpiece clamp or changing the weld angle of the gun.

2.3.0 Welding Cable

Cables used to carry welding current are designed for maximum strength and flexibility. The conductors inside the cable are made of fine strands of copper wire. The copper strands are covered with layers of rubber reinforced with nylon or Dacron® cord (*Figure 21*).

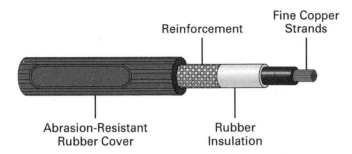

Figure 21 Cut-away section of welding cable.

The size of a welding cable is based on the number of copper strands it contains. Large-diameter cable has more copper strands and can carry more welding current. Typical welding cable sizes range from #4 to #3/0 (spoken as "3 aught"), as shown in *Table 1*.

TABLE 1 Welding Cable Sizes

Machine Size (A)	Duty Cycle (%)	Copper Cable Sizes for Combined Lengths of Electrodes Plus Workpiece Lead				
		Up to 50' (15 m)	50' to 100' (15 m to 30 m)	100' to 150' (30 m to 46 m)	150' to 200' (46 m to 61 m)	200' to 250' (61 m to 76 m)
100	20	#8	#4	#3	#2	#1
180	20	#5	#4	#3	#2	#1
180	30	#4	#4	#3	#2	#1
200	50	#3	#3	#2	#1	#1/0
200	60	#2	#2	#2	#1	#1/0
225	20	#4	#3	#2	#1	#1/0
250	30	#3	#3	#2	#1	#1/0
300	60	#1/0	#1/0	#1/0	#2/0	#3/0
400	60	#2/0	#2/0	#2/0	#3/0	#4/0
500	60	#2/0	#2/0	#3/0	#3/0	#4/0
600	60	#3/0	#3/0	#3/0	#4/0	**
650	60	#3/0	#3/0	#4/0	*	**

*Use double strand of #2/0
**Use double strand of #3/0

When selecting the welding cable size, the current flow (amperage) and the distance the current will travel must be considered. The longer the distance the current must travel, the larger the cable must be. Cable that is too small will become hot, and the voltage provided under load will drop due to the cable's electrical resistance. Excessive voltage drop may affect the performance of the welding machine. When selecting welding cable, use the rated capacity of the welding machine for the cable amperage requirement. For the distance, measure the electrode and workpiece leads and add the two lengths together. To identify the required welding cable size, refer to a recommended welding cable size table furnished by most welding cable manufacturers.

2.3.1 Welding Cable End Connections

Welding cables must be equipped with the proper end connections or terminals.

Lugs are used at the end of the welding cable to connect the cable to the welding machine terminals. The lugs come in various sizes to match the welding cable size and are mechanically secured to the welding cable.

Quick disconnects are also mechanically connected to the cable ends. They are insulated and serve as cable extensions for splicing two lengths of cable together. Quick disconnects are connected or disconnected with a half twist or similar motion. When using quick disconnects, care must be taken to ensure they are tightly connected to prevent overheating or arcing in the connector. *Figure 22* shows some typical quick disconnects and lugs used as welding cable connectors.

CAUTION

If the end connection is not tightly secured to the cable, the connection will overheat and oxidize. An overheated connection will cause variations in the welding current and permanent damage to the connector and/or cable. Check connections for tightness and repair loose or overheated connections. If quick disconnects are used, they must be properly sized to the cable, or overheating may occur.

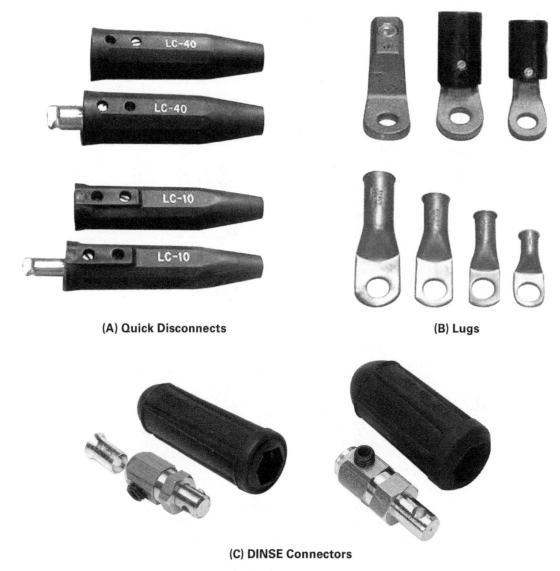

(A) Quick Disconnects

(B) Lugs

(C) DINSE Connectors

Figure 22 Quick disconnects, lugs, and DINSE connectors.
Source: Photo courtesy of The Lincoln Electric Company, Cleveland, OH, U.S.A. (22C)

The workpiece clamp (*Figure 23*) connects the end of the workpiece lead and the workpiece. Workpiece clamps are mechanically connected to the welding cable and come in a variety of shapes and sizes. The size of a workpiece clamp is the rated amperage that it can carry without overheating. Workpiece clamps must at least have the same rating as the rated capacity of the power source on which they will be used.

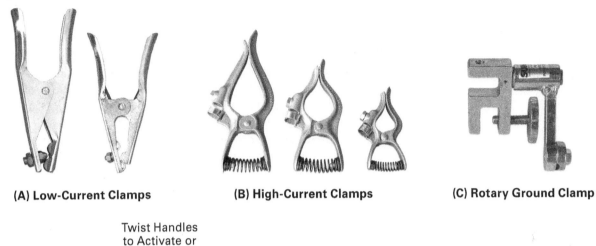

(A) Low-Current Clamps   **(B) High-Current Clamps**   **(C) Rotary Ground Clamp**

(D) Magnetic Grounding Clamps

Figure 23 Workpiece clamps.
Sources: Southwire (23C); Magswitch Technology Inc (23D)

2.4.0   External Wire Feeders

An external wire feeder pulls the electrode wire from a spool and pushes it through the gun cable and gun. It consists of a wire spool holder with a drag brake, an electric motor that drives either one or two sets of opposing slotted rollers, and various controls. The wire feeder also usually contains the shielding gas connections, the gas control solenoid, and (optionally) the cooling-water flow-control solenoid. Some of the controls that may be located on a wire feeder include the following:

- *Spool drag* — This is a manual adjustment used to adjust the drag on the wire spool to prevent the uncontrolled unwinding of the wire.
- *Wire feed speed control* — This is a variable control used to set the wire feed speed. Since GMAW welding machines are self-regulating (they produce the required current on demand), increasing or decreasing the wire feed speed will cause the welding machine to automatically increase or decrease its current output.
- *Voltage control* — This control is used to set the welding voltage.
- *Wire jog* — The manual wire jog switch is used to start and stop the wire feeder without energizing the welding current.
- *Wire meltback timer (burnback)* — This is an adjustable time delay that causes the welding current to continue after the trigger has been released. This consumes the length of wire that continues to feed after the trigger has been released so that the proper **stickout** is maintained to start the next weld.
- *Gas purge* — The manual gas purge switch is used to operate the gas solenoid to purge the shielding gas system prior to actual welding. It is also used to adjust the flowmeter.
- *Trigger hold* — This switch enables or disables a trigger hold function activated by the gun trigger. It enables welding without holding the trigger.
- *Gas preflow timer* — This is an adjustable time delay that prevents the welding current contactor from closing until the shielding gas is flowing.
- *Gas postflow timer* — This is an adjustable time delay that is used to keep shielding gas flowing for a short period after the gun trigger has been released.

Figure 24 shows some basic wire feeder units, while *Figure 25* shows other types of wire feeders.

Two critical properties of electrode wire must be considered before selecting the wire to be used with a wire feeder. Those two properties are the cast and helix (*Figure 26*). The cast and helix are determined by cutting a strand of the wire several feet long and laying it on a flat surface. The cast of the wire is the diameter of the circle formed by the strand of wire. The helix is the height, or vertical distance, of the end of the wire from the flat surface. The cast and helix of electrode wire primarily affect the wire feeder's ability to move the wire through the gun cable and gun. The cast and helix can also cause excessive wear to the contact tip and misguide the wire, causing it to be diverted away from the joint. These factors can affect the quality of the weld.

During operation, a wire feeder grips the wire between the grooves of opposing grooved rollers. There can be one or two sets of opposed rollers (*Figure 27*). The grooves can be V-shaped or U-shaped in their cross section.

V-grooves are generally used for hard wire. One roller slot in each V-groove set may be serrated for better gripping. U-groove rollers are used for soft or flux-cored wire, as they crush more easily. For extremely soft or flux-cored wire, cogged U-groove rollers are used.

Some rollers are made in two pieces with different-size half-grooves formed on the outside edges of each half-roller. The half-rollers are assembled to fit two different ranges of wire size. *Figure 28* shows several different groove styles and combination rollers.

Stickout: The length of electrode that projects from the contact tube, measured from the arc tip to the contact tube.

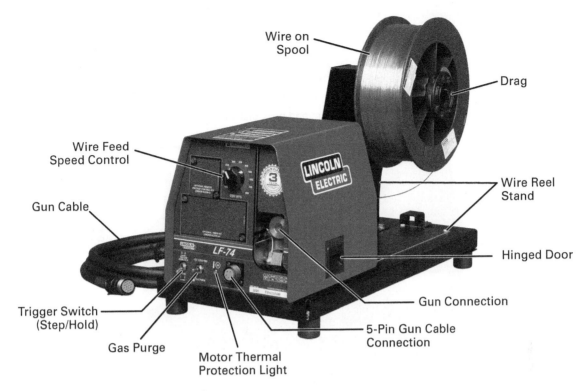

Figure 24 Basic GMAW/FCAW wire feeder units.
Source: Photo courtesy of The Lincoln Electric Company, Cleveland, OH, U.S.A.

(A) Double Reel Bench GMAW/FCAW Wire Feeder

(B) Single Reel Programmable Bench Wire Feeder for GMAW-P

(C) Weatherproof Portable GMAW/FCAW Feeder with FCAW-S Gun

(D) Programmable Portable Push-Pull Wire Feeder and Push-Pull Gun

Figure 25 Various types of wire feeders.
Sources: Photo courtesy of The Lincoln Electric Company, Cleveland, OH, U.S.A. (25A, 25B, 25D); © Miller Electric Mfg. LLC (25C)

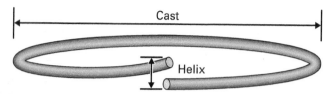

Figure 26 Cast and helix of a wire electrode.

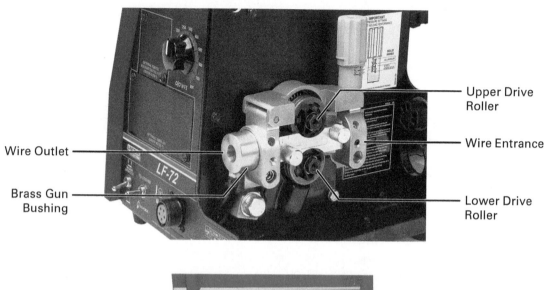

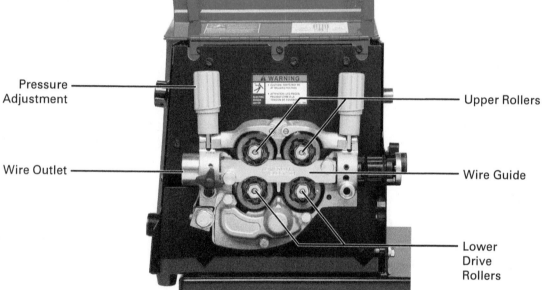

Figure 27 Wire feeder drive roll systems.
Source: Photo courtesy of The Lincoln Electric Company, Cleveland, OH, U.S.A.

When dust or dirt is a problem, a felt wiper can be added to the wire between the spool and the inlet guide to the feed rollers. The felt wipers are treated with a silicone lubricant that cleans and lubricates the wire. Special liquid silicone cleaner/lubricants can occasionally be added to the felt wiper.

U-Groove Roller   **V-Groove Roller**   **Knurled Roller (Core Wire Only)**

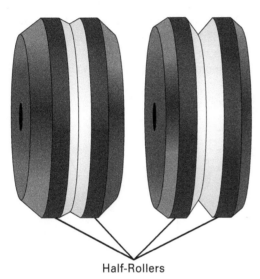

Half-Rollers

Combination Rollers

Figure 28 Groove styles and combination rollers.

Programmable Wire Feeders

Programmable wire feeders are available for many power sources. Custom welding programs can be created, or preset welding programs can be stored in the feeder or the power source. The program can be activated by the various controls and displays on the feeder.

Mode Selection and Customizable Process Display

Process Configuration Menus

Source: Photo courtesy of The Lincoln Electric Company, Cleveland, OH, U.S.A.

2.5.0   GMAW/FCAW-G Guns

The GMAW/FCAW-G gun (*Figure 29*) supports and guides the wire electrode and provides the electrical contact between the wire and welding machine electrode lead. The gun also contains a remote switch to start and stop the welding current and shielding gas flow. The tip of the gun has a nozzle to direct the shielding gas around the arc. Within the nozzle, the wire passes through a contact tip that provides the actual connection between the welding current and electrode wire. Nozzles and contact tips are subject to wear and electrical erosion, so both components are replaceable.

The duty cycle of GMAW guns is determined and rated with CO_2. The rating is reduced from 10% to 50% when mixed gases are used, depending on the gas mixture and welding parameters. The chosen transfer mode (short-circuiting, spray, or pulsed) may further derate their duty cycle.

Some types of guns contain motor-driven wire pullers within the gun handle (*Figure 30*). These work in conjunction with a push wire feeder located with the power source. This system was developed for feeding soft wire, such as aluminum, that tends to bend easily. With the wire being both pushed and pulled, neither feeder mechanism needs to grip the wire quite as tightly.

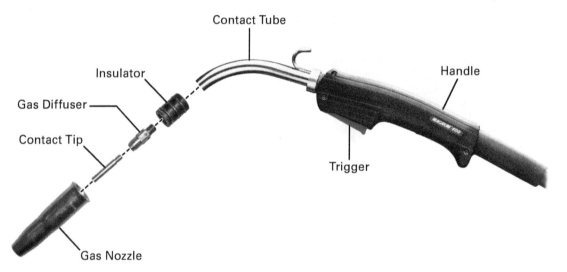

Figure 29 Construction of a typical GMAW/FCAW-G gun.
Source: Photo courtesy of The Lincoln Electric Company, Cleveland, OH, U.S.A.

(A) Air-Cooled   **(B) Water-Cooled**

Figure 30 Pull guns for push-pull wire feeders.
Source: Photo courtesy of The Lincoln Electric Company, Cleveland, OH, U.S.A.

In addition, there are spool guns designed for aluminum wire. These guns contain both a wire feeder and a small spool of wire. In some applications, an auxiliary gun control unit is required for proper operation. *Figure 31* shows one style of spool gun.

Light-duty guns are gas-cooled by the shielding gas flow. Heavy-duty guns are water-cooled. The cooling-water flow may be controlled by a solenoid in the wire feeder or by a separate dedicated cooling system. Cooling water is piped to and from the gun through flexible tubing integrated into the gun cable. Some cooling systems use domestic water to cool the gun. Others use a closed-loop system (*Figure 32*) that circulates demineralized water or special cooling fluid that will not corrode internal gun surfaces or plug the cooling passages with mineral scale.

Figure 31 Spool gun.
Source: Photo courtesy of The Lincoln Electric Company, Cleveland, OH, U.S.A.

Figure 32 Typical closed-loop cooling unit.
Source: © Miller Electric Mfg. LLC

2.5.1 FCAW-S Guns

FCAW-S guns are similar to GMAW/FCAW-G guns, but they are often designed to handle larger-diameter wire and do not provide shielding gas flow. The FCAW-S gun supports and guides the wire electrode. It also contains a remote switch for starting and stopping the welding current. The wire electrode passes through the contact tip at the end of the gun, which provides the electrical connection from the welding machine electrode lead to the electrode wire.

Guns designed for self-shielded FCAW do not have provisions for delivering shielding gas. This makes the gun nozzles smaller in diameter. To protect the contact tip, these guns have a fiber insulator that threads onto the end of the gun nozzle.

FCAW-S generates smoke, which can be removed at the gun tip with a special smoke extraction nozzle. The nozzle is connected to a vacuum pump or blower by a flexible tube that draws off smoke, which improves visibility.

2.5.2 Contact Tips and Nozzles

Contact tips and nozzles are a vital part of effective welding. The contact tip provides the connection between the welding current and the electrode wire. The nozzle directs shielding gas around the arc. Both components are available in many different shapes and sizes (*Figure 33*), depending on the application requirements. For that reason, several factors must be taken into consideration when selecting contact tips and nozzles. The chosen tip and nozzle must be compatible with the equipment in use.

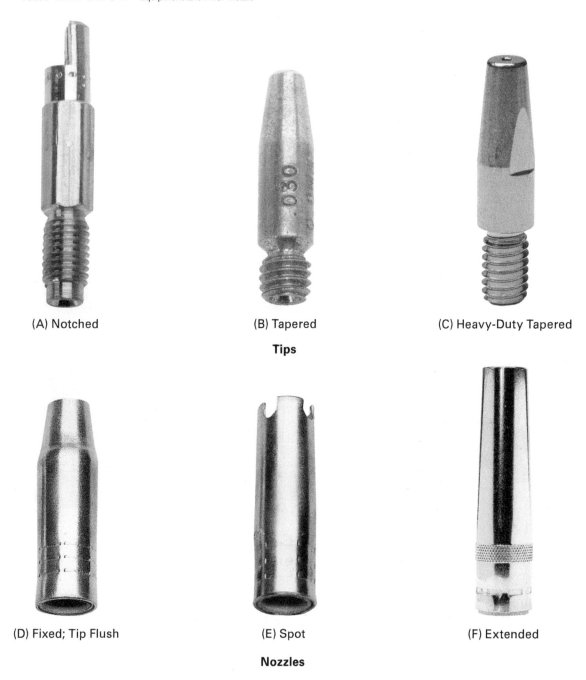

(A) Notched   (B) Tapered   (C) Heavy-Duty Tapered

Tips

(D) Fixed; Tip Flush   (E) Spot   (F) Extended

Nozzles

Figure 33 Various contact tips and nozzles.
Source: Photo courtesy of The Lincoln Electric Company, Cleveland, OH, U.S.A.

A contact tip should always correspond to the wire electrode size. For example, if a wire electrode with a diameter of 0.125" (3.2 mm) is being used, the contact tip should also be 0.125" (3.2 mm). If a contact tip is too small for the wire being used, the wire will jam. A contact tip that is larger than necessary may work, but it is not the ideal setup. If the contact tip is too large, poor electrical contact will cause damage from arcing between the tip and the wire.

Other factors to consider when selecting a contact tip include the following:

- *Tapered versus non-tapered* — Tapered tips are usually longer to allow for welding in areas with restricted access. Non-tapered tips have more mass and tend to last longer than tapered ones.
- *Threaded versus non-threaded* — Threaded tips are screwed into the gas diffuser. These are the most common contact tips. Non-threaded tips are dropped into the gas diffuser.
- *Standard versus heavy duty* — Contact tips can be small standard-duty types or larger heavy-duty types. Most low-amperage applications can be handled with a standard contact tip. High-amperage or extensive welding applications are best handled with a heavy-duty contact tip.

Like contact tips, there are a wide variety of nozzles. They can be tapered or non-tapered, threaded or non-threaded, have small or large inner diameters, and come in numerous shapes and sizes. Ideally, it is best to choose the largest nozzle possible that still allows access to the weld joint. This way, the greatest amount of shielding gas coverage is provided. If the welding joint has limited access, a smaller tapered nozzle will need to be used.

Nozzles are available in copper or brass. Copper nozzles can withstand extreme heat, so they are best suited for heavy-duty applications. Brass nozzles are subject to melting and burning if they are used in high-temperature applications, so they are more suited for light-duty applications. However, brass nozzles resist welding spatter better than copper nozzles.

2.6.0 Shielding Gases

Shielding gases are used to displace atmosphere from the weld zone and to prevent contamination of the weld puddle by oxygen, nitrogen, and moisture. The three principal shielding gases are argon, helium, and CO_2. These gases are often mixed in various proportions for specific applications.

The GMAW/FCAW-G process always uses shielding gas to protect the electrode and the weld from contamination. The shielding gas is directed around the end of the electrode and the weld zone by the nozzle of the gun.

Some FCAW core wires require the use of a shielding gas. The shielding gas is directed around the electrode end and the weld zone by the nozzle of the gun. The shielding gas used depends on the filler wire and base metal being welded. The most common shielding gases for FCAW carbon and low-alloy steels are CO_2 or an argon-CO_2 mixture. For FCAW high-alloy or stainless steels, an argon-CO_2 mixture is used. These gases are available separately or premixed in pressurized cylinders. They are also available separately in bulk liquid tanks, so they can be mixed at the point of use.

Each shielding gas has distinctive performance characteristics. Each gas performs differently in the various transfer modes and with different base metals. Mixtures of gases often have the best features of the individual constituents. The following sections explain some of the features of the principal gases and their primary uses.

Gas flow rate from the gun tip is important because it affects the quality and the cost of a weld. Too low a flow rate will not shield the weld zone adequately, which will result in a poor-quality weld. Excessive gas flow wastes expensive gases and can cause turbulence at the weld. The turbulence can pull atmosphere into the weld zone, causing porosity in the weld. Welding specifications indicate appropriate shielding gas flow rates.

CAUTION

Check your WPS or site quality standards for the type of shielding gas to use. If these are not available, refer to the filler metal supplier's specifications.

2.6.1 Argon

Argon (Ar) is the least expensive and most used inert shielding gas. Argon can be used alone or in combination with other shielding gases because of its current transfer and ionization properties. It forms a tight arc column with high current density, which concentrates the arc in a small area and produces deep penetration.

Pure argon is used with aluminum, nickel-based alloys, copper alloys, and the reactive metals zirconium, titanium, and tantalum. It provides excellent arc stability, penetration, and bead shape. For ferrous metals, argon is usually mixed with oxygen, helium, hydrogen, carbon dioxide, and/or nitrogen.

2.6.2 Helium

Helium is likely the most expensive of the inert gases, and supplies are diminishing. Because of thermal conductivity (heat transfer), helium (He) is used when deeper penetration and higher travel speed are required.

Arc stability with helium is not as good as it is with argon. The helium arc column is wider than the argon arc column, resulting in reduced current density. Helium is better for welding aluminum, magnesium, and copper alloys because its use results in better puddle fluidity and bead wetting. Some mixtures of helium and argon have the beneficial characteristics of both gases. Helium mixtures of 90% helium, 7.5% argon, and 2.5% CO_2 are widely used for short-circuiting GMAW of stainless steel in all positions.

2.6.3 Carbon Dioxide

CO_2 is commonly used to weld mild steel because of its weld performance and low cost. In the welding arc, it is not an inert gas because it breaks down into carbon monoxide (CO) and free oxygen. The oxygen combines with elements in the weld puddle to form slag. CO_2 does not support spray transfer, so short-circuiting transfer and globular transfer must be used. CO_2 provides good penetration. However, in GMAW, the weld surface is heavily oxidized and has high weld spatter. This is not true in FCAW. Normally, argon is mixed with no more than 25% CO_2 to improve the performance characteristics. *Figure 34* shows the effects of shielding gas on the welding arc in GMAW and FCAW.

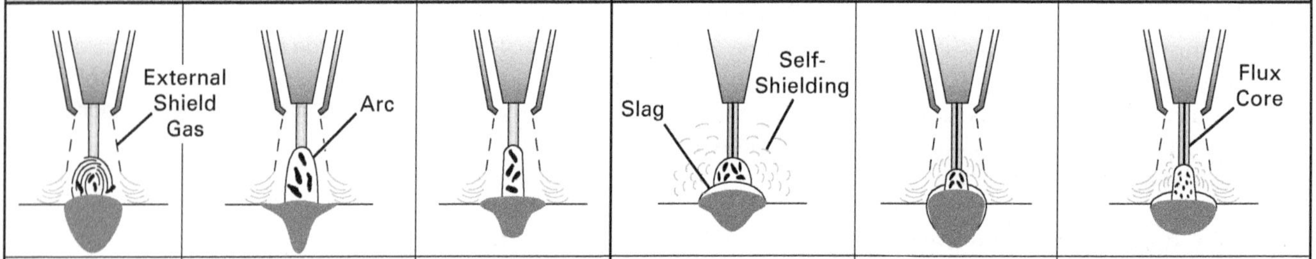

GMAW			FCAW		
CO_2	Ar-O_2	Ar-CO_2	Self-Shielding	Dual-Shielding with CO_2	Dual-Shielding with Ar-CO_2
Shorter, Colder Arc	Hotter Arc and Narrow Penetration	Eliminates Under Cutting	Better Arc Ionization and Provides Covering (Slag)	Increases Penetration	Increases Arc Length, Improves Spray, and Decreases Penetration

Figure 34 Effects of shield gas on the welding arc.

2.6.4 Gas Mixtures

The principal shielding gases are often mixed to improve their overall welding characteristics. For steel, small amounts of oxygen added to argon stabilize the arc, increase filler metal deposition rate, and improve wetting and bead shape. Small percentages of oxygen added to CO_2 improve arc characteristics. Helium is added to argon to increase the arc voltage and heat for welding nonferrous metals. Small amounts of nitrogen and oxygen are added to argon for welding stainless steel when the austenitic characteristics should be preserved. Helium and CO_2 are added to argon to improve arc stability, wetting, and bead profile. Some examples and uses of gas mixtures are shown in *Table 2*. Many other typical gas mixtures are available.

TABLE 2 Examples and Uses of Gas Mixtures for GMAW

Gas Mixture	Use
Argon with 1% oxygen	Spray transfer on stainless steels
Argon with 2% oxygen	Spray transfer on carbon steels, low-alloy steels, and stainless steels
Argon with 5% oxygen	General carbon steel welding using highly deoxidized wires
Argon with 8% to 12% oxygen	Single-pass spray-transfer welds with high puddle fluidity
Argon with 3% to 10% carbon dioxide	Spray transfer and short-circuiting transfer on carbon steels
Argon with 11% to 20% carbon dioxide	High-speed and out-of-position welds on carbon steels and low-alloy steels
Argon with 21% to 25% carbon dioxide	Short-circuiting transfer on mild steels
Argon with 25% helium	Sometimes used for aluminum where bead appearance is important
Argon with 75% helium	Welding heavy aluminum
Argon with 90% helium	Welding aluminum over 3" thick and copper over 1" thick
Argon with 5% to 10% carbon dioxide and 1% to 3% oxygen	Welding carbon steel, low-alloy steel, and stainless steel with all transfer modes
Argon with 10% to 30% helium and 5% to 15% carbon dioxide	Pulse spray transfer on heavy out-of-position carbon and low-alloy steel
60% to 70% helium with 20% to 35% argon and 4% to 5% carbon dioxide	Short-circuiting transfer of high-strength steels in all positions
90% helium with 7.5% argon and 2.5% carbon dioxide	Short-circuiting transfer of stainless steels in all positions
Argon, helium, carbon dioxide, and oxygen	High-rate deposition of low-alloy, high-tensile metals and mild steels

2.6.5 Shielding Gas Selection

Selecting the best shielding gas for a particular welding task involves considering many factors. The shielding gas selected will affect arc shape, arc density, arc temperature, arc stability, rate of filler metal transfer, degree of spatter, weld penetration, weld bead shape, weld bead appearance, weld porosity, weld chemistry, and weld quality.

Shielding gases affect the welding process in the following three ways:

- Gas thermal conductivity affects arc voltage and heat delivered to the weld. The thermal conductivity of helium and CO_2 is much higher than that of argon; both helium and CO_2 supply more heat to the weld. As a result, they also require more voltage and power to maintain a stable arc.

- Some gases can react with the filler and base metals and affect arc stability. CO_2 and most oxygen-bearing gases cannot be used with aluminum because they form oxides. However, they provide better fusion and arc stability when used with steels.

- Gases affect the mode of metal transfer and penetration depth. Mixtures containing more than 15% CO_2 do not allow true spray transfer.

Common short-circuiting shielding gases and their applications in short-circuiting transfer welding are shown in *Table 3*. Common shielding gases and their applications in spray-transfer welding are shown in *Table 4*.

TABLE 3 Short-Circuiting Shielding Gases and Their Applications for GMAW

Gas Mixture	Shielding Gas	Thickness	Advantages
Carbon steel	75% Argon + 25% Carbon Dioxide	Less than $\frac{1}{8}$" (3.2 mm)	High welding speeds without burn-through; minimum distortion and spatter
	75% Argon + 25% Carbon Dioxide	More than $\frac{1}{8}$" (3.2 mm)	Minimum spatter; clean weld appearance; good puddle control in vertical and overhead positions
	50% Argon + 50% Carbon Dioxide	—	Deeper penetration; faster welding speeds
Stainless steel	90% Helium + 7.5% Argon + 2.5% Carbon Dioxide	—	No effect on corrosion resistance; small HAZ; no undercutting; minimum distortion
Low-alloy steel	60% to 70% Helium + 25% to 35% Argon + 4.5% Carbon Dioxide	—	Minimum reactivity; excellent toughness; excellent arc stability, wetting characteristics, and bead contour; little spatter
	75% Argon + 25% Carbon Dioxide	—	Fair toughness; excellent arc stability, wetting characteristics, and bead contour; little spatter
Aluminum, copper, magnesium, nickel, and their alloys	Argon and Argon + Helium mixtures	Over $\frac{1}{8}$" (3.2 mm)	Argon satisfactory on sheet metal; argon-helium preferred base material

Source: AWS PHB-4:2000, Pages 7, 8, 9. Reproduced with permission from the American Welding Society (AWS), Miami, FL, USA

TABLE 4 Spray-Transfer Shielding Gases and Their Applications for GMAW

Gas Mixture	Shielding Gas	Thickness	Advantages
Aluminum	100% Argon	0" to 1" (0 mm to 25 mm)	Best metal transfer and arc stability; least splatter
	65% Argon + 35% Helium	1" to 3" (27 mm to 76 mm)	Higher heat input than straight argon; improved fusion characteristics with 5XXX series Al-Mg alloys
	75% Helium + 25% Argon	Over 3" (76 mm)	Highest heat input; minimizes porosity
Magnesium	100% Argon	—	Excellent cleaning action
Carbon steel	Argon + 3% to 5% Oxygen	—	Improves arc stability; produces a more fluid and controllable weld puddle; good coalescence and bead contour; minimizes undercutting; permits higher speeds than pure argon
	Argon + 8% to 10% Carbon Dioxide	—	High-speed mechanized welding; low-cost manual welding
Low-alloy steel	98% Argon + 2% Oxygen	—	Minimizes undercutting; provides good toughness
Stainless steel	99% Argon + 1% Oxygen	—	Improves arc stability; produces a more fluid and controllable weld puddle; good coalescence and bead contour; minimizes undercutting on heavier stainless steels
	98% Argon + 2% Oxygen	—	Provides better arc stability, coalescence, and welding speed than 1% oxygen mixture for thinner stainless steel materials
Nickel, copper, and their alloys	100% Argon	Up to $\frac{1}{8}$" (3.2 mm)	Provides good wetting; decreases fluidity of weld material
	Argon + Helium mixtures		Higher heat inputs of 50% and 75% helium mixtures offset high heat dissipation of heavier gauges
Titanium	100% Argon		Good arc stability; minimum weld contamination; inert gas backing is required to prevent air contamination on back of weld area

Source: AWS PHB-4:2000, Pages 7, 8, 9. Reproduced with permission from the American Welding Society (AWS), Miami, FL, USA

2.6.6 Cylinder Safety

Shielding gases may be supplied in bulk liquid tanks, in liquid cylinders, or in high-pressure cylinders of various sizes. The most common container is the high-pressure cylinder, which is portable and can be easily moved as needed.

When transporting and handling cylinders, always observe the following rules:

- Always install the safety cap over the valve, except when the cylinder is connected for use.

WARNING!

Do not remove the protective valve cap unless the cylinder is secured. If a cylinder falls over and the valve breaks off, the cylinder may rocket away uncontrollably, causing severe injury or death to anyone in its way.

- Secure the cylinder to prevent it from falling when it is in use. Chain or clamp it to the welding machine or to a post, beam, or pipe.
- Always use a cylinder cart to transport cylinders.
- Never hoist cylinders with a sling or magnets. Cylinders can slip out of the sling or fall from the magnets. Always use a hoisting basket or similar device.
- Always open the cylinder valve slowly. Once pressure has been applied to the system, open the valve completely to prevent leakage around the valve stem.

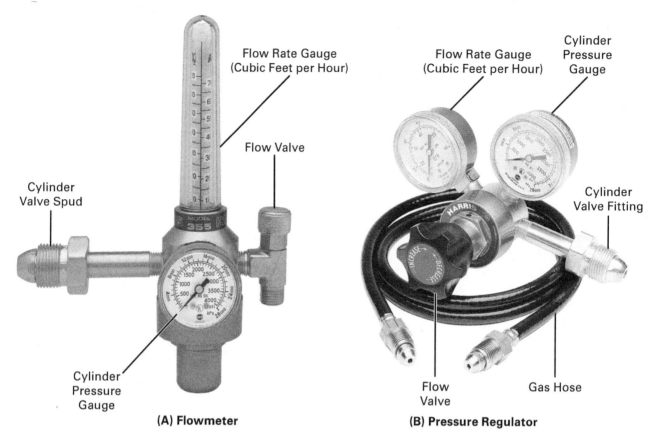

Figure 35 Shielding gas flowmeter and pressure regulator.
Source: Photo courtesy of The Lincoln Electric Company, Cleveland, OH, U.S.A.

2.6.7 Gas Regulators / Flowmeters

A gas pressure regulator and a flowmeter are required to meter the shielding gas to the gun at the proper flow rate and pressure (*Figure 35*). The two devices can be matched to each other in a single assembly or be separate devices. A typical cylinder pressure regulator includes two pressure gauges—one for cylinder pressure and another for outlet pressure. The flowmeter may also have a gauge. The flow rate indicator is a clear cylinder with a ball that floats freely inside the housing to indicate the actual flow rate. Increased flow rates cause the ball to rise higher in the housing. The metering, or flow, valve on the flowmeter is used to adjust the gas flow rate to the gun nozzle. The flow indicator shows the gas flow rate in cubic feet per hour (cfh) or liters per minute (lpm).

Gas flow to the gun is started and stopped by an electric solenoid valve that is usually located inside the control unit or the wire feeder unit. The solenoid is controlled by the operator with the trigger switch on the gun. The same trigger switch simultaneously starts and stops the wire feeder and welding current.

2.7.0   Welding Equipment Setup

To weld safely and efficiently, the welding equipment must be properly set up (*Figure 36*). The following sections explain the steps for setting up welding equipment.

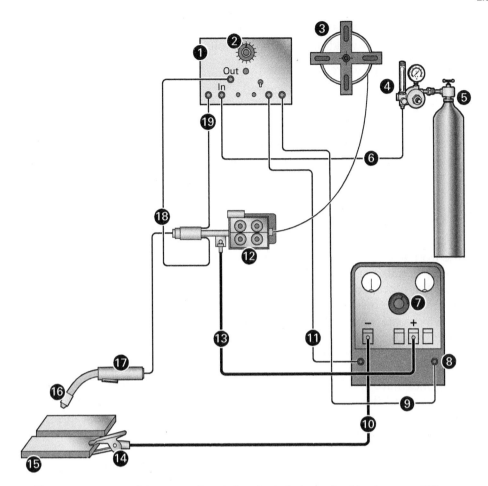

1	Control Unit
2	Wire Feed Speed Control
3	Electrode Wire Reel
4	Regulator/Flowmeter
5	Optional Shielding Gas Source
6	Gas-In Hose
7	Voltage Control
8	Power Source
9	Power to Control Unit
10	Workpiece Lead
11	Contactor Control Cable
12	Wire Feed Drive Assembly
13	Electrode Lead
14	Workpiece Clamp
15	Workpiece
16	Optional Gas Nozzle
17	GMAW or FCAW Gun
18	Gas-Out Hose
19	Gun Control Cable

Note: The polarity of the gun and workpiece leads is determined by the type of filler metal and application.

Figure 36 Diagram of a GMAW/FCAW system.

2.7.1 Selecting a GMAW Welding Machine

To select a GMAW welding machine, the following factors must be considered:

- A CV or CV/CC welding machine is required.
- A DCEP-type welding current is required.
- The maximum amperage required for the task must be available.
- The primary power requirements must be met. Ensure the proper voltage and circuit current capacity are available and that the welding machine plug and receptacle are compatible.

2.7.2 Selecting an FCAW Welding Machine

To select a FCAW welding machine, the following factors must be considered:

- A CV power source is required.
- A DC welding current is required.
- The required operating amperage ranges from 150 A with a 0.045" (1.2 mm) electrode to 650 A or more with a $\frac{1}{8}$" (3.2 mm) electrode; larger electrodes require even higher amperage.
- The primary power requirements must be met. Ensure the proper voltage and circuit current capacity are available and that the welding machine plug and receptacle are compatible.

2.7.3 Positioning the Equipment

Because the gun cables are short, the wire feeder must be positioned close to the work to be performed. The power source must also be located near the wire feeder because of the limited length of the contactor/control cable that runs from the power source to the wire feeder. Normally, the wire feeder is located on top of the power source, on the floor, or on a cart near the power source.

Select a site where the equipment will not be in the way but will be protected from welding, cutting, or grinding sparks. There should be good air circulation to keep the welding machine cool. The environment should be free from explosive or corrosive fumes and as free as possible from dust and dirt. Welding machines have internal cooling fans that will pull these materials into the welding machine if they are present. Also, dust and dirt can collect on the wire, causing weld contamination and clogging of the liner in the gun cable. The site should be free of standing water or water leaks. If an engine-driven unit is used, position it so that it can be easily refueled and serviced.

There should be easy access to the site so the equipment can be started, stopped, or adjusted as needed. If the machine will be plugged into an outlet, be sure that the outlet has been properly installed by a qualified electrician to ensure it is grounded. Also, be sure to identify the location of the electrical disconnect before plugging the welding machine into the outlet.

2.7.4 Moving Welding Machines

Large engine-driven generators are mounted on a trailer frame and can easily be moved by a pickup truck or tractor with a trailer hitch. Other types of welding machines may have a skid base or be mounted on steel or rubber wheels. Be careful when moving wheel-mounted welding machines by hand. Some machines are top-heavy and could fall over in a tight turn or if the floor or ground is uneven or soft.

> **WARNING!**
>
> Secure or remove the wire feeder and gas cylinder before attempting to move a welding machine. If a welding machine starts to fall over, do not attempt to hold it. Welding machines are very heavy, and severe crushing injuries can occur if a welding machine falls on you.

Most welding machines have a lifting eye used to move machines mounted on skids or lift any machine. Before lifting a welding machine, check the equipment specifications for the weight. Be sure the lifting device and tackle can handle the machine's weight. Always use a sling and/or a shackle. Never attempt to lift a machine by placing the lifting hook directly in the machine's lifting eye because the safety latch on the hook cannot be closed. Also, before lifting or moving a welding machine, always inspect the rigging equipment before each use and be sure the welding cables are secure. *Figure 37* shows the proper way to lift a welding machine.

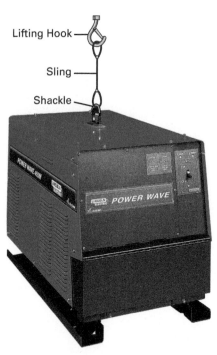

Figure 37 Lifting a welding machine.
Source: Photo courtesy of The Lincoln Electric Company, Cleveland, OH, U.S.A.

2.7.5 Connecting the Shielding Gas

The shielding gas connects to the wire feeder unit or the power source if the wire feeder is integrated with the power source.

Follow these steps to connect the shielding gas:

Step 1 Identify the required shielding gas by referring to the WPS or site quality standards.

Step 2 Locate a cylinder of the correct gas or mixture and secure it near the wire feeder. Be sure to secure the cylinder so that it cannot fall over.

Step 3 Remove the cylinder's protective cap, momentarily crack the cylinder valve open to blow out any dirt, and then close it.

> **WARNING!**
>
> Even though they are available, do not use an adapter to connect a regulator/flowmeter equipped with one type of Compressed Gas Association (CGA) connection to a gas cylinder with a different CGA connection. The CGA connections are specific to the types of gas and cylinder pressures permitted for the connection.

Step 4 Using a regulator/flowmeter with the correct CGA connection for the cylinder, install the pressure regulator/flowmeter on the cylinder.

Step 5 Connect the gas hose to the flowmeter and gas connection on the wire feeder (or welding machine).

Step 6 Check to be sure the flowmeter adjusting valve is closed.

> **WARNING!**
>
> Do not open the valve quickly, or the flowmeter gauges may break. If a gauge breaks, the glass may shatter, and the gas may escape, potentially injuring the operator.

Step 7 Slowly crack open the cylinder valve, then open it completely.

Step 8 Adjust the gas flow to the specified flow rate using the flowmeter valve.

Some flowmeters are equipped with several scales of different calibrations around the same sight tube for monitoring the flows of different gases. These scales differ due to the different densities of shielding gases. Be sure to rotate the scales or read the correct side for the type of gas being used.

2.7.6 Selecting and Installing Filler Wire

Follow these steps to select and install the filler wire:

Step 1 Identify the filler wire required by referring to the WPS or site quality standard.

Step 2 Locate a spool of the wire and mount it on the wire feeder spool holder. Lock the holder and set the drag brake for a slight drag. The brake control is usually located on the end of the spool axle.

Step 3 Check (and change if necessary) the wire feed drive wheels, liner, and gun contact tube to make sure they are the correct size for the wire selected. Adjust the wheel tension if necessary. For detailed instructions on changing or adjusting the wire feeder drive wheels or changing the gun's contact tube or liner, refer to the specific manufacturer's instructions.

Step 4 Feed the wire into the feeder wheels, and use the jog control to feed it through the gun cable and gun contact tube.

Adjusting Flowmeters

In most cases, the top of the floating flowmeter ball is used as the reference point for setting the gas flow. However, always check the manufacturer's instructions to determine the correct adjustment method. Use the lowest flow rate as often as possible to conserve gas and protect the environment.

Gas Flow Rates

More gas flow is not always better. Too much flow can deflect off the workpiece. The resultant swirling gas can create turbulence that draws in air, contaminating the weld and creating pinholes. If black flecks occur in the weld puddle or pinholes occur in the weld, try reducing the gas flow.

2.7.7 Placing the Workpiece Clamp

The workpiece clamp must be properly placed to prevent damage to surrounding equipment. If the electrical welding current travels through a bearing, seal, valve, or contacting surface, it could cause severe damage from heat and arcing, requiring these items be replaced. Carefully check the area to be welded and position the workpiece clamp so the welding current will not pass through any contacting surface. When in doubt, ask your supervisor for assistance before proceeding.

WARNING!

If welding is to be done near a battery, the battery must be removed. Batteries produce hydrogen gas, which is extremely explosive. A welding spark could cause the battery to explode, showering the area with battery acid.

Workpiece clamps must never be connected to pipes carrying flammable or corrosive materials. The welding current could cause overheating or sparks, resulting in an explosion or fire.

The workpiece clamp must make good electrical contact when it is connected. Dirt and paint will inhibit the connection and cause arcing, resulting in overheating of the workpiece clamp. Dirt and paint also make the welding current unstable and can cause defects in the weld. Clean the surface before connecting the workpiece clamp. If the workpiece clamp is damaged or does not close securely onto the surface, replace it.

2.7.8 Energizing the Power Source

Electrically powered welding machines are energized by plugging them into an electrical outlet. The electrical requirements will be located on the equipment specification tag prominently displayed on the machine. Most machines requiring single-phase, 240 VAC power have a three-pronged plug. Machines requiring three-phase, 208 VAC, or 460 VAC power have a four-pronged plug. Both types of plugs are shown in *Figure 38*.

If a welding machine does not have a power plug, an electrician must connect it. The electrician will add a plug or hard-wire the machine directly into an electrical box.

WARNING!

Never use a welding machine until you identify the location of the electrical disconnect switch for that specific machine or receptacle. In an emergency, you must be able to turn off the power quickly to the welding machine at the disconnect switch.

2.7.9 Starting Engine-Driven Generators

Before welding can take place with an engine-driven generator, the engine must be checked and started. As with a car engine, the engine powering the generator must also have routine maintenance performed.

Many sites have prestart checklists that must be completed and signed prior to starting or operating an engine-driven generator. Check with your supervisor. If your site has such a checklist, complete and sign it. If your site does not have a prestart checklist, perform the following checks before starting the engine:

- Check the oil using the engine oil dipstick. If the oil is low, add the appropriate grade of oil for the conditions according to the manufacturer's maintenance instructions.
- Check the coolant level in the radiator if the engine is liquid-cooled. If the coolant level is low, add the proper coolant.
- Check the fuel. The unit may have a fuel gauge or a dipstick. If the fuel is low, add the correct fuel (diesel or gasoline) to the fuel tank. The type of fuel required should be marked on the fuel tank. If it is not marked, contact your supervisor to verify the fuel required and have the tank marked.

CAUTION

Welding current passing through electrical or electronic equipment will cause severe damage to the affected equipment. Also, before welding on any type of equipment with a battery, the ground lead at the battery must be disconnected to protect the electrical system.

Figure 38 Typical three- and four-pronged plugs.

CAUTION

Do not add plain water to radiators that contain antifreeze. Antifreeze not only protects radiators from freezing in cold weather, but it also has rust inhibitors and additives to protect heat-transfer surfaces. If the antifreeze is diluted, it will not provide the proper protection.

CAUTION

Adding gasoline to a diesel engine or diesel to a gasoline engine will cause severe engine problems and may create a fire hazard. Always be sure to add the correct fuel to the fuel tank.

- Check the battery water level unless the battery is sealed. Add demineralized water if the battery water level is low.
- Check the electrode holder to be sure it is not grounded. If the electrode holder is grounded, it will arc and overheat the welding system when the welding machine is started. This may cause damage to the equipment.
- Open the fuel shutoff valve if the equipment has one. If there is a fuel shutoff valve, it should be located in the fuel line between the fuel tank and carburetor.
- Record the hours from the hour meter. An hour meter records the total number of hours that the engine runs. This information is used to determine when the engine needs to be serviced. The hours will be displayed on a gauge similar to an odometer.
- Clean the unit. Use a compressed air hose to blow off the engine and generator or alternator. Use a rag to remove heavier deposits that cannot be removed with the compressed air.

WARNING!

Always wear eye protection when using compressed air to blow dirt and debris from surfaces. Never point the compressed air nozzle at yourself or anyone else.

Most engines have an On/Off ignition switch and a starter. They may be combined into a key switch similar to the ignition on a car. To start the engine, turn on the ignition switch and press the starter. Release the starter when the engine starts. The engine speed is controlled by the governor. If the governor switch is set for idle, the engine will slow to idle after a few seconds. If the governor is set to welding speed, the engine will continue to run at the welding speed.

Small engine-driven generators may have an On/Off switch and a pull cord. These are started by turning on the ignition switch and pulling the cord, similar to how a lawn mower is started. Engine-driven generators should be started about 5 minutes to 10 minutes before they are needed for welding. This will allow the engine to warm up before a welding load is applied.

If no welding is required for 30 minutes or more, stop the engine by turning off the ignition switch. If you are finished with the welding machine for the day, close the fuel valve if there is one.

Engine-driven generators require regular preventive maintenance to keep the equipment operating properly. Most sites have a preventive maintenance schedule based on the hours that the engine operates. In severe conditions, such as in very dusty environments or cold weather, maintenance may have to be performed more frequently.

The responsibility for performing preventive maintenance varies by site. Check with your supervisor to determine who is responsible for performing preventive maintenance. Even though the welder may not be responsible for conducting preventive or periodic maintenance, all welders share an obligation to monitor and protect the equipment and to report problems. As is true for most crafts, a welder cannot produce welds without functional and reliable equipment.

When performing preventive maintenance, follow the manufacturer's guidelines in the equipment manual. Typical procedures to be performed as a part of preventive maintenance include the following:

- Changing the oil
- Changing the gas filter
- Changing the air filter
- Checking/changing the antifreeze
- Greasing the undercarriage
- Repacking the wheel bearings

CAUTION

To prevent damage to the equipment, perform preventive maintenance as recommended by the site procedures or by the manufacturer's maintenance schedule in the equipment manual.

Dust Removal

Blow out the welding machine and drive unit with compressed air occasionally to remove dust. A significant buildup of dust inside the machine can provide an alternate path for the electricity and could result in some of the components shorting out.

2.0.0 Section Review

1. The two types of voltage associated with welding current are _____.
   a. open-circuit voltage and operating voltage
   b. operating voltage and arc voltage
   c. closed-circuit voltage and welding voltage
   d. operating voltage and closed-circuit voltage

2. The deflection of an arc caused by the attraction or repulsion of the arc's magnetic field with the weld current's magnetic field in the base metal is known as _____.
   a. pinch effect
   b. slope curvature
   c. arc blow
   d. inductance bending

3. Quick disconnects used on welding cable are connected and disconnected with a _____.
   a. terminal lug
   b. screwing action
   c. pronged plug
   d. half twist

4. A device that can be used to start and stop the wire feeder without energizing the welding current is a _____.
   a. spool drag brake
   b. wire jog switch
   c. feed speed control
   d. wire meltback timer

5. The part of a GMAW/FCAW gun that provides the electrical connection from the welding machine electrode to the electrode wire is the _____.
   a. gun handle
   b. fiber insulator
   c. contact tip
   d. extraction nozzle

6. The *most* commonly used shielding gas for GMAW/FCAW-G processes is _____.
   a. argon (Ar)
   b. CO_2
   c. helium (He)
   d. carbon monoxide (CO)

7. What rigging device should never be placed into a welding machine's lifting eye when the machine is being lifted or moved?
   a. A synthetic sling
   b. The lifting hook
   c. A wire rope sling
   d. Any type of shackle

3.0.0 Filler Metals

Objective	Performance Task
Identify GMAW and FCAW filler metals. a. Identify various GMAW filler metals. b. Identify various FCAW filler metals.	1. Set up GMAW and FCAW equipment with appropriate shielding gases and filler metals.

GMAW and FCAW filler metals are explained in this section. The WPS specifies the filler metals to be used. The improper selection of filler metal can cause the weld to fail.

Regardless of the type of filler metal being used, there is often a need to cut and manipulate the wire. *Figure 39* shows an example of a special tool used for this purpose. Commonly called MIG welding pliers or wire feed pliers, this tool is designed for drawing out wire, cutting wire, removing and installing contact tips and nozzles, removing welding spatter from nozzles, and light hammering.

Filler Wire and Filler Metal

The terms *filler wire* and *filler metal* are both used to describe the electrode used in GMAW and FCAW. Filler metal is also referred to as *filler rod* when used for GTAW.

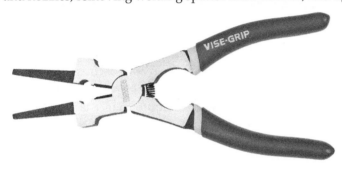

Figure 39 Wire feed pliers.
Source: Stanley Black & Decker, Inc.

3.1.0 GMAW Filler Metals

GMAW uses a continuous solid wire electrode for filler material. This filler wire is drawn from high-grade pure alloys compounded for different applications. During manufacture, the wire is drawn, cleaned, and inspected for defects several times. Standard wire diameters are 0.023" (0.6 mm), 0.030" (0.7 mm), 0.035" (0.9 mm), 0.045" (1.2 mm), 0.052" (1.4 mm), and 0.062" (1.5 mm). Note that the metric equivalents used here represent actual metric wire sizes and are not direct equivalents. The finished wire is wound on spools 4" (102 mm), 8" (204 mm), 12" (30 cm), or 14" (36 cm) in diameter. The most common spool is 12" (30 cm) in diameter and weighs 30 lb (13.6 kg). Reels or drums up to 30" (76 cm) in diameter and weighing up to 1,000 lb (454 kg) are also available for stationary production welding.

Several industry organizations and the United States government publish specification standards for welding wire. The most common are the AWS specifications. The purpose of the AWS specifications is to set standards that all manufacturers must follow when manufacturing welding consumables. This ensures consistency for the user regardless of who manufactured the product. The specifications set standards for the following:

- Classification system, identification, and marking
- Chemical composition of the deposited weld metal
- Mechanical properties of the deposited weld metal

NOTE

An AWS specification number is generally followed by a four-digit number that indicates the year the specification was last revised. When referring to an AWS specification on the job, always make sure to use the version called out by design and drawing documents rather than defaulting to the most current version.

Rod and Electrode Designations

The letter R that follows the E in an AWS classification number indicates that the classification is used as a metal filler rod as well as an electrode. GMAW and GTAW filler metals are labeled the same way when used for GTAW.

The following AWS specifications refer to GMAW wires and rods:

- *AWS A5.7, Specification for Copper and Copper-Alloy Bare Welding Rods and Electrodes*
- *AWS A5.9/A5.9M, Specification for Bare Stainless Steel Welding Electrodes and Rods*
- *AWS A5.10/A5.10M, Specification for Bare Aluminum and Aluminum-Alloy Welding Electrodes and Rods*
- *AWS A5.14/A5.14M, Specification for Nickel and Nickel-Alloy Bare Welding Electrodes and Rods*
- *AWS A5.16/A5.16M, Specification for Titanium and Titanium-Alloy Welding Electrodes and Rods*
- *AWS A5.18/A5.18M, Specification for Carbon Steel Electrodes and Rods for Gas Shielded Arc Welding*
- *AWS A5.19, Specification for Magnesium Alloy Welding Electrodes and Rods*
- *AWS A5.28/A5.28M, Specification for Low-Alloy Steel Electrodes and Rods for Gas Shielded Arc Welding*

Filler wire is graded for three major areas of use:

- *General use* — Wire meets specifications. No record of chemical composition, strength, or similar factors is supplied to the user with the wire purchase.
- *Rigid control fabrication* — A Certificate of Conformance is supplied with the wire at purchase. The stock is identified by code numbers located on the roll package.
- *Critical use* — A Certified Chemical Analysis report is supplied. Records are kept by the fabricator of welds and processes for later reference on aircraft, nuclear reactors, and pressure vessels.

The following factors affect the selection of a GMAW filler wire:

- Base metal chemical composition
- Base metal mechanical properties
- Weld joint design
- Service or specification requirements
- Shielding gas used

Commonly used filler wire types include the following:

- Carbon steel
- Low-alloy steel
- Stainless steel
- Aluminum and aluminum alloy
- Copper and copper alloy
- Nickel and nickel alloy
- Magnesium and magnesium alloy
- Titanium and titanium alloy

Settings for electrode wire of the same classification can vary widely depending upon the manufacturer. It is important to check the manufacturer's recommended settings for each specific wire used. This information can typically be found on the wire manufacturer's website or catalog.

3.1.1 Carbon Steel Filler Metals

Carbon steel filler metals are identified by *AWS A5.18*. The filler metal is available in wire reels and rod form. The wire classification number (*Figure 40*) is located on a label on the side of the reel or container.

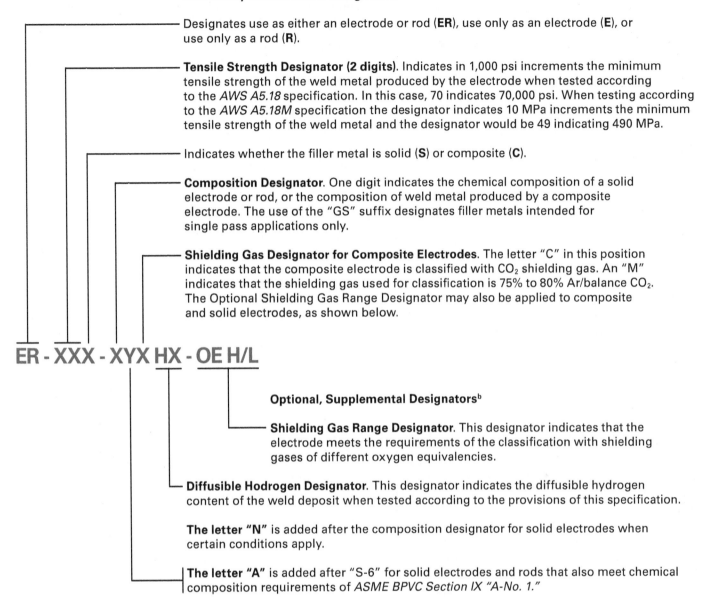

Mandatory Classification Designators[a]

Designates use as either an electrode or rod (**ER**), use only as an electrode (**E**), or use only as a rod (**R**).

Tensile Strength Designator (2 digits). Indicates in 1,000 psi increments the minimum tensile strength of the weld metal produced by the electrode when tested according to the *AWS A5.18* specification. In this case, 70 indicates 70,000 psi. When testing according to the *AWS A5.18M* specification the designator indicates 10 MPa increments the minimum tensile strength of the weld metal and the designator would be 49 indicating 490 MPa.

Indicates whether the filler metal is solid (**S**) or composite (**C**).

Composition Designator. One digit indicates the chemical composition of a solid electrode or rod, or the composition of weld metal produced by a composite electrode. The use of the "GS" suffix designates filler metals intended for single pass applications only.

Shielding Gas Designator for Composite Electrodes. The letter "C" in this position indicates that the composite electrode is classified with CO_2 shielding gas. An "M" indicates that the shielding gas used for classification is 75% to 80% Ar/balance CO_2. The Optional Shielding Gas Range Designator may also be applied to composite and solid electrodes, as shown below.

ER - XXX - XYX HX - OE H/L

Optional, Supplemental Designators[b]

Shielding Gas Range Designator. This designator indicates that the electrode meets the requirements of the classification with shielding gases of different oxygen equivalencies.

Diffusible Hodrogen Designator. This designator indicates the diffusible hydrogen content of the weld deposit when tested according to the provisions of this specification.

The letter "N" is added after the composition designator for solid electrodes when certain conditions apply.

The letter "A" is added after "S-6" for solid electrodes and rods that also meet chemical composition requirements of *ASME BPVC Section IX "A-No. 1."*

[a] The combination of these designators constitutes the electrode (or rod) classification.
[b] The designators are optional and do not constitute a part of the electrode (or rod) classification.

A5.18/A5.18M Classification System

Figure 40 AWS classification for carbon steel filler metal.
Source: AWS A5.18/A5.18M:2005, Figure A.1. Reproduced with permission from the American Welding Society (AWS), Miami, FL, USA

All carbon steel filler metals contain alloys such as silicon, manganese, aluminum, and carbon. Other alloys such as nickel, chromium, and molybdenum are also often added. The purpose of the alloys is as follows:

- *Silicon (Si)* — Concentrations of 0.40% to 1.00% are used to deoxidize the puddle and to strengthen the weld. Silicon above 1% may make the welds crack-sensitive. Silicon also lowers the melting temperature of the wire and promotes wetting.
- *Manganese (Mn)* — Concentrations of 1% to 2% are also employed as a deoxidizer and to strengthen the weld. Manganese also decreases hot crack sensitivity.
- *Aluminum (Al), titanium (Ti), and zirconium (Zr)* — One or more of these elements may be added in very small amounts for deoxidizing. These elements may also increase strength.
- *Carbon (C)* — Concentrations of 0.05% to 0.12% are used to add strength without adversely affecting ductility, porosity, or toughness.
- *Nickel (Ni), chromium (Cr), and molybdenum (Mo)* — These elements may be added in small amounts to improve corrosion resistance, strength, and toughness.

In many cases, carbon steel filler metal wire is copper coated. The copper coating protects the filler metal wire from corrosion and makes electrical contact easier as the wire passes through the copper contact tube at the end of a gun.

AWS carbon steel filler metal wire classifications and their uses are listed in *Table 5*.

TABLE 5 AWS Carbon Steel Wire Classifications and Uses

Wire Classification	Use
ER70S-2 ER48S-2	These wires weld all grades of carbon steel, including killed (highly deoxidized by adding aluminum powder or ferrosilicon), semi-killed, and rimmed (slightly deoxidized by adding a small amount of aluminum powder). Also welds carbon steel with rusted surfaces. Acceptable shielding gases include Ar-O_2, Ar-CO_2, and CO_2. The viscous weld puddle is well suited to out-of-position welding. Wetting is helped by keeping the oxygen or CO_2 percentage relatively high.
ER70S-3 ER48S-3	These wires weld killed and semi-killed steels (used in appliances, automobiles, and farm equipment) with CO_2, Ar-O_2 or Ar-CO_2 shielding gases.
ER70S-4 ER48S-4	These wires weld semi-killed and rimmed steels (used in boilers and pressure vessels, piping, ship steels, structural steels, and A515 grades 55-70 steels) with CO_2, Ar-O_2, or Ar-CO_2 shielding gases.
ER70S-6 ER48S-6	These wires weld many carbon steels with CO_2 and high welding current. High-speed welding can be done with Ar-O_2 with 5% or more O_2. Either spray transfer or short-circuiting transfer can be used.
ER70S-7 ER48S-7	These wires weld a wide variety of steels with superior weld quality and appearance. Porosity and weld defects caused by mill scale or rust are minimized.
ER70S-G ER48S-G E70C-G E48C-G	Filler metals are not included in the preceding classes that only have certain mechanical properties specified. Absence of a C or M suffix means that the manufacturer should be consulted for the recommended shielding gas and intended applications.
E70C-GS E48C-GS	These wires are composite stranded or metal-cored electrodes intended for only single-pass applications. These electrodes may have higher alloy content, and the manufacturer should be consulted for specific applications.
E70C-3 E48C-3 E70C-6 E48C-6	These wires are composite stranded or metal-cored electrodes intended for both single-pass and multipass applications on steel. They are characterized by a spray arc and excellent bead wash characteristics.

3.1.2 Low-Alloy Steel Filler Metals

Low-alloy steel filler metals are identified by *AWS A5.28*. The filler metal is available in wire reel and rod form. The wire classification number is located on a label on the side of the reel or container. The classification is similar to the carbon steel filler metal classification except that there is a letter in the suffix at the end of the classification (*Figure 41*). The letter suffix indicates that the filler metal contains alloys (nickel, manganese, molybdenum, etc.) that give the deposited weld metal special properties. Also, the tensile strength of low-alloy filler metals is often higher than that of carbon steel filler metals. For specific information on low-alloy filler metal composition and intended use, refer to the manufacturer's specifications.

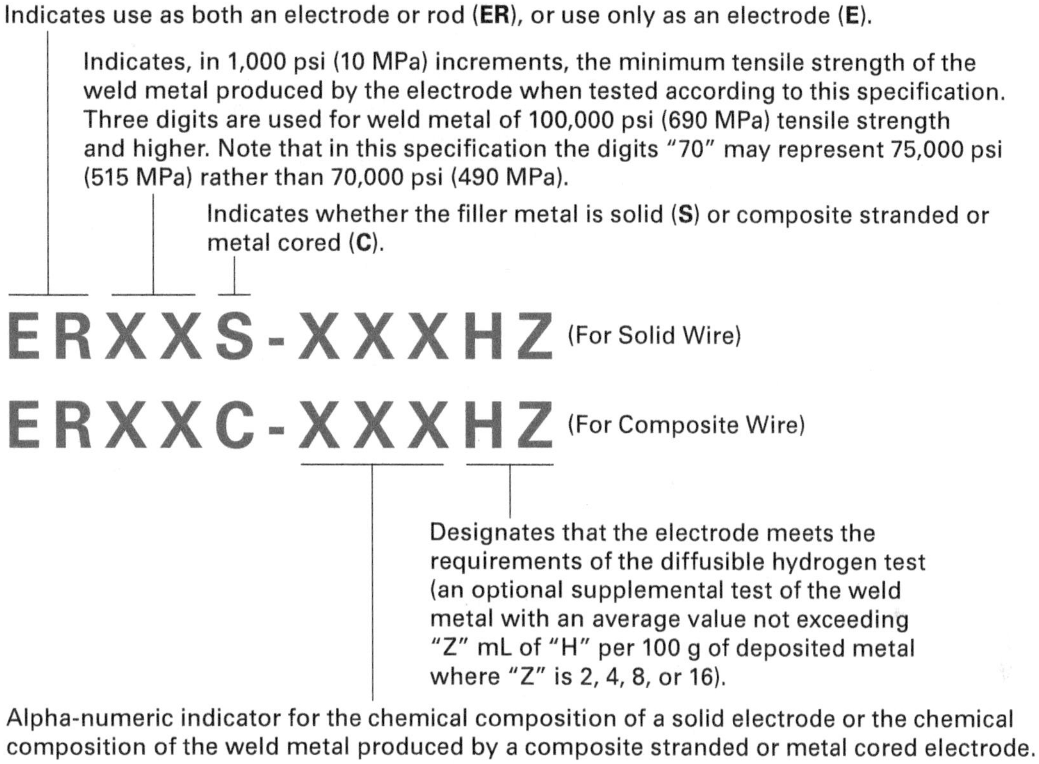

Indicates use as both an electrode or rod (**ER**), or use only as an electrode (**E**).

Indicates, in 1,000 psi (10 MPa) increments, the minimum tensile strength of the weld metal produced by the electrode when tested according to this specification. Three digits are used for weld metal of 100,000 psi (690 MPa) tensile strength and higher. Note that in this specification the digits "70" may represent 75,000 psi (515 MPa) rather than 70,000 psi (490 MPa).

Indicates whether the filler metal is solid (**S**) or composite stranded or metal cored (**C**).

ERXXS-XXXHZ (For Solid Wire)

ERXXC-XXXHZ (For Composite Wire)

Designates that the electrode meets the requirements of the diffusible hydrogen test (an optional supplemental test of the weld metal with an average value not exceeding "Z" mL of "H" per 100 g of deposited metal where "Z" is 2, 4, 8, or 16).

Alpha-numeric indicator for the chemical composition of a solid electrode or the chemical composition of the weld metal produced by a composite stranded or metal cored electrode.

Figure 41 AWS classification for low-alloy steel filler metal.

3.1.3 Stainless Steel Filler Metals

Stainless steel filler metals are identified by *AWS A5.9* as shown in *Figure 42*. The filler metal is available in wire reel, flat strip, and rod form.

Stainless steel filler metal should be selected to closely match the alloy composition of the base metal. Stainless steel filler metals also require specific shielding gases or gas mixtures.

3.1.4 Aluminum and Aluminum Alloy Filler Metals

Aluminum filler metals are covered by *AWS A5.10*. Aluminum filler metals usually contain magnesium, manganese, zinc, silicon, or copper for increased strength. Corrosion resistance and weldability are also considerations. Aluminum filler metals are designed to weld specific types of aluminum and should be selected for compatibility. The most widely used aluminum filler metals are ER4043 (contains silicon) and ER5356 (contains magnesium). The filler metal classification number (*Figure 43*) is located on a label on the side of the reel or container.

3.1.5 Copper and Copper Alloy Filler Metals

Copper and copper alloy filler metals are covered by *AWS A5.7*. Most copper filler metals contain other elements that increase strength, deoxidize the weld metal, and match the base metal composition. *Figure 44* shows the AWS classification for copper filler metal.

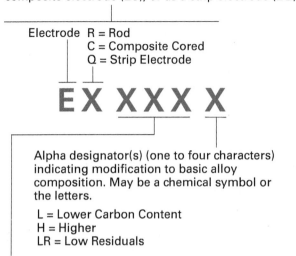

Figure 42 AWS classification for stainless steel filler metal.

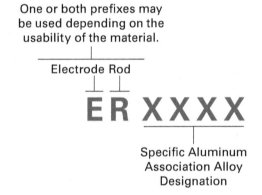

Figure 43 AWS classification for aluminum filler metal.

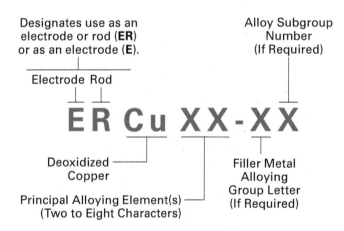

Figure 44 AWS classification for copper filler metal.

3.1.6 Nickel and Nickel Alloy Filler Metals

Nickel-based filler metals (*Figure 45*) are covered by *AWS A5.14*. These filler metals contain other elements to match base metal applications and to increase the strength and quality of the weld metal. For GMAW, DCEP is used. Argon shielding gas is often used, as are mixtures of argon and helium.

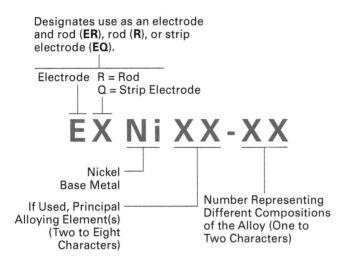

Figure 45 AWS classification for nickel filler metal.

3.1.7 Magnesium Alloy Filler Metals

Magnesium alloy filler metals (*Figure 46*) are covered by *AWS A5.19*. These filler metals are usually used with GTAW, GMAW, and plasma arc welding (PAW) processes. Oxyfuel welding should be used only for temporary repair work. GMAW of magnesium alloys is the same as for other metals. Argon is generally used for shielding; however, mixtures of argon and helium are occasionally used. Pulsed and short-circuiting GMAW are used for magnesium alloys, and spray transfer without pulsing is also used; however, globular transfer is not suitable. Magnesium fines are flammable, so don't allow them to accumulate.

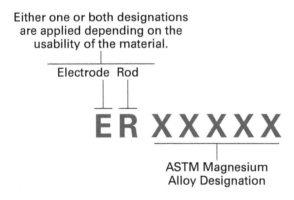

Figure 46 AWS classification for magnesium filler metal.

3.1.8 Titanium and Titanium Alloy Filler Metals

Titanium and titanium alloy filler metals (*Figure 47*) are covered by *AWS A5.16*. These filler metals are generally used with GTAW, GMAW, submerged arc welding (SAW), and PAW processes. Titanium is sensitive to embrittlement by oxygen, nitrogen, and hydrogen at temperatures above 500°F (260°C). Like aluminum, titanium requires weld cleaning, high-purity gas shielding, and an especially adequate post-flow shielding to avoid embrittlement. Titanium can be successfully fusion welded to zirconium, tantalum, niobium, and vanadium. Titanium should not be welded to copper, iron, nickel, or aluminum. Like magnesium, titanium fines are flammable.

Magnesium Fires

While the ignition of magnesium is a very remote possibility when welding, the fire will cease when the heat source is removed. Ignition of the weld pool is also prevented by the gas shielding used in GTAW, GMAW, and PAW processes. Most magnesium fires occur when fines from grinding or filing, or chips from machining, are allowed to accumulate. Take care to prevent the accumulation of fines on clothing. Graphite-based or salt-based powders recommended for extinguishing magnesium fires should be stored in the work area. If large amounts of fines are produced, collect them in a water-wash dust collector designed for use with magnesium. Follow special precautions for the handling of wet magnesium fines.

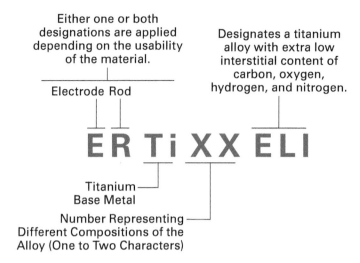

Figure 47 AWS classification for titanium filler metal.

3.2.0 | FCAW Filler Metals

FCAW of Structural Steel

For large structural welding projects such as bridges and skyscrapers, many construction companies prefer carbon or low-alloy steel FCAW to meet the toughness requirements for new construction using the newer self-shielding E71T-8 or gas-shielded E71T-1 electrodes. Many ironworkers who use SMAW are reluctant to use FCAW, and those who have used FCAW predominantly relied upon E71T-11 (a limited-thickness electrode) for structural welds because it was easy to use in all positions. However, welds with T-11 sometimes had cracking problems and had to be rewelded. When welders are properly retrained for FCAW using the newer electrodes, the cracking problems are eliminated, and the deposition rates are higher because the newer electrodes run hotter and faster than the T-11 electrodes or SMAW electrodes.

FCAW is primarily used for the welding of ferrous metals. These include low- and medium-carbon steels, some low-alloy steels, cast irons, and stainless steels.

The FCAW electrode is a continuous flux-cored wire electrode. The wire is made by forming a thin filler metal strip into a U-shape, filling the U with the flux material, squeezing the strip shut around the flux, and then drawing the tube through dies to size the wire and compact the flux. During manufacture, the wire is carefully inspected for defects. Flux-cored wires are manufactured in a range of sizes beginning at 0.023" (0.6 mm). The finished flux-cored wire is coiled on all the standard spool sizes and forms. Alternatively, some FCAW wires are welded closed or seamless-type wires. The use of seamless wires in welds has been associated with fewer problems of hydrogen-induced cracking.

Keep in mind that all flux-cored wires are generally considered to be low hydrogen. For this reason, proper electrode control procedures must be followed to keep the wire dry.

By varying the power settings on the welding machine, many smaller sizes of electrode wire can be used for thin metals and multipass welds on metals of unlimited thickness in all positions.

FCAW flux-cored electrodes are now classified by the AWS either as a fixed or an open classification system. The fixed classification system has been carried over from an earlier system that provided limited descriptors for the welding characteristics of the electrode. As an example, *Figure 48* shows the AWS fixed classification for mild steel and low-alloy steel FCAW electrodes.

The open classification system allows more designators to be defined, thus providing more information to personnel for selecting electrodes based on process, welding characteristics, strengths, and weld positions.

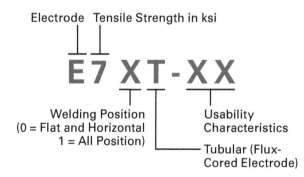

Figure 48 AWS fixed classification for mild steel and low-alloy steel FCAW electrodes.

3.2.1 Carbon Steel Flux-Cored Electrodes

The carbon steel flux-cored electrode welding characteristics, as indicated by the AWS classification, are shown in *Table 6A* and *Table 6B*. Carbon steel flux-cored electrode classes and uses are shown in *Table 7*.

TABLE 6A Carbon Steel Flux-Cored Electrode Welding Position, Shielding, Polarity, and Application Requirements (1 of 2)

Usability Designator	AWS Classification		Position of Welding[a,b]	External Shielding[c]	Polarity[d]	Application[e]
	A5.20	**A5.20M**				
1	E70T-1C	E490T-1C	H, F	CO_2	DCEP	M
	E70T-1M	E490T-1M	H, F	75% to 80% Ar/ bal CO_2		
	E71T-1C	E491T-1C	H, F, VU, OH	CO_2		
	E71T-1M	E491T-1M	H, F, VU, OH	75% to 80% Ar/ bal CO_2		
2	E70T-2C	E490T-2C	H, F	CO_2	DCEP	S
	E70T-2M	E490T-2M	H, F	75% to 80% Ar/ bal CO_2		
	E71T-2C	E491T-2C	H, F, VU, OH	CO_2		
	E71T-2M	E491T-2M	H, F, VU, OH	75% to 80% Ar/ bal CO_2		
3	E70T-3	E490T-3	H, F	None	DCEP	S
4	E70T-4	E490T-4	H, F	None	DCEP	M
5	E70T-5C	E490T-5C	H, F	CO_2	DCEP	M
	E70T-5M	E490T-5M	H, F	75% to 80% Ar/ bal CO_2		
	E70T-5C	E491T-5C	H, F, VU, OH	CO_2	DCEP or DCEN[f]	
	E70T-5M	E491T-5M	H, F, VU, OH	75% to 80% Ar/ bal CO_2		
6	E70T-6	E490T-6	H, F	None	DCEP	M
7	E70T-7	E490T-7	H, F	None	DCEN	M
	E71T-7	E491T-7	H, F, VU, OH			
8	E70T-8	E490T-8	H, F	None	DCEN	M
	E71T-8	E491T-8	H, F, VU, OH			
9	E70T-9C	E490T-9C	H, F	CO_2	DCEP	M
	E70T-9M	E490T-9M	H, F	75% to 80% Ar/ bal CO_2		
	E71T-9C	E491T-9C	H, F, VU, OH	CO_2		
	E71T-9M	E491T-9M	H, F, VU, OH	75% to 80% Ar/ bal CO_2		
10	E70T-10	E490T-10	H, F	None	DCEN	S

Source: AWS A5.20/A5.20M:2005, Table 2. Reproduced with permission from the American Welding Society (AWS), Miami, FL, USA

TABLE 6B Carbon Steel Flux-Cored Electrode Welding Position, Shielding, Polarity, and Application Requirements (2 of 2)

Usability Designator	AWS Classification		Position of Welding[a,b]	External Shielding[c]	Polarity[d]	Application[e]
	A5.20	A5.20M				
11	E70T-11	E490T-11	H, F	None	DCEN	M
	E71T-11	E491T-11	H, F, VU, OH			
12	E70T-12C	E490T-12C	H, F	CO_2	DCEP	M
	E70T-12M	E490T-12M	H, F	75% to 80% Ar/bal CO_2		
	E71T-12C	E491T-12C	H, F, VU, OH	CO_2		
	E71T-12M	E491T-12M	H, F, VU, OH	75% to 80% Ar/bal CO_2		
13	E61T-13	E431T-13	H, F, VU, OH	None	DCEN	S
	E71T-13	E491T-13				
14	E71T-14	E491T-14	H, F, VU, OH	None	DCEN	S
	E60T-G E70T-G	E430T-G E490T-G	H, F	Not Specified	Not Specified	M
	E61T-G E71T-G	E431T-G E491T-G	H, F, VD or VU, OH	Not Specified	Not Specified	M
G	E60T-GS E70T-GS	E430T-GS E490T-GS	H, F	Not Specified	Not Specified	S
	E61T-GS E71T-GS	E431T-GS E491T-GS	H, F, VD or VU, OH	Not Specified	Not Specified	S
	EX0T-G[g]	EX0T-G[g]	H, F	Not Specified	Not Specified	S
	EX1T-G[g]	EX1T-G[g]	H, F, VD or VU, OH	Not Specified	Not Specified	
	EX0T-GS[g]	EX0T-GS[g]	H, F	Not Specified	Not Specified	
	EX1T-GS[g]	EX1T-GS[g]	H, F, VD or VU, OH	Not Specified	Not Specified	

Notes:

[a]H = horizontal position; F = flat position; OH = overhead position; VD = vertical position with downward progression; VU = vertical position with upward progression.

[b]Electrode sizes suitable for out-of-position welding, i.e., welding positions other than flat or horizontal, usually sizes smaller than the $3/32$" (2.4 mm) size. For that reason, electrodes meeting the requirements for the groove weld tests and the fillet weld tests may be classified as EX1T-X or EX1T-XM (where X represents the tensile strength and usability designators) regardless of their size.

[c]Properties of weld metal from electrodes used with external gas shielding (EXXT-1, EXXT-1M, EXXT-2, EXXT-2M, EXXT-5, EXXT-5M, EXXT-9, EXXT-9M, EXXT-12, and EXXT-12M) vary according to the shielding gas employed. Electrodes classified with the specified shielding gas should not be used with other shielding gases without first consulting the manufacturer of the electrode.

[d]The term "DCEP" refers to direct current electrode positive (DC, reverse polarity). The term "DCEN" refers to direct current electrode negative (DC, straight polarity).

[e]M = single or multipass; S = single pass only. (Single-pass-only classes are highly oxidizing and are used on metals that are rusty or have mill scale.)

[f]Some E71T-5 and E71T-5M electrodes may be recommended for use on DCEN for improved out-of-position welding. Consult the manufacturer for the recommended polarity.

[g]The letter "X" can be replaced with either a "6" or "7" to designate the weld metal's tensile strength.

Source: AWS A5.20/A5.20M:2005, Table 2. Reproduced with permission from the American Welding Society (AWS), Miami, FL, USA

TABLE 7 Carbon Steel Flux-Cored Electrode Classes and Uses

AWS Classification	Intended Uses
EXXT-1 EXXT-1M	Electrodes designated as -1 use CO_2 shielding. However, other gas mixtures can be used to improve arc stability if recommended by the manufacturer. Electrodes designated as -1M use an argon-CO_2 mixture to increase manganese and silicon in the weld metal to increase yield and tensile strengths. Reducing argon in a gas mixture or using CO_2 shielding only with -1M electrodes may cause deterioration of the arc, out-of-position welding characteristics, and loss of strength. Both electrodes have spray transfer characteristics and low spatter. Most electrodes of this class have rutile-base slag and produce high disposition rates.
EXXT-2	These electrodes are similar to -1 and -1M types with higher manganese, silicon, or both. The higher levels of deoxidizers allow single-pass welding of heavily oxidized or rimmed steel. The -2M electrodes have an even higher manganese content and can be used on metals with heavier mill scale, rust, or foreign matter that cannot be tolerated by some -1 or -1M electrodes.
EXXT-3	These self-shielding electrodes are designed for single-pass, very high-speed welding and use spray transfer. Besides the listed positions, they can be used on 20° inclined downhill welds of sheet metal. They are not recommended for T or lap joints in materials thicker than $\frac{3}{16}$" or for butt, edge, or corner joints in materials thicker than $\frac{1}{4}$".
EXXT-4	These self-shielding electrodes use globular-type transfer and are designed for very high deposition rates. They produce low-sulfur welds that resist hot cracking. They are also designed for low penetration beyond the root of the weld because of poor joint fit-up.
EXXT-5 EXXT-5M	These electrodes have a globular transfer and a lime-fluoride base slag that produces welds with better impact strength and hot and/or cold cracking resistance than other similar classes with rutile-based slags. The -5 electrodes are used with a CO_2 shielding gas; however, argon-CO_2 mixes can be used to reduce spatter if recommended by the manufacturer. Reducing argon in a gas mixture or using only CO_2 shielding with -5M electrodes may cause deterioration of the arc, out-of-position welding characteristics, and loss of strength. Operator appeal for both types of electrodes is not as good as other similar classes with rutile slag.
EXXT-6	These self-shielding electrodes use spray-type transfer. The slag system is designed for good penetration into the root of the weld and excellent slag removal, even in deep grooves.
EXXT-7	These self-shielding electrodes use a droplet/spray-type transfer and produce a low-sulfur weld that is very resistant to cracking. The slag system is designed to produce high deposition rates for larger size electrodes in the flat or horizontal positions.
EXXT-8	These self-shielded electrodes have small droplet/spray-type transfer. They are suitable for all positions. The weld metal has good toughness, low temperature performance, and crack resistance.
EXXT-9 EXXT-9M	These electrodes are essentially the same as the -1 and -1M electrodes, except for the deposit weld metal with improved impact properties. Some electrodes in this class require joints that are relatively clean of oil, excess oxide, and scale.
EXXT-10	These self-shielding electrodes have small droplet transfer. They are used for high-speed, single-pass welds on any thickness material in the flat, horizontal, and 20° inclined vertical positions.
EXXT-11	These self-shielding electrodes have a smooth spray-type transfer and are general-purpose electrodes for all positions. Their use on material thicknesses greater than $\frac{3}{4}$" is not generally recommended unless preheat and interpass temperatures are maintained. Consult the manufacturer for specific recommendations.
EXXT-12 EXXT-12M	These electrodes are essentially the same as -1 and -1M electrodes, except they are modified to improve impact toughness and to lower manganese requirements of the A-1 Analysis Group in the *American Society of Mechanical Engineers (ASME) Boiler and Pressure Vessel Code, Section IX*. As a result, tensile strength and hardness will be decreased from -1 and -1M electrodes. Users should check hardness when a required hardness level is specified or required.
EXXT-13	These self-shielding electrodes are usually welded with a short-circuiting transfer. The slag system allows an all-position root pass on pipe welds. They are recommended only for the root pass and are not generally recommended for multiple passes.
EXXT-14	These self-shielding electrodes have a smooth spray-type transfer. The slag system allows high-speed welds in all positions. They are used to make welds on sheet metal up to $\frac{3}{16}$" thick and are often designed especially for galvanized, aluminized, or other coated steels. They are generally not recommended for T or lap joints in materials thicker than $\frac{3}{16}$" or butt, edge, or corner joints in materials thicker than $\frac{1}{4}$". The manufacturer should be consulted for specific recommendations.
EXXT-G	These are multipass electrodes not covered by other classifications. Except for requirements to assure carbon steel deposits and tensile strengths, other requirements are agreed upon by the purchaser and manufacturer.
EXXT-GS	These are single-pass electrodes not covered by other classifications. Except for tensile strength, other requirements are agreed upon by the purchaser and the manufacturer.

3.2.2 Low-Alloy Steel Flux-Cored Electrodes

Figure 49 shows the *AWS A5.29* specification format for low-alloy steel flux-cored electrodes.

This designator indicates that the electrode is a flux-cored electrode.

This designator is either 0 or 1. It indicates the positions of welding for which the electrode is intended. 0 is for flat and horizontal positions only; 1 is for all positions.

This designator is some number: 1, 4, 5, 6, 7, 8, 11, or the letter "G". The number refers to the usability of the electrode. The "G" indicates that the slag system and shielding gas are not specified.

This designator indicates the minimum tensile strength (in ksi ⊠ 10) of the weld metal when the weld metal is made in the manner prescribed by this specification. (Two digits may be required.)

Designates the chemical composition of the deposited weld. Specific chemical compositions are not always identified with specific mechanical properties in the specification. A supplier is required by the specification to include the mechanical properties appropriate for a particular electrode in classification of that electrode. Thus, for example, a complete designation is E80T5-Ni3, EXXT5-Ni3 is not a complete classification. The letter "G" indicates that the chemical composition is not specified.

Designates an electrode.

An "M" designator in this position indicates that the electrode is classified using 75% to 80% argon/balance CO_2 shielding gas. When the "M" designator does not appear, it signifies that either the shielding gas used for classification is CO_2 or that the electrode is a self-shielded electrode.

E X X T X - X M J H Z

Designates that the electrode meets the requirements for improved toughness by meeting a requirement of 20' × lbf (27J) at a test temperature of 20°F (11°C) lower than the temperature shown for the classification. Absence of the "J" indicates normal impact requirements.

If present, designates that the electrode meets the requirements of the diffusible hydrogen test (an optional supplemental test of the weld metal with an average value not exceeding "Z" mL of "H" per 100 g of deposited metal where "Z" is 4, 8, or 16).

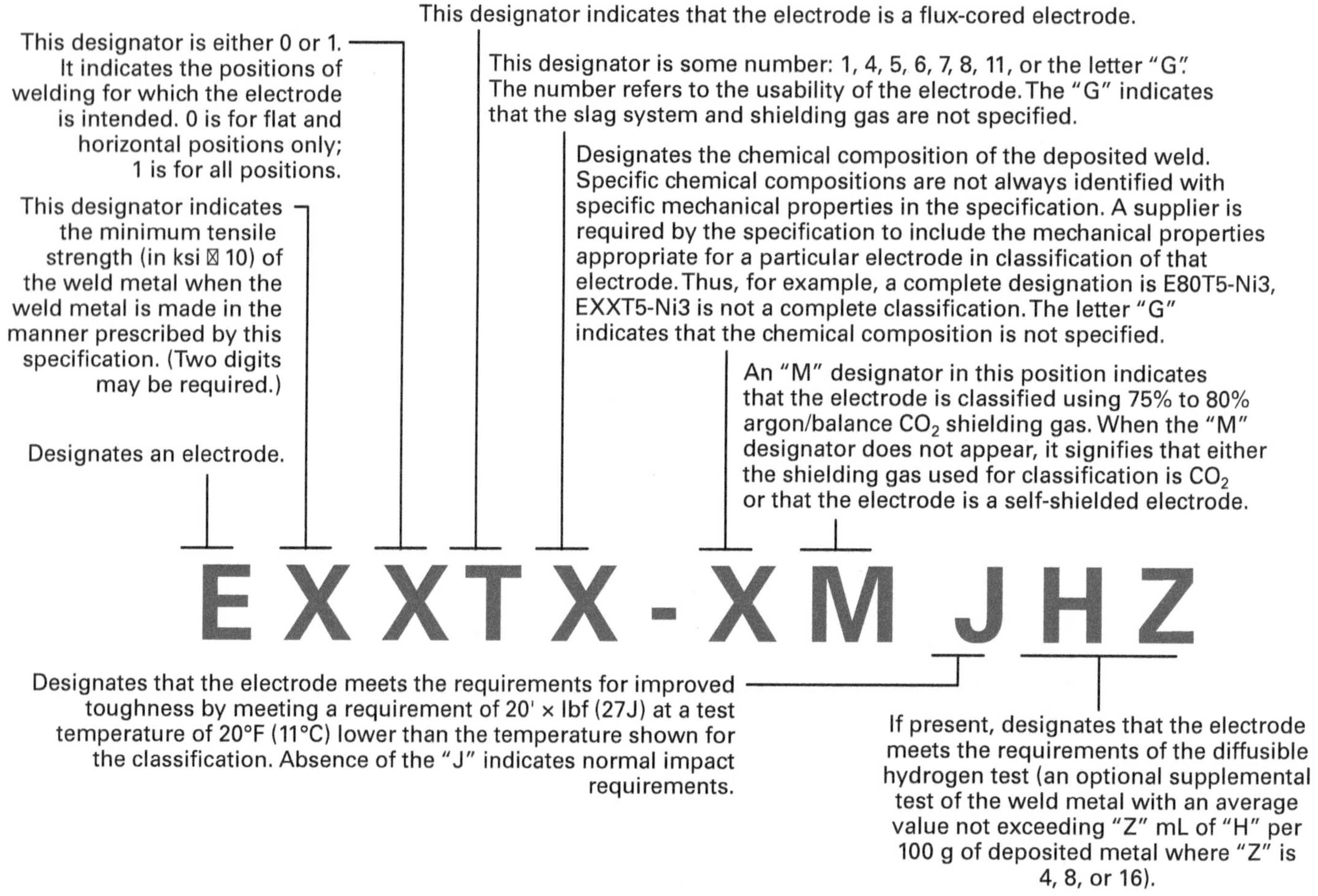

Figure 49 AWS classification for low-alloy steel flux-cored electrodes.
Source: AWS A5.20/A5.20M:2005, Figure A.1. Reproduced with permission from the American Welding Society (AWS), Miami, FL, USA

3.0.0 Section Review

1. Which of these filler metals is sensitive to embrittlement by oxygen, nitrogen, and hydrogen at temperatures above 500°F (260°C)?
   a. Nickel
   b. Magnesium
   c. Titanium
   d. Aluminum

2. In FCAW operations, it is possible to use a smaller size electrode wire for thin metals and multipass welds by varying the welding machine's _____.
   a. pinch
   b. polarity cycling
   c. feed drag
   d. power settings

1. Areas where oil, grease, and other petroleum products are present should be kept clear of accumulated _____.
   a. oxygen
   b. helium
   c. argon
   d. nitrogen

2. In GMAW, the arc is stabilized, and the molten filler and base metals are protected from oxidation using a(n) _____.
   a. differential pressure
   b. self-fluxing alloy
   c. intense heat
   d. shielding gas

3. Of the two basic FCAW processes, one is self-shielding while the other _____.
   a. requires no shielding gas
   b. uses ferrous core vapors
   c. uses an external shielding gas
   d. utilizes an oxygen generator

4. Which metal transfer mode uses a high welding voltage that ultimately causes fine droplets of electrode wire to be carried axially to the molten weld puddle?
   a. Globular transfer
   b. Inverter transfer
   c. Short-circuiting transfer
   d. Spray transfer

5. Short-circuiting metal transfer is the *best* mode to use for welding _____.
   a. cast iron blocks
   b. light-gauge sheet metal
   c. heavy steel beams
   d. aluminum plates

6. FCAW welds made with CO_2 shielding gas produce _____.
   a. deep penetration
   b. a slag-free weld
   c. shallow penetration
   d. a lot of spatter

7. Depending on the range selected, for every 100 A of current, the arc voltage of a welding machine is typically lower than the open-circuit voltage by _____.
   a. 0 V to 1 V
   b. 1 V to 2 V
   c. 2 V to 3 V
   d. 4 V to 5 V

8. The welding arc is established between the workpiece and the _____.
   a. beginning of the wire electrode
   b. end of the wire electrode
   c. middle of the wire electrode
   d. clamp of the wire electrode

9. One advantage of using engine-driven generators to power welders is that the equipment is _____.
    a. less expensive than single-phase welders
    b. easier to operate than all other welders
    c. free of most maintenance requirements
    d. portable so it can be used in the field

10. Which welding power source increases the frequency of the incoming primary power and is well suited for applications where space is limited?
    a. Inverter
    b. Engine-driven generator
    c. Axial
    d. Transformer-rectifier

11. When plotted on a graph, the general angle showing the voltage-to-amperage relationship for GMAW is known as the _____.
    a. pinch
    b. arc blow
    c. slope
    d. hypotenuse

12. Two factors that *must* be considered when selecting the size of a welding cable are the amperage load and the _____.
    a. type of copper strands in the cable
    b. distance the current will travel
    c. speed of the wire electrode feed
    d. pinch effect imparted on the cable

13. The duty cycle of a GMAW gun is determined and rated with _____.
    a. carbon dioxide (CO_2)
    b. helium (He)
    c. argon (Ar)
    d. oxygen (O)

14. Light-duty GMAW guns are cooled by _____.
    a. domestic water flow
    b. special cooling fluid
    c. shielding gas flow
    d. demineralized water

15. Because of its current transfer and ionization properties, the *most* common shielding gas that can be used alone or in combination with other gases is _____.
    a. carbon dioxide (CO_2)
    b. helium (He)
    c. nitrogen (N)
    d. argon (Ar)

16. What is required to meter shielding gas to a welding gun at the proper flow rate?
    a. A gas mixer / stabilizer
    b. A variable vacuum tip
    c. A gas regulator / flowmeter
    d. An auxiliary gun controller

17. After checking the equipment specifications to determine the weight of a welding machine, what is the proper way to lift and move the machine?
    a. Use a wire rope basket hitch under the welder.
    b. Attach a sling and/or shackle to the welder lift eye.
    c. Use numerous synthetic slings attached to the corners of the welder.
    d. Attach the lifting hook from the lifting device to the welder lift eye.

18. A tool designed for cutting and manipulating wire, removing and installing contact tips and nozzles, and removing welding spatter from nozzles is a pair of _____.
    a. cast and helix straightener bars
    b. knurled combination rollers
    c. push-pull wire straighteners
    d. MIG welding or wire feed pliers

19. Many types of GMAW carbon steel filler wire are protected from corrosion by a(n) _____.
    a. copper coating
    b. aluminum core
    c. mercury alloy
    d. nonferrous skin

20. The primary use of FCAW is for the welding of _____.
    a. nonferrous metals
    b. aluminum
    c. ferrous metals
    d. titanium

Answers to odd-numbered questions are found in the Review Question Answer Keys at the back of this book.

Answers to Section Review Questions

Answer	Section Reference	Objective
Section 1.0.0		
1. b	1.1.3	1a
2. c	1.2.0	1b
3. d	1.3.3	1c
4. b	1.4.0	1d
Section 2.0.0		
1. a	2.1.1	2a
2. c	2.2.3	2b
3. d	2.3.1	2c
4. b	2.4.0	2d
5. c	2.5.2	2e
6. a	2.6.1	2f
7. b	2.7.4	2g
Section 3.0.0		
1. c	3.1.8	3a
2. d	3.2.0	3b

User Update

Did you find an error? Submit a correction by visiting **https://www.nccer.org/olf** or by scanning the QR code using your mobile device.

GMAW – Plate

Source: Photo courtesy of The Lincoln Electric Company, Cleveland, OH, U.S.A.

Objectives

Successful completion of this module prepares you to do the following:

1. Identify GMAW-related safety practices and explain how to prepare for welding.
   a. Identify GMAW-related safety practices.
   b. Describe basic GMAW processes.
   c. Explain how to safely set up the equipment and work area for welding.
2. Describe equipment control and welding procedures for GMAW and explain how to produce basic weld beads.
   a. Describe equipment control and welding techniques related to GMAW.
   b. Explain how to produce basic GMAW weld beads.
3. Describe the welding procedures needed to produce proper fillet and V-groove welds using GMAW welding techniques.
   a. Describe the welding procedures needed to produce proper fillet welds using GMAW welding techniques.
   b. Describe the welding procedures needed to produce proper V-groove welds using GMAW welding techniques.

Performance Tasks

Under supervision, you should be able to do the following:

1. Make multipass GMAW-S (short-circuit) fillet welds on carbon steel plate coupons in all four 1F through 4F positions, using solid or composite electrode.
2. Make multipass GMAW-S (short-circuit) V-groove welds on carbon steel plate coupons in all four 1G through 4G positions, with or without backing, using solid or composite electrode.
3. Make multipass GMAW spray-transfer fillet welds on carbon steel plate coupons in both the 1F and 2F positions, using solid or composite electrode.
4. Make multipass GMAW spray-transfer V-groove welds on carbon steel plate coupons in the 1G position, with backing, using solid or composite electrode.

Overview

Gas metal arc welding (GMAW) is one of the most common welding processes. GMAW equipment is available in many different sizes and varieties; however, the basic operating principles of GMAW apply to all makes and models of equipment. This module describes how GMAW equipment is used, identifies GMAW-related safety practices, and examines different techniques used to produce various types of basic weld beads.

Digital Resources for Welding

Scan this code using the camera on your phone or mobile device to view the digital resources related to this craft.

1.0.0 GMAW Welding

Performance Tasks

1. Make multipass GMAW-S (short-circuit) fillet welds on carbon steel plate coupons in all four 1F through 4F positions, using solid or composite electrode.

2. Make multipass GMAW-S (short-circuit) V-groove welds on carbon steel plate coupons in all four 1G through 4G positions, with or without backing, using solid or composite electrode.

3. Make multipass GMAW spray-transfer fillet welds on carbon steel plate coupons in both the 1F and 2F positions, using solid or composite electrode.

4. Make multipass GMAW spray-transfer V-groove welds on carbon steel plate coupons in the 1G position, with backing, using solid or composite electrode.

Objective

Identify GMAW-related safety practices and explain how to prepare for welding.

a. Identify GMAW-related safety practices.
b. Describe basic GMAW processes.
c. Explain how to safely set up the equipment and work area for welding.

Gas metal arc welding (GMAW) is often called *metal inert gas (MIG) welding*. As the names imply, GMAW joins together two metals using a filler metal electrode and an inert shielding gas. The following sections describes GMAW safety practices, GMAW equipment and how it works, and examines how to prepare for GMAW work.

1.1.0 Safety Summary

The following sections summarize the safety procedures and practices that welders must observe when cutting or welding. Complete safety coverage is provided in NCCER Module 29101, *Welding Safety*, which should be completed before continuing with this module. The safety procedures and practices are in place to prevent potentially life-threatening injuries from welding accidents. Above all, be sure to wear appropriate protective clothing and equipment when welding or cutting.

1.1.1 Protective Clothing and Equipment

Welding activities can cause injuries unless you wear all the protective clothing and equipment designed specifically for the welding industry. The following safety guidelines about protective clothing and equipment should be followed to prevent injuries:

- Wear a face shield over snug-fitting cutting goggles or safety glasses for gas welding or cutting. The welding hood should be equipped with an approved shade for the application. A welding hood equipped with a properly tinted lens is best for all forms of welding.

- Wear proper protective leather and/or flame-retardant clothing along with welding gloves that protect the welder from flying sparks, molten metal, and heat.

- Wear high-top safety shoes or boots. Make sure the tongue and lace area of the footwear will be covered by a pant leg. If the tongue and lace area is exposed or the footwear must be protected from burn marks, wear leather spats under the pants or chaps over the front and top of the footwear.

- Wear earmuffs or, at least earplugs, to protect ear canals from sparks.

- Wear a 100% cotton cap with no mesh material included in its construction. The bill of the cap points to the rear. If a hard hat is required for the environment, use one that allows the attachment of rear deflector material and a face shield. A hard hat with a rear deflector is generally preferred when working overhead and may be required by some employers and jobsites.

WARNING!

Using proper personal protective equipment (PPE) for the hands and eyes is particularly important. The most common injuries that welders experience during GMAW operations are injuries to the fingers and eyes.

1.1.2 Fire/Explosion Prevention

Welding activities usually involve the use of fire or extreme heat to melt metal. Whenever fire is used in a weld, the fire must be controlled and contained. Welding or cutting activities are often performed on vessels that may have once contained flammable or explosive materials. Residues from those materials can catch fire or explode when a welder begins work on such a vessel. The following fire and explosion prevention guidelines associated with welding contribute to a safe work zone:

- Never carry matches or gas-filled lighters in your pockets. Sparks can cause the matches to ignite or the lighter to explode, causing serious injury.

- Never use oxygen to blow dust or dirt from clothing. The oxygen can remain trapped in the fabric for a time. If a spark hits clothing during this time, the clothing can burn rapidly and violently out of control.

- Make sure any flammable material in the work area is moved or shielded by a fire-resistant covering. Approved fire extinguishers must be available before attempting any heating, welding, or cutting operations. If a hot work permit and a fire watch are required, be sure those items are in place before beginning and that all site requirements have been observed.

- Never release a large amount of oxygen or use oxygen as compressed air. The presence of oxygen around flammable materials or sparks can cause rapid and uncontrolled combustion. Keep oxygen away from oil, grease, and other petroleum products.

- Never release a large amount of fuel gas, especially acetylene. Propane tends to concentrate in and along low areas and can ignite at a considerable distance from the release point. Acetylene is lighter than air but is even more dangerous than propane. When mixed with air or oxygen, acetylene will explode at much lower concentrations than any other fuel.

- To prevent fires, maintain a neat and clean work area, and make sure that any metal scrap or slag is cold before disposing of it.

- Before cutting containers such as tanks or barrels, check to see if they have contained any explosive, hazardous, or flammable materials, including petroleum products, citrus products, or chemicals that decompose into toxic fumes when heated. As a standard practice, always clean and then fill any tanks or barrels with water or purge them with a flow of inert gas such as nitrogen to displace any oxygen. Containers must be cleaned by steam cleaning, flushing with water, or washing with detergent until all traces of the material have been removed.

WARNING!

Welding or cutting must never be performed on drums, barrels, tanks, vessels, or other containers until they have been emptied and cleaned thoroughly, eliminating all flammable materials and all substances (such as detergents, solvents, greases, tars, or acids) that might produce flammable, toxic, or explosive vapors when heated. Clean containers only in well-ventilated areas, as vapors can accumulate during cleaning, causing explosions or injury.

Proper procedures for cutting or welding hazardous containers are described in the *American Welding Society (AWS) F4.1, Safe Practices for the Preparation of Containers and Piping for Welding, Cutting, and Allied Processes.*

1.1.3 Work Area Ventilation

Vapors and fumes tend to rise in the air from their sources. Welders must often work above the welding area where the fumes are being created. Welding fumes can cause personal injuries including long-term respiratory harm. Good work area ventilation helps to remove the vapors and protect the welder. The following is a list of work area ventilation guidelines to consider before and during welding activities:

- Follow confined space procedures before conducting any welding or cutting in a confined space.
- Never use oxygen for ventilation in confined spaces.
- Always perform cutting or welding operations in a well-ventilated area. Cutting or welding operations involving zinc or cadmium materials or coatings result in toxic fumes. For long-term cutting or welding of such materials, always wear an approved full-face, supplied-air respirator that uses breathing air supplied from outside of the work area. For occasional, very short-term exposure, a high-efficiency particulate arresting-rated (also called HEPA-rated) or metal fume filter may be used on a standard respirator.
- Make sure confined spaces are properly ventilated for cutting or welding purposes. Use powered extraction systems when available. *Figure 1* shows mobile extraction equipment used to remove cutting or welding fumes and vapors from work areas.

> **WARNING!**
>
> Backup flux may contain silica, fluorides, or other toxic materials. Alcohol and acetone used for the mixture are flammable. Weld only in well-ventilated spaces. Failure to provide proper ventilation could result in personal injury or death.

NOTE

The use of welding processes that use shielding gases is a particular concern in confined spaces, as they may displace the air available, risking suffocation. Confined spaces can be hazardous since they're more difficult to exit quickly. Most are poorly ventilated and have limited oxygen capacity. They can trap fumes, particles, and gases. Some are small enough that they can trap workers. Working in a confined space requires careful thought and planning to avoid unnecessary risks. Before working in a confined space, workers must get a confined space entry permit.

Figure 1 Mobile and portable extraction equipment.
Source: © Miller Electric Mfg. LLC

1.2.0 | Welding Processes

Arc welding processes use the intense heat produced by an electrical arc between a filler metal electrode and a workpiece to melt the metals and cause them to merge or coalesce. GMAW is one common form of arc welding used extensively in the welding industry. Performing GMAW safely and effectively requires a good understanding of how the process and equipment work and the critical safety practices.

1.2.1 The GMAW Process

In the GMAW process, a continuous, consumable, solid or composite metal-cored electrode is used for the filler metal while shielding gas is used to protect the weld zone. Composite metal-cored electrodes and solid electrodes are subsets of GMAW carbon steel electrodes as defined in *AWS Specification A5.18*. Keep in mind that the proper term for the consumable material used for filler metal is electrode, although it is commonly referred to as wire.

GMAW is a fast and effective method of making high-quality welds. The GMAW process is commonly used to make welds on carbon steel plate as well as on aluminum and other metals.

1.2.2 GMAW Equipment

The equipment used for gas-shielded GMAW includes guns equipped with a gas nozzle, gas hoses, a gas control solenoid valve, a gas regulator/flowmeter, and a source of shielding gas. *Figure 2* is an example of typical GMAW equipment in operation.

> **NOTE**
>
> Reference *Appendix 29209A* for Performance Accreditation Tasks (PATs) designed to evaluate your ability to run fillet and groove welds with GMAW equipment.

Figure 2 Gas-shielded GMAW equipment.
Source: Photo courtesy of The Lincoln Electric Company, Cleveland, OH, U.S.A.

1.3.0 GMAW Welding Equipment Setup

Before welding can take place, the area must be made ready, the welding equipment must be set up, and the metal to be welded must be prepared. The following sections explain how to set up equipment for welding.

1.3.1 Preparing the Welding Area

To practice welding, a welding table, bench, or stand is needed. The welding surface must be steel, and provisions must be made for placing weld coupons out of position (*Figure 3*).

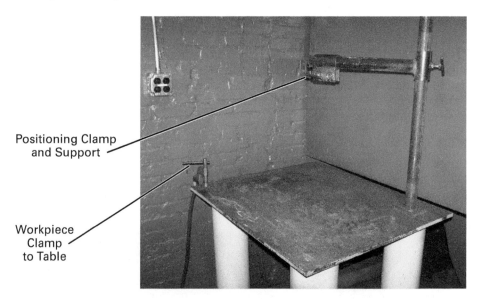

Figure 3 Welding station.
Source: Terry Lowe

Follow these steps to set up the area for welding:

Step 1 Make sure that the area is properly ventilated. Make use of doors, windows, and fans.

Step 2 Check the area for fire hazards. Remove any flammable materials before proceeding.

Step 3 Know the location of the nearest fire extinguisher. Know how to use the fire extinguisher, and do not proceed unless the extinguisher is charged.

Step 4 Set up welding curtains around the welding area.

Machine Bevels and Root Faces for GMAW

For V-groove joints made in thick materials, machined or flame-cut bevels and root faces provide the cleanest and most uniform way to prepare materials for satisfactory GMAW welds. While SMAW, FCAW, and GTAW can tolerate some poor fit-up for root passes, controlling penetration of root passes with poor fit-up using GMAW is difficult.

1.3.2 Preparing the Practice Weld Coupons

If possible, the practice weld coupons should be $\frac{3}{8}$" (10 mm) thick carbon steel. If this size is not readily available, $\frac{1}{4}$" to $\frac{3}{4}$" (6 mm to 20 mm) thick steel can be used to practice welding. Use a wire brush or grinder to remove heavy mill scale or corrosion.

The following outlines how to prepare weld coupons for various methods of practice:

- *Running beads* — The coupons can be any size or shape that can be easily handled.

- *Overlapping beads* — The coupons can be any size or shape that can be easily handled.

- *Fillet welds* — Cut the metal into 2" × 6" (51 mm × 152 mm) rectangles for the base and 1" × 6" (25 mm × 152 mm) rectangles for the web. *Figure 4* shows the weld coupons for fillet welding.

- *Open V-groove welds* — As shown in *Figure 5,* cut the metal into 3" × 7" (76 mm × 178 mm) rectangles with one (or both) of the lengths beveled at 30° to 37.5° for carbon steel or low-alloy plate. Note that the bevel angle for stainless steel coupons is 22.5°. Grind up to a $\frac{1}{8}$" (3.2 mm) root face on the bevel as chosen and directed by the instructor.

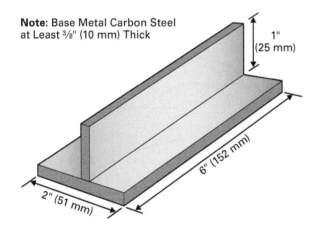

Note: Base Metal Carbon Steel at Least $\frac{3}{8}$" (10 mm) Thick

1" (25 mm)

6" (152 mm)

2" (51 mm)

Figure 4 Fillet weld coupons.

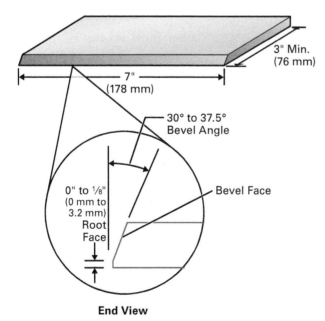

3" Min. (76 mm)

7" (178 mm)

30° to 37.5° Bevel Angle

Bevel Face

0" to $\frac{1}{8}$" (0 mm to 3.2 mm) Root Face

End View

Note: Base Metal Carbon Steel at Least $\frac{3}{8}$" (10 mm) Thick

Figure 5 Metal cut for open V-groove weld coupons.

Follow these steps to prepare each V-groove weld coupon:

Step 1 Check the bevel face of each coupon. There should be no dross and a 0" to ⅛" (0 mm to 3.2 mm) root face. The bevel angle should be 30° to 37.5°, as shown in *Figure 6*.

Step 2 Align the beveled coupons on a flat surface with a root opening determined by the instructor. Place tack welds on the back side of the joint. Use from three to four ¼" (6.4 mm) tack welds.

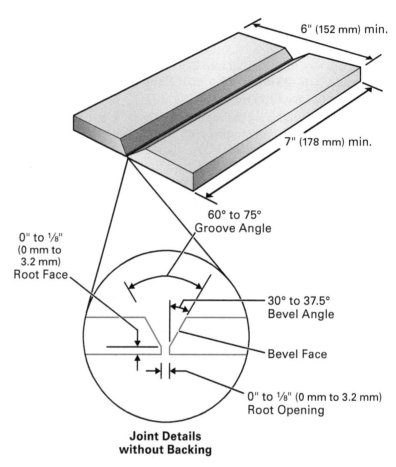

Note: Base Metal Carbon Steel, at Least ⅜" (10 mm) Thick

Figure 6 Open V-groove weld coupon.

Going Green

Recycling Metal

When metals are no longer usable in the welding shop or on the job, the metals should be collected and sold to a recycling company rather than sent to a landfill. Selling the scrap metal can make money for the company or welding school and protect the environment from resource depletion. Steel is the most recycled material in the world.

1.3.3 The Welding Machine

Identify the proper welding machine for GMAW use and follow these steps to set it up for welding:

Step 1　Verify that the welding machine can be used for GMAW.

Step 2　Verify the location of the primary disconnect.

Step 3　Configure the welding machine for GMAW as directed by the instructor (*Figure 7*). Configure the gun polarity and equip the gun with the correct liner material and contact tip for the diameter of the electrode being used and the correct nozzle for the application.

Step 4　In accordance with the manufacturer's instructions, load the wire feeder and gun with a solid or composite electrode of the proper diameter as directed by the instructor.

Step 5　Connect the proper shielding gas for the application as specified by the electrode manufacturer, the Welding Procedure Specification (WPS), the site quality standards, or the instructor.

Step 6　Connect the clamp of the workpiece lead to the workpiece.

Step 7　Turn on the welding machine and, if necessary, purge the gun as directed by the gun manufacturer's instructions.

Step 8　Set the initial welding voltage and wire feed speed for the type and size of electrode as recommended by the manufacturer, directed by the instructor, or provided in the WPS.

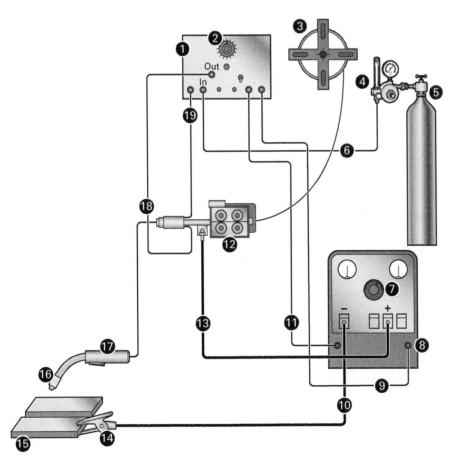

1. Control Unit
2. Wire Feed Speed Control
3. Electrode Wire Reel
4. Regulator/Flowmeter
5. Shielding Gas Source
6. Gas-In Hose
7. Voltage Control
8. Power Source
9. Power to Control Unit
10. Workpiece Lead
11. Contactor Control Cable
12. Wire Feed Drive Assembly
13. Electrode Lead
14. Workpiece Clamp
15. Workpiece
16. Gas Nozzle
17. GMAW Gun
18. Gas-Out Hose
19. Gun Control Cable

Note: The polarity of the gun and workpiece leads is determined by the type of filler metal and application.

Figure 7 Configuration diagram of a typical GMAW welding machine.

1.0.0 Section Review

1. In the GMAW process, the shielding gas that is used to protect the weld zone comes from a _____.
   a. flux-cored electrode
   b. shielding gas cylinder or other gas source
   c. solid-cored wire
   d. gaseous filler metal

2. The best procedure for safely cutting into a previously used vessel is to first fill the container with _____.
   a. sand
   b. air
   c. oxygen
   d. water

3. When setting up GMAW equipment for practice welding, the table, bench, or stand that will be used must have a _____.
   a. steel welding surface
   b. secondary disconnect
   c. shielding gas liner
   d. dry, wooden pedestal

2.0.0 Welding Procedures and Basic Beads

Performance Tasks

1. Make multipass GMAW-S (short-circuit) fillet welds on carbon steel plate coupons in all four 1F through 4F positions, using solid or composite electrode.

2. Make multipass GMAW-S (short-circuit) V-groove welds on carbon steel plate coupons in all four 1G through 4G positions, with or without backing, using solid or composite electrode.

3. Make multipass GMAW spray-transfer fillet welds on carbon steel plate coupons in both the 1F and 2F positions, using solid or composite electrode.

4. Make multipass GMAW spray-transfer V-groove welds on carbon steel plate coupons in the 1G position, with backing, using solid or composite electrode.

Objective

Describe equipment control and welding procedures for GMAW and explain how to produce basic weld beads.

a. Describe equipment control and welding techniques related to GMAW.

b. Explain how to produce basic GMAW weld beads.

Effective welding requires the welder to properly control the equipment being used and employ various welding techniques to produce strong and acceptable weld beads. This section examines factors that affect GMAW and how these factors are controlled. The process for producing basic GMAW weld beads is also described.

2.1.0   Welding Procedures

Welding can be affected by numerous factors, including the welding voltage, the welding amperage, the position and travel speed of the gun, and other aspects of the welding equipment and process.

2.1.1 Welding Voltage

Arc length is determined by voltage, which is set at the power source. Arc length is the distance from the electrode tip to the base metal or molten pool at the base metal. If voltage is set too high, the arc will be too long, which could cause the electrode to fuse to the tip. Too high a voltage also causes porosity and excessive spatter. Voltage and wire feed speed are correlated, and fine adjustments of either can be necessary to achieve the desired weld bead. Voltage may be increased or decreased as wire feed speed is increased or decreased. Set the voltage so that the arc length is just above the surface of the base metal (*Figure 8*).

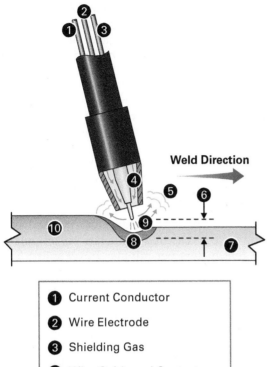

Weld Direction

1. Current Conductor
2. Wire Electrode
3. Shielding Gas
4. Wire Guide and Contact
5. Shielding Gas
6. Arc Length
7. Base Metal
8. Molten Weld Metal
9. Arc
10. Solidified Weld Material

Figure 8 Arc length.

Synergic welding machines control voltage, current, and wire feed at any given moment using a program that's customized for each welding situation. These machines will have a trim control that works similarly to the voltage control on other machines. Essentially, it overrides the program.

Synergic: Welding technology that controls voltage, current, and wire feed using a program optimized for each welding situation.

2.1.2 Welding Amperage

With a standard welding machine, the wire feed speed control on the power source controls the welding amperage. The welding machine provides the amperage necessary to melt the electrode while maintaining the selected welding voltage. Within limits, when the wire feed speed is increased, the welding amperage and deposition rate also increase. This results in higher welding heat, deeper penetration, and higher beads. When the wire feed speed is decreased, the welding amperage automatically decreases. With lower welding amperage and less heat, the deposition rate drops, and the weld beads are smaller and have less penetration.

Note that some constant-voltage welding machines used for GMAW provide varying degrees of current modification, such as slope or induction adjustments. Power sources with a slope adjustment allow the amount of amperage change to vary in relation to the arc length of the weld. This allows the correct short-circuiting current to be established for certain welding applications and electrode requirements. Standard constant-voltage GMAW units have a current slope that is fixed by the manufacturer for general welding applications and conditions. Other units allow adjustment of the response time for short-circuit current by using an induction control.

Trim: A control found on synergic welding machines that is similar to the voltage control on regular machines and acts as a kind of override to the program.

Arc Voltage

A minimum arc voltage is needed to maintain spray transfer. However, penetration is not directly related to voltage. Penetration will increase with voltage for a time, but beyond the optimum voltage, penetration will actually decrease as the voltage increases further.

In an instantaneous short-circuit current draw mode of operation, the machine draws a high level of current as soon as the electrode touches the workpiece. In the adjustable induction mode of operation, the machine gradually increases the welding current after the initial contact of the electrode to the workpiece. This smooths out the arc characteristics during short-circuiting transfer.

Pulse-transfer welding machines allow the adjustment of the peak current for the pulse and the background current between pulses to match specific welding applications and electrode requirements. In other cases, where a voltage-sensing wire feeder is used with a constant-current machine, the current is adjustable, and the wire feeder varies the wire feed speed to maintain the proper arc voltage at the electrode tip.

2.1.3 Welding Travel Speed

Weld travel speed is the speed of the electrode tip across the base metal in the direction of the weld, measured in inches per minute (IPM) or a metric equivalent such as centimeters per minute. Travel speed has a great effect on penetration and bead size. Slower travel speeds build higher beads with deeper penetration, until the arc no longer contacts the base metal at the front of the puddle, and then can actually interfere with proper fusion. Faster travel speeds build smaller beads with less penetration. Ideally, the welding parameters should be adjusted so that the electrode tip is positioned at the leading edge of the weld puddle during travel.

Inexperienced welders may be tempted to turn down the wire feed speed if they experience difficulty controlling the weld puddle. Electrodes must be run at certain balanced parameters that cannot be changed individually. Voltage, wire feed speed, and travel speed are adjusted and balanced together to control the weld puddle. The voltage selected is usually determined by the thickness of the base metal or its heat-absorption capacity.

2.1.4 Gun Position

Gun position for GMAW also influences weld penetration along with voltage and travel speed. The gun travel angle (*Figure 9*), is an angle less than 90° between the electrode axis and a line perpendicular to the weld axis at the point of electrode contact in a plane determined by the electrode axis and the weld axis. For pipe, the plane is determined by the electrode axis and a line tangent to the pipe surface at the same point. The following list describes different types of travel angles:

- *Neutral angle* — A neutral angle is the travel angle when the center line of the gun electrode is exactly perpendicular (90°) to the welding surface. The term is nonstandard for a 0° travel angle. Weld penetration is moderate when the gun electrode is used in this position.

- *Push angle* — A push angle is generally used for welding thin materials or when shallow penetration is required. A push angle is the travel angle created when the welding gun is tilted back so that the electrode tip precedes the gun in the direction of the weld. In this position, the electrode tip and shielding gas are being directed ahead of the weld bead. Welding with this angle produces a wider weld bead with less penetration. Push angles greater than 15° are not recommended for manual welding.

- *Drag angle* — A drag angle is the travel angle created when the gun is tilted forward so that the electrode tip trails the gun in the direction of the weld. In this position, the electrode tip and shielding gas are being directed back toward the weld bead. Welding with this angle produces a higher and narrower weld bead with more penetration. Maximum penetration is achieved with a drag angle of approximately 25°.

The gun work angle, also shown in *Figure 9*, is an angle less than 90° between a line perpendicular to the major workpiece surface at the point of electrode contact and a plane determined by the electrode **axis** and weld axis. For a T-joint or corner joint, the line is 45° to the non-butting member. For pipe, the plane is determined by the electrode axis and a line tangent to the pipe surface at the same point.

Axis: The centerline of the weld or prepared base metal.

Arc Blow

When current flows, magnetic fields are created. The magnetic fields tend to concentrate in corners, deep grooves, or the ends of the base metal. When the arc approaches these concentrated magnetic fields, it is deflected from its intended path. Arc blow can cause defects such as excessive weld spatter and porosity. If arc blow occurs, try one or more of these methods to control it:

- Change the position of the workpiece clamp. This will change the flow of welding current, affecting the way the magnetic fields are created.
- Change the angle of the gun. Reducing or even reversing the angle of the gun can compensate for the arc blow.

GMAW of Stainless Steel

For stainless steel, a drag angle is normally used for fillet welds and a push angle for square or grooved butt joints. Stainless steel is usually welded in the flat and horizontal positions with spray transfer. Vertical and overhead welds usually employ short-circuiting transfer for at least the root passes, if not the fill and cover passes as well. Pulsed-spray transfer can also be used for fill and cover passes. Remember that all open joints in stainless steel materials may be made with a copper, stainless steel, or ceramic backing to prevent air from reaching the weld puddle until the puddle solidifies. When practicing welding using stainless steel electrodes and carbon steel coupons, use the travel angles and transfer methods outlined, along with carbon steel backing strips or a backing fixture for open or butt joints, when required. When welding plate that is thicker than $1/4$" (6.4 mm), the gun should be moved back and forth in the travel direction, with a slight side-to-side motion. On thinner materials, only the back-and-forth motion is necessary.

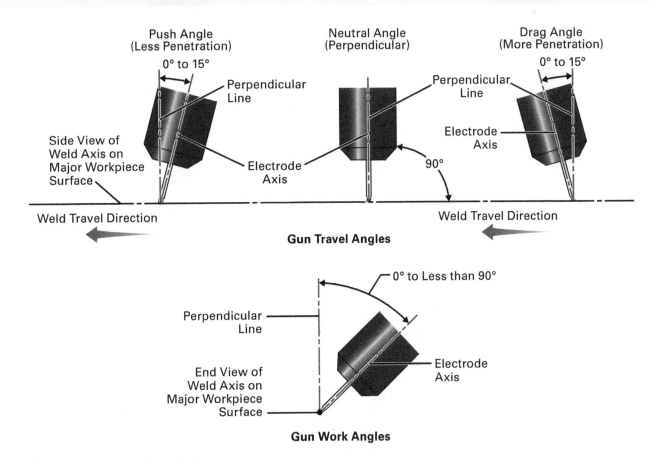

Figure 9 Gun travel and work angles.

CAUTION

Overuse of electrode extension is a common mistake for new welders. Too much electrode extension will reduce amperage and potentially risk weld quality and adherence to welding specifications. Paying attention to the amperage lost with every interval of electrode extension will help new welders to gain a sense of the relationship.

2.1.5 Electrode Extension, Stickout, and Standoff Distance

Electrode extension is the length of the electrode that extends beyond the contact tip of the welding gun. Relatedly, the **contact tip to work distance (CTWD)** is the distance between the contact tip of the electrode and the base metal to be welded. Changing the extension influences the welding current for low-conductivity metal electrodes because it changes the preheating of the electrode caused by the resistance of the electrode to current flow. As the extension increases, preheating of the electrode increases.

Since the welding power source is self-regulating, it does not have to furnish as much welding current to melt the electrode, so the current output automatically decreases. This results in less penetration and increased deposition rates. Increasing extension is useful for bridging gaps and compensating for mismatch, but it can cause overlap or lack of fusion and a ropy bead appearance. When the extension is decreased, the power source is forced to increase its current output to burn off the electrode. Too little extension can cause the electrode to weld to the contact tip or develop porosity in the weld. For metal electrodes that are highly conductive, the preheating effect of electrode resistance is minimal, and electrode speed and voltage settings have a more direct effect.

GMAW extensions vary between $\frac{1}{4}$" and 1" (6.4 mm and 25.4 mm), depending on the transfer mode. *Figure 10* shows typical electrode extensions for GMAW gun configurations.

Additional gun nomenclature is also defined in *Figure 10*. Stickout is the distance from the gas nozzle or insulating nozzle to the end of the electrode. Standoff distance is the distance from the gas nozzle or insulating nozzle to the workpiece. Contact tip extension or setback is usually dependent on the transfer mode for the GMAW application.

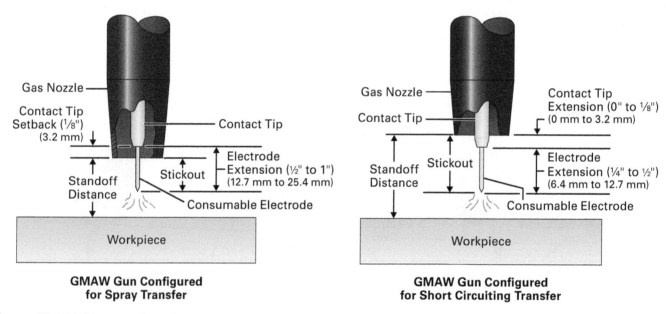

GMAW Gun Configured for Spray Transfer

GMAW Gun Configured for Short Circuiting Transfer

Figure 10 GMAW gun configurations.

2.1.6 Gas Nozzle Cleaning

As a GMAW welding machine is being used, weld spatter accumulates on the gas nozzle and contact tip. If the gas nozzle is not properly cleaned, it will restrict the shielding gas flow and cause porosity in the weld. When making out-of-position welds, accumulated spatter in the nozzle can result in a short circuit between the gas diffuser and the nozzle shell, causing the nozzle to stick to the workpiece.

Clean the gas nozzle with a reamer or round file, the tang of a standard file, or the jaw of the MIG pliers. After cleaning, the nozzle can be sprayed or dipped in a special anti-spatter compound. The anti-spatter compound helps prevent the spatter from sticking to the nozzle. Ensure that any anti-spatter product used is compatible with the welding application and equipment in use.

CAUTION

When using anti-spatter, be sure not to let any of the compound come into contact with the welding wire. If contact does occur, trim off the contaminated wire. If the contaminated wire is not trimmed off, the compound on the wire could cause weld defects or discontinues.

2.2.0 GMAW Bead Types

The following sections explain how to create weld beads using GMAW equipment. The two basic bead types are stringer beads and weave beads.

2.2.1 Stringer Beads

Stringer beads are made with little or no side-to-side motion of the electrode. Practice running stringer beads in the flat position. Experiment with different voltages, travel angles, and stickouts. Change the wire feed speed and observe the bead size. Try different travel angles and observe the bead shape and size. Use solid or composite electrodes and the appropriate shielding gas as directed by the instructor. If the practice coupon gets too hot between passes, it may have to be cooled in water.

CAUTION

The width of stringer beads is specified in the welding code or WPS being used at a site. Do not exceed the widths specified for the site.

WARNING!

Use pliers to handle the hot practice coupons. Wear gloves when placing the practice coupon in water. Steam will rise off the coupon and can burn or scald unprotected hands and arms.

Follow these steps to run stringer beads:

Step 1 Hold the gun at the desired travel angle (drag or push), with the electrode tip directly over the point where the weld is to begin and pull the gun trigger.

Step 2 Hold the arc in place until the weld puddle begins to form.

Step 3 Slowly advance the arc while maintaining the gun angle. Stay in the leading edge of the puddle to prevent incomplete fusion.

Step 4 Continue to weld until a bead about 2" to 6" (50 mm to 152 mm) long is formed.

Step 5 Stop, raise the gun to a neutral (0°) travel angle, pause until the crater is filled, and then release the trigger.

Step 6 Inspect the bead for the following:

- Straightness of the bead
- Uniform appearance of the bead face
- Smooth, flat transition with complete fusion at the toes of the weld
- No excessive porosity with no pores larger than $3/32$" (2.38 mm)
- No undercut greater than $1/32$" (0.8 mm) deep or 10% of the base metal thickness, whichever is less
- No overlap
- No lack of fusion
- Complete penetration
- No pinholes (fisheyes)
- Filled crater
- No cracks

Step 7 Continue practicing stringer beads until acceptable welds are made every time.

CAUTION

Cooling with water (quenching) is done only on practice coupons. Never cool test coupons or on-the-job welds with water. Cooling with water can cause weld cracks and affect the mechanical properties of the base metal.

Going Green

Conserving Material

Steel for practice welding is expensive and difficult to obtain. Every effort should be made to conserve the available material. Reuse weld coupons until all surfaces and joints have been used for welding, then cut the weld coupon apart and reuse the pieces. Use material that cannot be cut into weld coupons to practice running beads.

Anti-Spatter Materials

Use only anti-spatter material specifically designed for GMAW gas nozzles. Water-based products may cause hydrogen embrittlement in hydrogen-sensitive steel. Always refer to the SDS/MSDS for specific safety considerations. Caution should be taken not to use too much anti-spatter, as using more anti-spatter than needed can cause porosity or contaminate the weld. Anti-spatter sprays and gels are usually water- or petrolatum-based. However, always check the SDS for the specific type of anti-spatter being used to ensure you know the materials you are working with. Additionally, some of these products are prohibited for use in code welding, so check the WPS before using them. For explosion and flammability reasons, be sure that all aerosol cans are a safe distance from the welding area.

Source: Photo courtesy of The Lincoln Electric Company, Cleveland, OH, U.S.A.

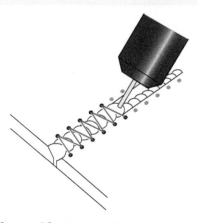

Figure 12 Weave motion.

2.2.2 Weave Beads

Weave beads are made with wide, side-to-side motions of the electrode. The width of a weave bead is determined by the amount of side-to-side motion. *Figure 11* shows an example of a weave bead.

When making a weave bead, use care at the toes to be sure there is proper tie-in to the base metal. To ensure proper tie-in at the toes, slow down or pause slightly at the edges. The pause at the edges will also flatten out the weld, giving it the proper profile.

Practice running weave beads about $5/8$" (16 mm) wide in the flat position. Experiment with different weave motions, travel angles, and stickouts. Change the wire feed speed and observe the bead size. Try different gun angles and observe the bead shape and size. Use solid or composite electrodes and appropriate shielding gas as directed by the instructor.

Figure 11 Close-up view of weave beads.
Source: Dan Sterry, TIC

Follow these steps to run weave beads:

Step 1 Hold the gun at the desired travel angle with the electrode tip directly over the point where the weld is to begin.

Step 2 Pull the gun trigger.

Step 3 Hold the arc in place until the weld puddle begins to form.

Step 4 Slowly advance the arc in a weaving motion while maintaining the gun angle (*Figure 12*).

Step 5 Continue to weld until a bead about 2" to 6" (50 mm to 152 mm) long is formed.

Step 6 Stop, raise the gun to a neutral (0°) travel angle, pause until the crater is filled, and then release the trigger.

Step 7 Inspect the bead for the following:

- Straightness of the bead
- Uniform appearance of the bead face
- Smooth, flat transition with complete fusion at the toes of the weld
- No excessive porosity with no pores larger than $3/32$" (2.38 mm)
- No undercut greater than $1/32$" (0.8 mm) deep or 10% of the base metal thickness, whichever is less
- No overlap
- No lack of fusion
- Complete penetration
- No pinholes (fisheyes)
- Filled crater
- No cracks

Step 8 Continue practicing weave beads until acceptable welds are made every time.

2.2.3 Weld Restarts

A restart is the junction where a new weld connects to and continues the bead of a previous weld. Restarts are important because an improperly made restart creates a weld discontinuity. One advantage of the semiautomatic GMAW process is that restarts can be minimized. Plan the weld to eliminate restarts, if possible. If a restart must be made, it must be made so that it blends smoothly with the previous weld and does not stand out. The technique for making a restart is the same for both stringer and weave beads. Practice weld restarts using GMAW as directed by the instructor.

Follow these steps to make a restart:

Step 1 With the weld clean, hold the gun at the proper travel angle while restarting the arc directly over the center of the crater. Remember that welding codes do not allow arc strikes outside the area of permanent welds.

Step 2 Pull the gun trigger.

Step 3 Move the electrode tip over the crater in a small circular motion to fill the crater with a molten puddle.

Step 4 Move to the leading edge of the puddle; as soon as the puddle fills the crater, continue the stringer or weave bead pattern being used.

Step 5 Inspect the restart. A properly made restart will blend into the bead, making it hard to detect. If the restart has undercut, not enough time was spent in the crater to fill it. If the undercut is on one side or the other, use more of a side-to-side motion while moving back into the crater. If the restart has a lump, it was overfilled; too much time was spent in the crater before resuming the forward motion.

Step 6 Continue to practice restarts until they are correct.

2.2.4 Weld Terminations

A weld termination is made at the end of a weld. A termination normally leaves a crater. When making a termination, the welding codes require that the crater must be filled to the full cross section of the weld. This can be difficult, since most terminations are at the edge of a plate, where welding heat tends to build up, making filling the crater more difficult. *Figure 13* shows the proper technique for terminating a weld.

Follow these steps to make a weld termination:

Step 1 Start to bring the gun up to a 0° travel angle and slow the forward travel as the end of the weld is approached.

Step 2 Stop the forward movement about $\frac{1}{8}$" (3.2 mm) from the end of the plate.

Step 3 With the gun at a 0° travel angle, allow the crater to fill, and then move about $\frac{1}{8}$" (3.2 mm) toward the start of the weld.

Step 4 Release the trigger when the crater is completely filled. Hold the gun in place until the gas post-flow cools the weld metal enough to prevent oxidation damage.

Step 5 Inspect the termination. The crater should be filled to the full cross section of the weld.

Step 6 Continue to practice terminations until they are correct.

CAUTION

Do not remove the gun from the weld until the puddle has solidifed and cooled. This continued flow of gas will protect the molten weld metal, preventing crater porosity and cracks.

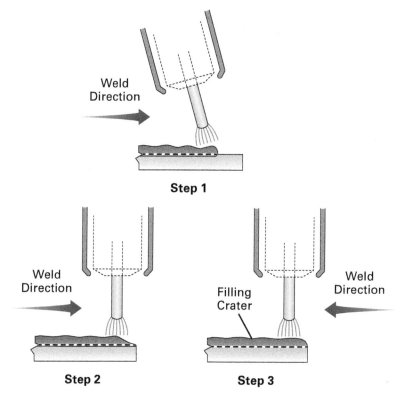

Figure 13 Terminating a weld.

2.2.5 Overlapping Beads

Overlapping beads are made by depositing connective weld beads parallel to one another. The parallel beads overlap, forming a flat surface. This is also called *padding*. Overlapping beads are used to build up a surface and to make multipass welds. Both stringer and weave beads can be overlapped. When viewed from the end, properly overlapped beads will have a relatively flat surface. *Figure 14* shows proper and improper overlapping string and weave beads.

Follow these steps to weld overlapping stringer or weave beads, using a specific diameter solid or composite electrode and the proper shielding gas as directed by the instructor.

Step 1 Mark out a square on a piece of steel.

Step 2 Weld a stringer or weave bead along one edge at a 0° work angle.

Step 3 Clean the weld.

Step 4 Position the gun at a work angle of 10° to 15° toward the side of the previous bead to get proper tie-in (*Figure 15*) and pull the trigger.

Step 5 Continue running overlapping stringer or weave beads until the square is covered. Clean each bead before running the new bead.

Step 6 Continue building layers of stringer or weave beads, one on top of the other, until perfecting the technique.

Build a pad using stringer beads (*Figure 16*). Repeat building the pad using weave beads.

A Few Suggestions

To improve your technique, practice with both hands in opposing directions when welding in the 1G, 2G, and 4G positions. Also, the first weld bead can be done up the center of the pad, with the remaining beads being worked from the center out. This tactic gives you practice overlapping beads on the pad as you might do in a groove weld.

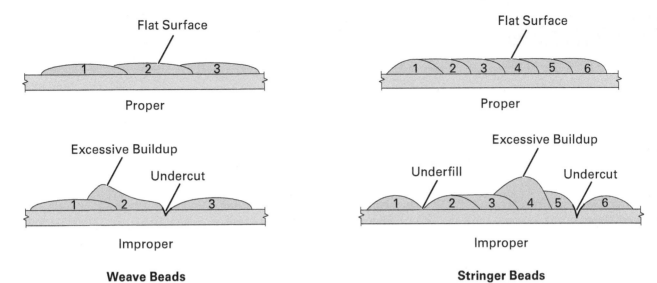

Figure 14 Proper and improper overlapping beads.

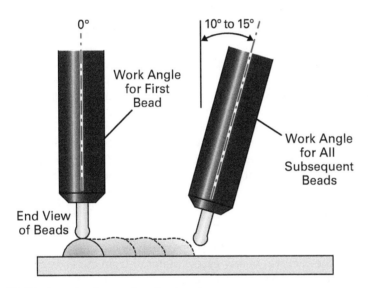

Figure 15 Work angles for overlapping beads.

Figure 16 Pad built with stringer beads.
Source: Photo courtesy of The Lincoln Electric Company, Cleveland, OH, U.S.A.

2.0.0 Section Review

1. Depending on the transfer mode, GMAW electrode extensions vary between _____.
   a. $^1/_{16}$" and $^1/_4$" (1.6 and 6.4 mm)
   b. $^1/_8$" and $^1/_2$" (3.2 and 12.7 mm)
   c. $^1/_4$" and 1" (6.4 and 25.4 mm)
   d. $^1/_2$" and $1^1/_2$" (12.7 and 38 mm)

2. The width of a weave bead is determined by the amount of _____.
   a. forward and backward gun movement
   b. stringer bead padding overlap
   c. filler material in the flux core
   d. side-to-side motion of the electrode

3.0.0 GMAW Fillet and V-Groove Welds

Performance Tasks

1. Make multipass GMAW-S (short-circuit) fillet welds on carbon steel plate coupons in all four 1F through 4F positions, using solid or composite electrode.

2. Make multipass GMAW-S (short-circuit) V-groove welds on carbon steel plate coupons in all four 1G through 4G positions, with or without backing, using solid or composite electrode.

3. Make multipass GMAW spray-transfer fillet welds on carbon steel plate coupons in both the 1F and 2F positions, using solid or composite electrode.

4. Make multipass GMAW spray-transfer V-groove welds on carbon steel plate coupons in the 1G position, with backing, using solid or composite electrode.

Objective

Describe the welding procedures needed to produce proper fillet and V-groove welds using GMAW welding techniques.

a. Describe the welding procedures needed to produce proper fillet welds using GMAW welding techniques.

b. Describe the welding procedures needed to produce proper V-groove welds using GMAW welding techniques.

Fillet welds and V-groove welds are two common types of welds produced with GMAW. This section examines the procedures used to make each of these welds in four different positions: flat, horizontal, vertical, and overhead.

3.1.0 Fillet Welds

Fillet welds require little base metal preparation, except for the cleaning of the weld area and the removal of any excess material from cut surfaces. Any dross from cutting will cause porosity in the weld. For this reason, the codes require that this material is entirely removed prior to welding.

The most common fillet welds are made in lap and T-joints. The weld position for plate is determined by the weld axis and the orientation of the workpiece. The positions for fillet welding on plate are flat (1F, where F stands for fillet), horizontal (2F), vertical (3F), and overhead (4F) as shown in *Figure 17*. In the 1F and 2F positions, the weld axis can be inclined up to 15°. Any weld axis inclination for the other positions varies with the rotational position of the weld face as specified in AWS standards.

Fillet welds can be concave or convex, depending on the WPS or site quality standards. The welding codes require a fillet weld to have a uniform concave or convex face, although a slightly nonuniform face is acceptable. Convexity is the distance the weld extends above a line drawn between the toes of the weld. The convexity of a fillet weld or individual surface bead must not exceed that permitted by the applicable code or standard. In single-pass, weave-bead fillet welds where two workpieces are being joined at an angle (not lap joints), flat or slightly convex faces are usually preferred because weld stresses are more uniformly distributed through the fillet weld and workpieces. A fillet weld must be repaired if the profile has defects.

A fillet weld is unacceptable and must be repaired if the profile has insufficient throat, excessive convexity, undercut, overlap, insufficient leg, or incomplete fusion. *Figure 18* shows ideal, acceptable, and unacceptable fillet weld profiles.

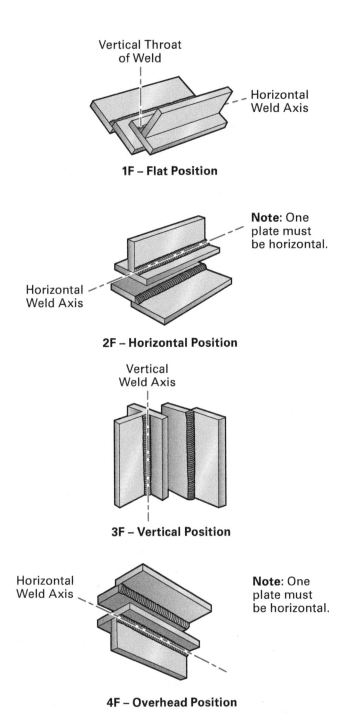

Figure 17 Fillet weld positions.

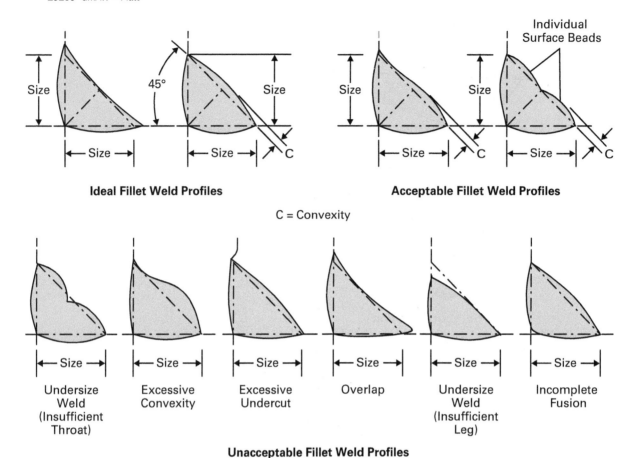

Ideal Fillet Weld Profiles   **Acceptable Fillet Weld Profiles**

C = Convexity

Unacceptable Fillet Weld Profiles

Figure 18 Ideal, acceptable, and unacceptable fillet weld profiles.

3.1.1 Practicing Flat (1F) Position Fillet Welds

Practice flat (1F) position fillet welds using GMAW spray and/or short-circuiting transfer as directed by the instructor. Weld multipass (six-pass) convex fillet welds in a T-joint. Use an instructor-specified size of solid or composite electrode with appropriate shielding gas. When making flat fillet welds, pay close attention to the work angle and travel speed. For the first bead, the work angle is vertical (45° to both plate surfaces). The angle is adjusted for all subsequent beads. Increase or decrease the travel speed to control the amount of weld metal buildup. *Figure 19* shows the work angles for 1F position fillet welds.

Follow these steps to make a flat fillet weld. Be sure to clean all weld beads thoroughly before beginning the next bead.

Step 1 Tack two plates together to form a T-joint for the fillet weld coupon (*Figure 20*). Clean the tack welds.

Step 2 Clamp or tack-weld the coupon in the flat position.

Step 3 Run the first bead along the root of the joint, using a work angle of 45° and a 5° to 10° drag angle. Use a slight side-to-side motion (oscillation).

Step 4 Run the second bead along a toe of the first weld, overlapping about 75% of the first bead. Alter the work angle as shown in *Figure 19* and use a 5° to 10° drag angle with a slight oscillation.

Step 5 Run the third bead along the other toe of the first weld, filling the groove created when the second bead was run. Use the work angle shown in *Figure 19* and a 5° to 10° drag angle with a slight oscillation.

Step 6 Run the fourth bead along the outside toe of the second weld, overlapping about half the second bead. Use the work angle shown in *Figure 19* and a 5° to 10° drag angle with a slight oscillation.

Step 7 Run the fifth bead along the inside toe of the fourth weld, overlapping about half the fourth bead that was run. Use the work angle shown in *Figure 19* and a 5° to 10° drag angle with a slight oscillation.

Step 8 Run the sixth bead along the toe of the fifth weld, filling the groove created when the fifth bead was run. Use the work angle shown in *Figure 19* and a 5° to 10° drag angle with a slight oscillation.

Step 9 Inspect the weld. The weld is acceptable if it has the following features:

- Uniform rippled appearance of the bead face
- Craters and restarts filled to the full cross section of the weld
- Uniform weld size $\pm^1/_{16}$" (1.6 mm)
- Acceptable weld profile in accordance with the applicable code or standard
- Smooth, flat transition with complete fusion at the toes of the weld
- No excessive porosity with no pores larger than $^3/_{32}$" (2.38 mm)
- No undercut greater than $^1/_{32}$" (0.8 mm) deep or 10% of the base metal thickness, whichever is less
- No overlap
- No cracks

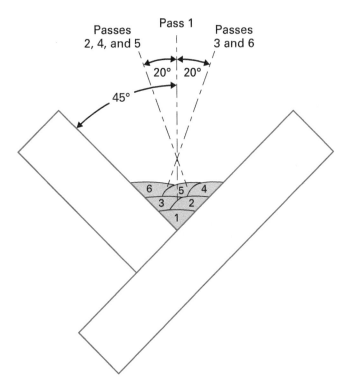

Figure 19 Multipass 1F bead sequence and work angles.

Figure 20 Fillet weld coupon.
Source: Holley Thomas, SME

3.1.2 Practicing Horizontal (2F) Position Fillet Welds

Practice horizontal (2F) fillet welding using GMAW spray and/or short-circuiting transfer as directed by the instructor. Place multipass fillet welds in a T-joint using solid or composite electrodes and the appropriate shielding gas as directed by the instructor. When making horizontal fillet welds, pay close attention to the gun angles and travel speed. For the first bead, the electrode work angle is 45°. The work angle is adjusted for all other welds. Increase or decrease the travel speed to control the amount of weld metal buildup.

Follow these steps to make a horizontal fillet weld:

Step 1 Tack two plates together to form a T-joint for the fillet weld coupon. Clean the tack welds.

Step 2 Clamp or tack-weld the coupon in the horizontal position.

Step 3 Run the first bead along the root of the joint using a work angle of approximately 45° with a 5° to 10° drag angle and a slight oscillation (*Figure 21*).

Step 4 Clean the weld.

Step 5 Run the remaining passes at the appropriate work angles using a 5° to 10° drag angle and a slight oscillation at the appropriate work angles as shown in *Figure 21*. Overlap each previous pass and clean the weld after each pass.

Step 6 Have the instructor inspect the weld. The weld is acceptable if it has the following features:

- Uniform rippled appearance of the bead face
- Craters and restarts filled to the full cross section of the weld
- Uniform weld size $\pm\frac{1}{16}$" (1.6 mm)
- Acceptable weld profile in accordance with the applicable code or standard
- Smooth, flat transition with complete fusion at the toes of the weld
- No excessive porosity with no pores larger than $\frac{3}{32}$" (2.38 mm)
- No undercut greater than $\frac{1}{32}$" (0.8 mm) deep or 10% of the base metal thickness, whichever is less
- No overlap
- No cracks

T-Joint Heat Dissipation

In T-joints, the welding heat dissipates more rapidly in the thicker or non-butting member. On various bead passes, the arc may have to be concentrated slightly more on the thicker or the non-butting member to compensate for the heat loss.

T-Joint Vertical Plate Undercut

The most common defect for T-joints is an undercut on the vertical plate of the joint. If this problem occurs, angling the arc slightly more toward the vertical plate and the bead at the top of the weave will force more metal into the bead at the top edge of the weld.

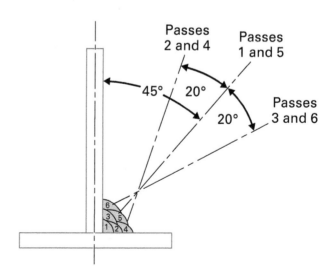

Figure 21 Multipass 2F bead sequence and work angles.

3.1.3 Practicing Vertical (3F) Position Fillet Welds

Practice vertical (3F) fillet welding using GMAW as directed by the instructor. Place multipass fillet welds in a T-joint using solid or composite electrodes and appropriate shielding gas as directed by the instructor. Normally, vertical welds are made welding uphill (also called *vertical-up welding*) from the bottom to the top, using a gun push angle (up-angle). However, downhill welding (also called *vertical-down welding*) can also be accomplished using a slight drag angle. When vertical welding, either stringer or weave beads can usually be used. Check with the instructor to see if stringer or weave beads, or both techniques, should be practiced. However, weave beads are avoided on stainless steel or in downhill welding. On the job, the site WPS or quality standard will specify which technique to use.

When making vertical fillet welds, pay close attention to the gun angles and travel speed. Increase or decrease the travel speed to control the amount of weld metal buildup.

Follow these steps to make an uphill fillet weld with weave beads:

Step 1 Tack two plates together to form a T-joint for the fillet weld coupon.

Step 2 Clamp or tack-weld the coupon in the vertical position.

Step 3 Starting at the bottom, run the first bead along the root of the joint using a work angle of approximately 45° and a 10° to 15° push angle. Pause in the weld puddle to fill the crater. Ensure it's filled to the full cross section at the end of the pass.

Step 4 Clean the weld.

Step 5 Run the remaining passes using a 5° to 10° push angle and a side-to-side weave technique with a 45° work angle (*Figure 22*). Use a slow motion across the face of the weld, pausing at each toe for penetration and to fill the crater. Adjust the travel speed across the face of the weld to control the buildup. Overlap each previous pass as required and clean the weld after each pass.

Step 6 Have the instructor inspect the weld. The weld is acceptable if it has the following features:

- Uniform rippled appearance of the bead face
- Craters and restarts filled to the full cross section of the weld
- Uniform weld size $\pm^1/_{16}$" (1.6 mm)
- Acceptable weld profile in accordance with the applicable code or standard
- Smooth, flat transition with complete fusion at the toes of the weld
- No excessive porosity with no pores larger than $^3/_{32}$" (2.38 mm)
- No undercut greater than $^1/_{32}$" (0.8 mm) deep or 10% of the base metal thickness, whichever is less
- No overlap
- No cracks

Repeat vertical fillet (3F) welding using stringer beads. Use a slight oscillation and side-to-side motion, pausing slightly at each toe to prevent undercut. For stringer beads, use a 0° to 10° push angle and the work angles shown in *Figure 22* as required.

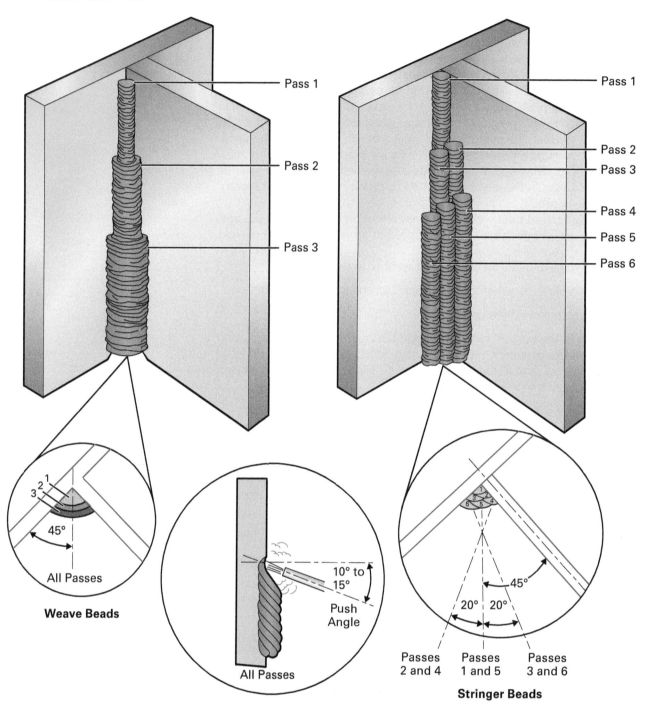

Figure 22 Multipass 3F bead sequences and work angles for weave and stringer beads.

Vertical Weave Bead

To help control undercut on a root pass, an alternate pattern such as a small triangular weave or an inverted-V motion can be used in vertical welds. By pausing the arc at the edges, the previous undercut can be filled. This action also creates undercut at the existing weld pool, but that will be filled in by the next weave. This technique should only be applied to the root pass.

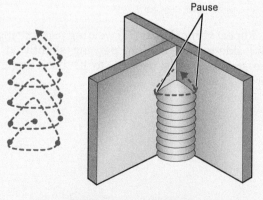

Triangular Weave Pattern

3.1.4 Practicing Overhead (4F) Position Fillet Welds

Practice overhead (4F) fillet welding using GMAW as directed by the instructor. Place multipass fillet welds in a T-joint using solid or composite electrodes and appropriate shielding gas as directed by the instructor. When making overhead fillet welds, pay close attention to the gun angles and travel speed. For the first bead, the work angle is approximately 45°. The work angle is adjusted for all other welds. Increase or decrease the travel speed to control the amount of weld metal buildup.

Follow these steps to make an overhead fillet weld:

Step 1 Tack two plates together to form a T-joint for the fillet weld coupon.

Step 2 Clamp or tack-weld the coupon so it is in the overhead position.

Step 3 Run the first bead along the root of the joint using a work angle of approximately 45° with a 5° to 10° drag angle. Use a slight oscillation to tie in the weld at the toes.

Step 4 Clean the weld.

Step 5 Using a slight oscillation, run the remaining passes using a 5° to 10° drag angle and the work angles shown in *Figure 23*. Overlap each previous pass and clean the weld after each pass.

Step 6 Have the instructor inspect the weld. The weld is acceptable if it has the following features:

- Uniform rippled appearance of the bead face
- Craters and restarts filled to the full cross section of the weld
- Uniform weld size ±$\frac{1}{16}$" (1.6 mm)
- Acceptable weld profile in accordance with the applicable code or standard
- Smooth, flat transition with complete fusion at the toes of the weld
- No excessive porosity with no pores larger than $\frac{3}{32}$" (2.38 mm)
- No undercut greater than $\frac{1}{32}$" (0.8 mm) deep or 10% of the base metal thickness, whichever is less
- No overlap
- No cracks

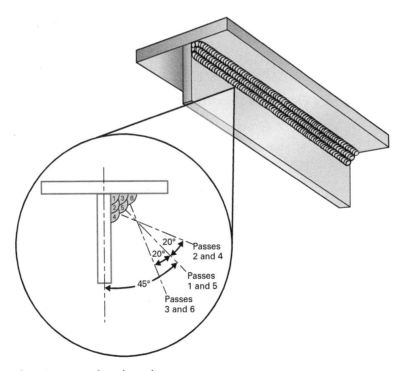

Figure 23 Multipass 4F bead sequence and work angles.

3.2.0   Open V-Groove Welds

The open V-groove weld is a common groove weld normally made on plate and pipe. Practicing the open V-groove weld on plate will help with making the more difficult pipe welds covered in *Welding Level Three*.

3.2.1 Root Pass

The most difficult part of making an open V-groove weld is the root pass. The root pass is made from the V-groove side of the joint and must have complete penetration, but not an excessive amount of root reinforcement. The penetration is controlled with a side-to-side motion or oscillation.

After the root pass has been run, it should be cleaned and inspected. Inspect the face of the weld for excess buildup or undercut (*Figure 24*). Remove excess buildup or undercut with a hand grinder by grinding the face of the root pass with the edge of a grinding disk. Use care not to grind through the root pass or widen the groove.

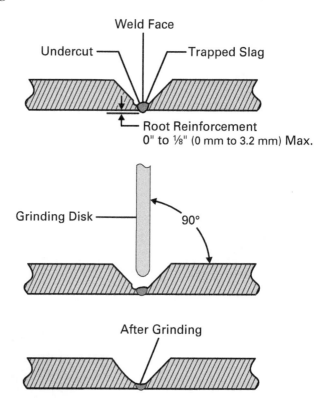

Figure 24 Grinding a root pass.

3.2.2 Groove Weld Positions

As shown in *Figure 25*, groove welds can be made in all positions. The weld position for plate is determined by the axis of the weld and the orientation of the workpiece. Groove weld positions for plate are flat (1G, where G stands for groove), horizontal (2G), vertical (3G), and overhead (4G).

In the 1G and 2G positions, the weld axis can be inclined up to 15°. Any weld axis inclination for the other positions varies with the rotational position of the weld face as specified in the WPS.

3.2.3 Acceptable and Unacceptable Groove Weld Profiles

Groove welds should be made with slight reinforcement and a gradual transition to the base metal at each toe. Groove welds must not have excessive reinforcement, undercut, underfill, or overlap (*Figure 26*). If a groove weld has any of these defects, it must be repaired.

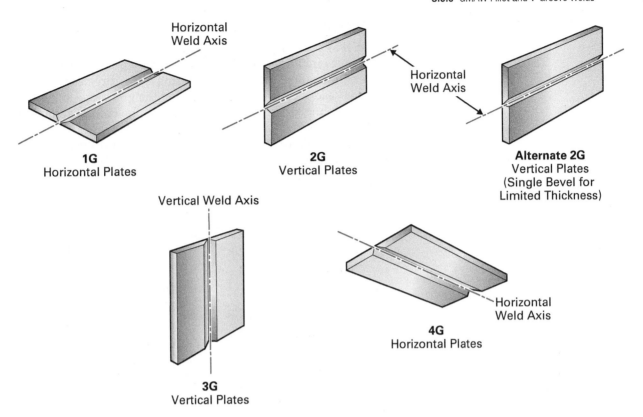

Figure 25 Groove weld positions for plate.

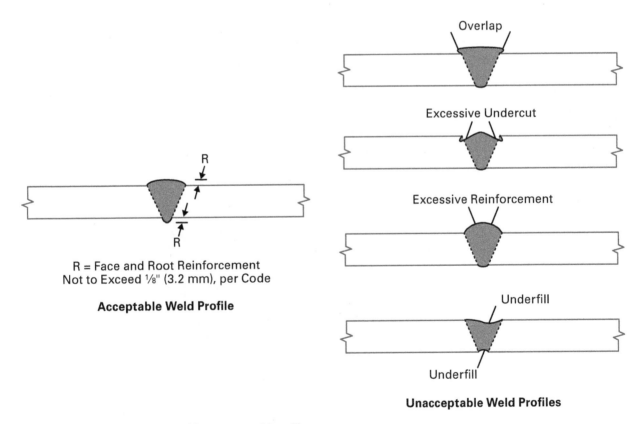

Figure 26 Acceptable and unacceptable groove weld profiles.

3.2.4 Practicing Flat (1G) Position Open V-Groove Welds

Practice flat (1G) open V-groove welds, as shown in *Figure 27* using GMAW (spray and short-circuiting transfer) as directed by the instructor. Use solid or composite electrodes and appropriate shielding gas as directed by the instructor. Use a stringer bead for the root pass, keeping the gun angle at 90° to the plate surface (0° work angle) with a 5° to 10° drag angle. For fill and cover passes, stringer or weave beads may be used. When using stringer beads, use a 5° to 10° drag angle, but adjust the work angle to tie in and fill the bead. For weave beads, also use a 5° to 10° drag angle, but keep the gun angle at 90° to the plate surface (0° work angle). Pay special attention at the termination of the weld to fill the crater.

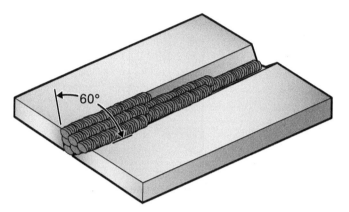

Note: The actual number of weld beads will vary depending on the plate thickness.

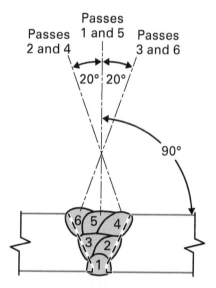

Stringer Bead Sequence

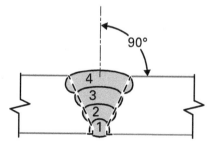

Weave Bead Sequence

Figure 27 Multipass 1G bead sequences and work angles.

Follow these steps to practice open V-groove welds in the flat position:

Step 1 Tack-weld the practice coupon together as explained earlier.

Step 2 Clamp or tack-weld the weld coupon in the flat position above the welding table surface.

Step 3 Run the root pass using a 5° to 10° drag angle and a slight side-to-side motion to control penetration.

Step 4 Clean the root pass and grind if required.

Step 5 Run the remaining passes using a 5° to 10° drag angle and applicable work angles. Overlap each previous pass and clean the weld after each pass.

Step 6 Inspect the weld. The weld is acceptable if it has the following:

- Uniform rippled appearance of the bead face
- Craters and restarts filled to the full cross section of the weld
- Uniform weld width $\pm^1/_{16}$" (1.6 mm)
- Acceptable weld profile in accordance with the applicable code or standard
- Smooth, flat transition with complete fusion at the toes of the weld
- Complete uniform root reinforcement at least flush with the base metal to a maximum buildup of $^1/_8$" (3.2 mm)
- No excessive porosity with no pores larger than $^3/_{32}$" (2.38 mm)
- No undercut greater than $^1/_{32}$" (0.8 mm) deep or 10% of the base metal thickness, whichever is less
- No cracks

NOTE

Avoid using weave beads for stainless steel.

Root Pass Welding

In practice, open V-groove root passes are usually made on various metals using GTAW, SMAW, or GMAW methods. The fill and cover passes are then made using FCAW or GMAW.

Whiskers

When making a root pass, make sure that the electrode does not get ahead of the leading edge of the weld puddle. If it does, a portion of the electrode may penetrate through the joint and be melted off and fused into the back side of the weld. This will cause pieces of electrode, sometimes called *whiskers*, to protrude from the back of the weld. Strategies to avoid the formation of whiskers include reducing the speed of the wire feed and using a keyhole welding technique. The keyhole technique in welding focuses a heat source on a workpiece to form a hole (keyhole) that is then filled with a weld bead made of molten metal.

3.2.5 Practicing Horizontal (2G) Position Open V-Groove Welds

A horizontal (2G) open V-groove weld is shown in *Figure 28*. Practice horizontal open V-groove welds using solid or composite electrodes and appropriate shielding gas as directed by the instructor. For the gun angles, use a 5° to 10° drag angle and adjust the work angle as required. Pay particular attention to filling the crater at the termination of the weld.

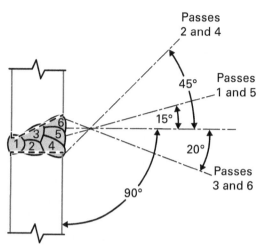

Alternate Joint Representation

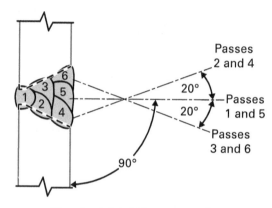

Standard Joint Representation

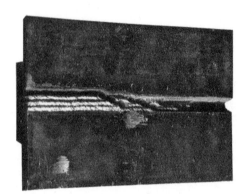

Note: The actual number of weld beads will vary depending on the plate thickness.

Figure 28 Multipass 2G bead sequences and work angles.
Source: Terry Lowe (photo)

Follow these steps to practice welding open V-groove welds in the horizontal position:

Step 1 Tack-weld the practice coupon together as explained earlier. Use the standard or alternate horizontal weld coupon as directed by the instructor.

Step 2 Tack-weld the flat plate weld coupon into the horizontal position.

Step 3 Run the root pass using a 5° to 15° drag angle and the appropriate work angle. Use a slight side-to-side motion to control penetration.

Step 4 Clean the weld.

Step 5 Run the remaining passes using a 5° to 10° drag angle and appropriate work angles. Overlap each previous pass and clean the weld after each pass.

Step 6 Inspect the weld. The weld is acceptable if it has the following:

- Uniform rippled appearance of the bead face
- Craters and restarts filled to the full cross section of the weld

- Uniform weld width ±$\frac{1}{16}$" (1.6 mm)
- Acceptable weld profile in accordance with the applicable code or standard
- Smooth, flat transition with complete fusion at the toes of the weld
- Complete uniform root reinforcement at least flush with the base metal to a maximum buildup of $\frac{1}{8}$" (3.2 mm)
- No excessive porosity with no pores larger than $\frac{3}{32}$" (2.38 mm)
- No undercut greater than $\frac{1}{32}$" (0.8 mm) deep or 10% of the base metal thickness, whichever is less
- No cracks

3.2.6 Practicing Vertical (3G) Position Open V-Groove Welds

Practice vertical (3G) multipass groove welds using stringer beads or weave beads. Set the welding parameters at the lower ends of the suggested ranges. Out-of-position welding, such as vertical and overhead welding, is done at lower welding voltages and currents than flat or horizontal welds to prevent the molten weld metal from sagging or running.

When welding vertically, either stringer or weave beads can be used. However, weave beads are not normally used on stainless steel. On the job, the WPS or site quality standards will specify which technique to use. Check with the instructor to see if stringer or weave beads, or both techniques, should be practiced. The root, fill, and cover beads can be run either uphill or downhill. However, downhill weave beads are not normally used. Use the direction preferred at the site or as specified in the WPS or site quality standards.

Run the root bead using a work angle of 0° (nozzle centerline bisects the V-groove) and a 5° to 15° push angle. The work angle is adjusted for all other welds.

Practice 3G open V-groove welds using GMAW as directed by the instructor. Use solid or composite electrodes and appropriate shielding gas as directed by the instructor.

Follow these steps to make a vertical groove weld with stringer beads:

Step 1 Tack-weld the practice coupon together as explained earlier.

Step 2 Clamp or tack-weld the coupon in the vertical position.

Step 3 Run the root pass uphill with a work angle of 0° (nozzle centerline bisects the V-groove) and a 5° to 15° push angle.

Step 4 Clean the weld.

Step 5 Run the remaining passes at the work angles shown in *Figure 29*, using a 5° to 15° push angle. Overlap each previous pass and use a slight oscillation to tie in the welds at the toes. Clean the weld after each pass.

Step 6 Inspect the weld. The weld is acceptable if it has the following features:

- Uniform rippled appearance of the bead face
- Craters and restarts filled to the full cross section of the weld
- Uniform weld width ±$\frac{1}{16}$" (1.6 mm)
- Acceptable weld profile in accordance with the applicable code or standard
- Smooth, flat transition with complete fusion at the toes of the weld
- Complete uniform root reinforcement at least flush with the base metal to a maximum buildup of $\frac{1}{8}$" (3.2 mm)
- No excessive porosity with no pores larger than $\frac{3}{32}$" (2.38 mm)
- No undercut greater than $\frac{1}{32}$" (0.8 mm) deep or 10% of the base metal thickness, whichever is less
- No cracks

Note: The actual number of weld beads will vary with the metal thickness.

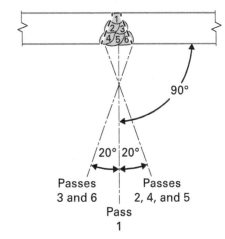

Figure 29 Multipass 3G bead sequence and work angles for stringer beads.
Source: Terry Lowe (photo)

Follow these steps to make a vertical groove weld with weave beads:

Step 1 Tack-weld the practice coupon together as explained earlier.

Step 2 Clamp or tack-weld the coupon in the vertical position.

Step 3 Run the root pass uphill with a work angle of 0° (90° from the plate) and a 5° to 15° push angle. Use a slight oscillation to control the penetration.

Step 4 Clean the weld.

Step 5 Run the remaining passes as uphill weave beads across the root bead at a 0° work angle. Clean the weld after each pass. Use a zigzag weave pattern across the root bead and pause at each toe for good fusion. Use a 5° drag angle to avoid cold wire. Increase or decrease the travel speed to control the amount of metal buildup. *Figure 30* shows the zigzag pattern and bead sequence for weave beads.

Step 6 Inspect the weld. The weld is acceptable if it has the following features:

- Uniform rippled appearance of the bead face
- Craters and restarts filled to the full cross section of the weld
- Uniform weld width $\pm\frac{1}{8}$" (3.2 mm)
- Acceptable weld profile in accordance with the applicable code or standard
- Smooth, flat transition with complete fusion at the toes of the weld
- Complete uniform root reinforcement at least flush with the base metal to a maximum buildup of $\frac{1}{8}$" (3.2 mm)
- No excessive porosity with no pores larger than $\frac{3}{32}$" (2.38 mm)
- No undercut greater than $\frac{1}{32}$" (0.8 mm) deep or 10% of the base metal thickness, whichever is less
- No cracks

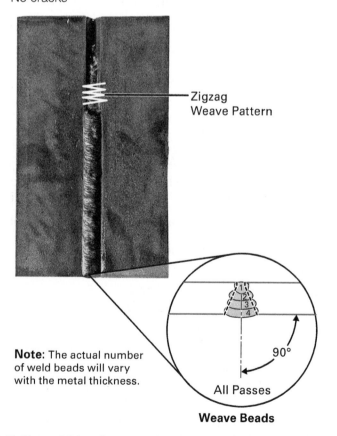

Note: The actual number of weld beads will vary with the metal thickness.

Figure 30 Multipass 3G bead sequence and work angle for weave beads.
Source: Terry Lowe (photo)

3.2.7 Practicing Overhead (4G) Position Open V-Groove Welds

Practice overhead (4G) open V-groove welds using GMAW as directed by the instructor. Use solid or composite electrodes and appropriate shielding gas as directed by the instructor. To reduce heat and prevent the molten weld metal from sagging or running, GMAW overhead welds are made with lower welding voltages and currents than flat or horizontal welds. Use a work angle of 0° (90° to plate) and a 5° to 15° drag angle for the root bead.

Follow these steps to make an overhead groove weld with stringer beads:

Step 1 Tack-weld the practice coupon together as explained earlier.

Step 2 Clamp or tack-weld the coupon in the overhead position.

Step 3 Run the root pass using a work angle of 0° and a 5° to 15° drag angle. Use a slight oscillation to control the penetration.

Step 4 Clean the weld.

Step 5 Run the remaining passes at the work angles shown in *Figure 31* using a 5° to 10° drag angle. Overlap each previous pass and use a slight side-to-side motion to tie in the welds at the toes. Each bead should be cleaned before running the next bead.

Step 6 Inspect the weld. The weld is acceptable if it has the following features:

- Uniform rippled appearance of the bead face
- Craters and restarts filled to the full cross section of the weld
- Uniform weld width ±$\frac{1}{16}$" (1.6 mm)
- Acceptable weld profile in accordance with the applicable code or standard
- Smooth, flat transition with complete fusion at the toes of the weld
- Complete uniform root reinforcement at least flush with the base metal to a maximum buildup of $\frac{1}{8}$" (3.2 mm)
- No excessive porosity with no pores larger than $\frac{3}{32}$" (2.38 mm)
- No undercut greater than $\frac{1}{32}$" (0.8 mm) deep or 10% of the base metal thickness, whichever is less
- No cracks

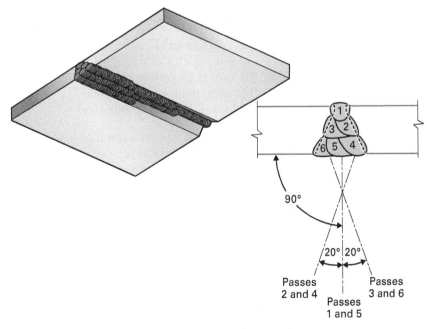

Stringer Bead Sequence

Figure 31 Multipass 4G bead sequence and work angles.

3.0.0 Section Review

1. The *most* common fillet welds are made in _____.
   a. horizontal piping
   b. lap and T-joints
   c. root passes
   d. vertical piping

2. The position of a 3G groove weld is _____.
   a. vertical
   b. overhead
   c. flat
   d. horizontal

Module 29209   Review Questions

1. The filler metal in a GMAW process is provided by _____.
   a. individual rods, or fillets, of steel
   b. a solid- or composite electrode
   c. the expansion of molten base metal
   d. the material in the flux-cored wiring

2. Typical protective clothing and equipment for welding includes welding hood, high-top safety shoes or boots, protective leather clothing, and _____.
   a. sunglasses
   b. mesh cap
   c. earmuffs
   d. safety shoelaces

3. A gas that is commonly used to purge areas or containers prior to GMAW operations is _____.
   a. nitrogen
   b. propane
   c. oxygen
   d. acetylene

4. If possible, practice weld coupons for GMAW should be _____.
   a. $\frac{1}{4}$" (6.4 mm) thick stainless steel
   b. $\frac{3}{8}$" (10 mm) thick carbon steel
   c. $\frac{3}{4}$" (19.1 mm) thick low-alloy plate
   d. $\frac{7}{8}$" (22.2 mm) thick composite steel

5. Setting up a GMAW machine requires loading the wire feeder and gun with the proper electrode as well as _____.
   a. locking out and tagging out the primary disconnect
   b. removing any liner material from the feeder and gun
   c. attaching the ground lead to the gun handle
   d. connecting the proper shielding gas for the application

6. In GMAW, the distance from the contact tip to the base metal to be welded is known as _____.
   a. contact tip to work distance
   b. strikeout distance
   c. stickout length
   d. slope reach

7. When weave beads are made, the motion of the electrode is _____.
   a. forward and backward
   b. up and down
   c. side to side
   d. circular and forward

8. If a weld restart has undercut, it is because _____.
   a. there was a previous weld termination
   b. not enough time was spent filling in the crater
   c. there was too little fillet in the electrode
   d. too much time was spent filling in the crater

9. Unless the proper technique is used, a weld termination normally leaves _____.
   a. a crack
   b. porosity
   c. spatter
   d. a crater

10. The process of depositing connective weld beads parallel to one another is known as _____.
    a. padding
    b. stringing
    c. pooling
    d. weaving

11. When flat (1F) fillet welds are being made, the travel speed can be increased or decreased to control the amount of _____.
    a. base metal fusion
    b. root penetration
    c. weld metal buildup
    d. flux-core shielding

12. What work angle should be used for the first bead in a horizontal (2F) fillet weld?
    a. 25°
    b. 45°
    c. 60°
    d. 90°

13. With GMAW, groove welds can be made in _____.
    a. only the 1G and 2G positions
    b. only the 2G and 3G positions
    c. only the 1G, 2G, and 3G positions
    d. all positions

14. When making a 3G open V-groove weld with weave beads, use a zigzag pattern across the root bead and pause at each toe _____.
    a. for complete penetration
    b. to prevent excessive spatter
    c. for good fusion
    d. to fill in the crater

15. One of the criteria for an acceptable overhead open V-groove weld with stringer beads is a uniform weld width of _____.
    a. $\pm^1/_{16}$" (1.6 mm)
    b. $\pm^1/_8$" (3.2 mm)
    c. $\pm^3/_8$" (9.5 mm)
    d. $\pm^1/_2$" (12.7 mm)

Answers to odd-numbered questions are found in the Review Question Answer Keys at the back of this book.

Answers to Section Review Questions

Answer	Section Reference	Objective
Section 1.0.0		
1. b	1.2.2	1b
2. d	1.1.2	1a
3. a	1.3.1	1c
Section 2.0.0		
1. c	2.1.5	2a
2. d	2.2.2	2b
Section 3.0.0		
1. b	3.1.0	3a
2. a	3.2.2	3b

User Update

Did you find an error? Submit a correction by visiting **https://www.nccer.org/olf** or by scanning the QR code using your mobile device.

FCAW – Plate

Source: © Miller Electric Mfg. LLC

Objectives

Successful completion of this module prepares you to do the following:

1. Identify FCAW-related safety practices and explain how to prepare for welding.
   a. Describe FCAW-related safety practices.
   b. Describe basic FCAW processes.
   c. Explain how to safely set up the equipment and work area for welding.
2. Describe equipment control and welding procedures for FCAW and explain how to produce basic weld beads.
   a. Describe equipment control and welding techniques related to FCAW.
   b. Explain how to produce basic FCAW weld beads.
3. Describe the welding procedures needed to produce proper fillet and V-groove welds using FCAW welding techniques.
   a. Explain how to produce proper fillet welds using FCAW welding techniques.
   b. Explain how to produce proper V-groove welds using FCAW welding techniques.

Performance Tasks

Under supervision, you should be able to do the following:

1. Make multipass FCAW-G/GM (gas-shielded) and/or FCAW-S (self-shielded) fillet welds on carbon steel plate coupons in all four 1F through 4F positions.
2. Make multipass FCAW-G/GM (gas-shielded) and/or FCAW-S (self-shielded) V-groove welds on carbon steel plate coupons in all four 1G through 4G positions, with or without backing.

Overview

Flux-cored arc welding (FCAW) is a common welding process used in industries throughout the world. FCAW equipment is available in different sizes and varieties; however, the basic operating principles of FCAW should apply to all makes and models. This module describes how FCAW equipment is used, identifies FCAW-related safety practices, and examines different techniques used to produce various types of weld beads.

Digital Resources for Welding

Scan this code using the camera on your phone or mobile device to view the digital resources related to this craft.

1.0.0 FCAW Welding

Performance Tasks

1. Make multipass FCAW-G/GM (gas-shielded) and/or FCAW-S (self-shielded) fillet welds on carbon steel plate coupons in all four 1F through 4F positions.

2. Make multipass FCAW-G/GM (gas-shielded) and/or FCAW-S (self-shielded) V-groove welds on carbon steel plate coupons in all four 1G through 4G positions, with or without backing.

Objective

Identify FCAW-related safety practices and explain how to prepare for welding.
a. Describe FCAW-related safety practices.
b. Describe basic FCAW processes.
c. Explain how to safely set up the equipment and work area for welding.

Flux-cored arc welding (FCAW) is based on the use of a flux-cored metal electrode and, in some cases, a shielding gas. This section describes FCAW equipment and how it works, identifies safety practices related to FCAW, and examines how to prepare for FCAW.

1.1.0 Safety Summary

The following sections summarize the safety procedures and practices that welders must observe when cutting or welding. Complete safety coverage is provided in NCCER Module 29101, *Welding Safety*, which should be completed before continuing with this module. The safety procedures and practices are in place to prevent potentially life-threatening injuries from welding accidents. Above all, be sure to wear appropriate protective clothing and equipment when welding or cutting.

1.1.1 Protective Clothing and Equipment

Welding activities can cause injuries unless you wear all the protective clothing and equipment designed specifically for the welding industry. The following safety guidelines about protective clothing and equipment should be followed to prevent injuries:

- Wear a face shield over snug-fitting cutting goggles or safety glasses for gas welding or cutting. The welding hood should be equipped with an approved shade for the application. A welding hood equipped with a properly tinted lens is best for all forms of welding.

- Wear proper protective leather and/or flame-retardant clothing along with welding gloves that protect the welder from flying sparks, molten metal, and heat.

- Wear high-top safety shoes or boots. Make sure the tongue and lace area of the footwear will be covered by a pant leg. If the tongue and lace area is exposed or the footwear must be protected from burn marks, wear leather spats under the pants or chaps over the front and top of the footwear.

- Wear a 100% cotton cap with no mesh material included in its construction. The bill of the cap points to the rear. If a hard hat is required for the environment, use one that allows the attachment of rear deflector material and a face shield. A hard hat with a rear deflector is generally preferred when working overhead and may be required by some employers and jobsites.

- Wear earmuffs, or at least earplugs, to protect your ear canals from sparks.

WARNING!

Using proper personal protective equipment (PPE) for the hands and eyes is particularly important. The most common injuries that welders experience during FCAW operations are injuries to the fingers and eyes.

1.1.2 Fire/Explosion Prevention

Welding activities usually involve the use of fire or extreme heat to melt metal. Whenever fire is used in a weld, the fire must be controlled and contained. Welding or cutting activities are often performed on vessels that may have once contained flammable or explosive materials. Residues from those materials can catch fire or explode when a welder begins work on such a vessel. The following fire and explosion prevention guidelines associated with welding contribute to a safe work zone:

- Never carry matches or gas-filled lighters in your pockets. Sparks can cause the matches to ignite or the lighter to explode, causing serious injury.

- Never use oxygen to blow dust or dirt from clothing. The oxygen can remain trapped in the fabric for a time. If a spark hits clothing during this time, the clothing can burn rapidly and violently out of control.

- Make sure any flammable material in the work area is moved or shielded by a fire-resistant covering. Approved fire extinguishers must be available before attempting any heating, welding, or cutting operations. If a hot work permit and a fire watch are required, be sure those items are in place before beginning and that all site requirements have been observed.

- Never release a large amount of oxygen or use oxygen as compressed air. The presence of oxygen around flammable materials or sparks can cause rapid and uncontrolled combustion. Keep oxygen away from oil, grease, and other petroleum products.

- Never release a large amount of fuel gas, especially acetylene. Propane tends to concentrate in and along low areas and can ignite at a considerable distance from the release point. Acetylene is lighter than air but is even more dangerous than propane. When mixed with air or oxygen, acetylene will explode at much lower concentrations than any other fuel.

- To prevent fires, maintain a neat and clean work area, and make sure that any metal scrap or slag is cold before disposing of it.

- Before cutting containers such as tanks or barrels, check to see if they have contained any explosive, hazardous, or flammable materials, including petroleum products, citrus products, or chemicals that decompose into toxic fumes when heated. As a standard practice, always clean and then fill any tanks or barrels with water or purge them with a flow of inert gas such as nitrogen to displace any oxygen. Containers must be cleaned by steam cleaning, flushing with water, or washing with detergent until all traces of the material have been removed.

WARNING!

Welding or cutting must never be performed on drums, barrels, tanks, vessels, or other containers until they have been emptied and cleaned thoroughly, eliminating all flammable materials and all substances (such as detergents, solvents, greases, tars, or acids) that might produce flammable, toxic, or explosive vapors when heated. Clean containers only in well-ventilated areas, as vapors can accumulate during cleaning, causing explosions or injury.

Proper procedures for cutting or welding hazardous containers are described in the *American Welding Society (AWS) F4.1, Safe Practices for the Preparation of Containers and Piping for Welding, Cutting, and Allied Processes.*

1.1.3 Work Area Ventilation

Vapors and fumes tend to rise in the air from their sources. Welders must often work above the welding area where the fumes are being created. Welding fumes can cause personal injuries including long-term respiratory harm. Good work area ventilation helps to remove the vapors and protect the welder. The following is a list of work area ventilation guidelines to consider before and during welding activities:

- Follow confined space procedures before conducting any welding or cutting in a confined space.

- Never use oxygen for ventilation in confined spaces.

- Always perform cutting or welding operations in a well-ventilated area. Cutting or welding operations involving zinc or cadmium materials or coatings result in toxic fumes. For long-term cutting or welding of such materials, always wear an approved full-face, supplied-air respirator that uses breathing air supplied from outside of the work area. For occasional, very short-term exposure, a high-efficiency particulate arresting-rated (also called HEPA-rated) or metal fume filter may be used on a standard respirator.

- Make sure confined spaces are properly ventilated for cutting or welding purposes. Use powered extraction systems when available. *Figure 1* shows mobile extraction equipment that removes cutting or welding fumes and vapors from work areas.

WARNING!

Backup flux may contain silica, fluorides, or other toxic materials. Alcohol and acetone used for the mixture are flammable. Weld only in well-ventilated spaces. Failure to provide proper ventilation could result in personal injury or death.

Figure 1 Mobile extraction equipment.
Source: © Miller Electric Mfg. LLC

Welding Processes

FCAW processes use the intense heat produced by an electrical arc between a filler metal electrode and a workpiece to melt the metals and cause them to merge or coalesce. FCAW is one common form of arc welding used extensively in industry. Performing FCAW safely and effectively requires a good understanding of how the process and equipment work. It is also critical to understand the safety practices necessary during an FCAW operation.

1.2.1 The FCAW Process

In the FCAW process, a continuous, consumable, flux-cored electrode is used as a filler metal to form a weld (*Figure 2*). Keep in mind that the proper term for the consumable material used for this filler metal is "electrode," although it is commonly referred to as wire. A shielding gas that protects the molten weld metal from contamination by the atmosphere can be generated entirely from the flux in the electrode core (self-shielded FCAW or FCAW-S) or from both the flux in the electrode core and an external shielding gas (gas-shielded FCAW or FCAW-G).

NOTE

Reference *Appendix 29210A* for Performance Accreditation Tasks (PATs) designed to evaluate your ability to run fillet and groove welds with FCAW equipment.

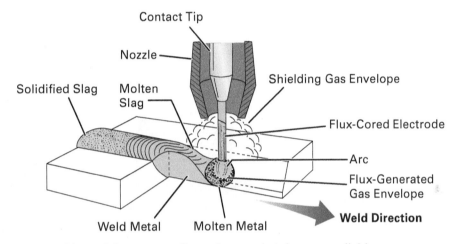

Note: Other gun configurations and styles are available.

Gas-Shielded FCAW

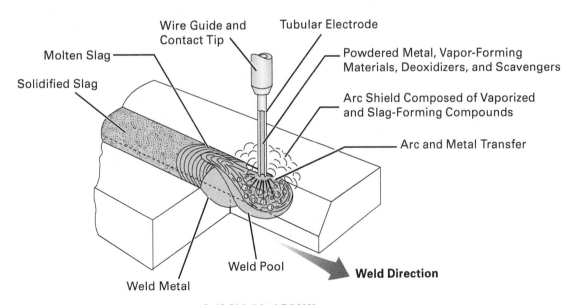

Self-Shielded FCAW

Figure 2 FCAW process.

Degassing: The escape of gases formed during the welding process that can cause imperfections in the weld.

In all types of flux-cored electrodes, the flux contains agents for **degassing** the weld, stabilizing the arc, and forming a protective slag over the weld bead. Other elements are often added to the flux-cored electrode to alloy with the base metal and improve its chemical and mechanical characteristics. The FCAW process is basically limited to welding low-carbon steels, some low-alloy steels, and some stainless steels. FCAW is an excellent process for making fillet and groove welds on carbon steel plate.

1.2.2 FCAW Equipment

The equipment used for FCAW-G or FCAW-S is basically the same except for the components needed for the gas used in the gas-shielding process. This additional equipment includes guns equipped with a gas nozzle, gas hoses, a gas control solenoid valve, a gas regulator/flowmeter, and a shielding gas cylinder or other source. *Figure 3* is an example of typical FCAW-S equipment.

FCAW-S is particularly useful in situations with high winds because the shielding gas is created inside the welding arc as the flux vaporizes. The molten puddle has a constant supply of shield gas displacing the air from the arc outward, and the puddle is immediately covered by the residual slag, similar to shielded metal arc welding (SMAW). An external shield gas, in contrast, attempts to force gas down from the nozzle to displace the air, and the gas can be blown away from the weld puddle by wind.

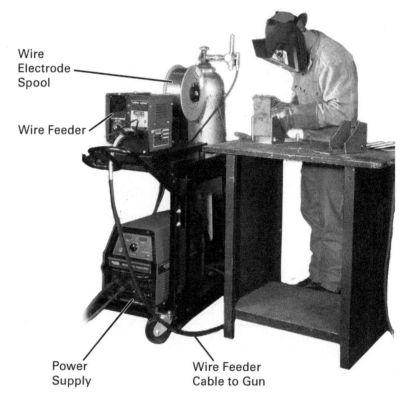

Figure 3 FCAW-S equipment.
Source: Photo courtesy of The Lincoln Electric Company, Cleveland, OH, U.S.A.

1.3.0   FCAW Welding Equipment Setup

Before welding can take place, the area must be made ready, the welding equipment must be set up, and the metal to be welded must be prepared. The following sections explain how to set up equipment for welding.

1.3.1 Preparing the Welding Area

To practice welding, a welding table, bench, or stand is needed. The welding surface must be steel, and provisions must be made for placing weld coupons out of position (*Figure 4*).

Positioning Clamp
and Support

Workpiece Clamp to Table

Figure 4 Welding station.
Source: Terry Lowe

Follow these steps to set up an area for welding:

Step 1 Make sure the area is properly ventilated. Make use of doors, windows, and fans.

Step 2 Check the area for fire hazards. Remove any flammable materials before proceeding.

Step 3 Know the location of the nearest fire extinguisher and how to use it. Don't proceed unless the extinguisher is charged.

Step 4 Set up flash shields around the welding area.

1.3.2 Preparing the Practice Weld Coupons

If possible, the practice weld coupons should be $\frac{3}{8}$" (10 mm) thick carbon steel. If this size is not readily available, $\frac{1}{4}$" to $\frac{3}{4}$" (6 mm to 20 mm) thick steel can be used to practice welding. Use a wire brush or grinder to remove heavy mill scale or corrosion.

The following outlines how to prepare weld coupons for various methods of practice:

- *Running beads* — The coupons can be any size or shape that can be easily handled.

- *Overlapping beads* — The coupons can be any size or shape that can be easily handled.

- *Fillet welds* — Cut the metal into 2" × 6" (51 mm × 152 mm) rectangles for the base and 1" × 6" (25 mm × 152 mm) rectangles for the web. *Figure 5* shows the weld coupons for fillet welding.

- *V-groove welds* — As shown in *Figure 6*, cut the metal into 3" × 7" (76 mm × 178 mm) rectangles with one (or both) of the lengths beveled at 30° to 37.5° for carbon steel or low-alloy plate (*Figure 6*). Note that the bevel angle for stainless steel coupons is 22.5°. Grind up to a $\frac{1}{8}$" (3.2 mm) root face on the bevel as chosen and directed by the instructor.

Machine Bevels and Root Faces for FCAW

For V-groove joints made in thick materials, machined bevels and root faces provide the cleanest and most uniform way to prepare materials for satisfactory FCAW welds. SMAW, FCAW, and gas tungsten arc welding (GTAW) can tolerate some poor fit-up for root passes, but the better the preparation, the better the result.

NOTE

Open V-groove welds are rarely done using the FCAW process. It is assumed that the root pass will be done using another process better suited for open-root passes, such as GMAW, GTAW, or SMAW. If the root pass is to be done using the FCAW process, a backing bar should be added or back gouging followed by a back weld.

Follow these steps to prepare each V-groove weld coupon:

Step 1 Check the bevel face of each coupon. There should be no dross and a 0" to ⅛" (0 mm to 3.2 mm) root face. The bevel angle should be 30° to 37.5°. An example of this is shown in *Figure 7*.

Step 2 Align the beveled coupons on a flat surface with a root opening determined by the instructor. Place tack welds on the back side of the joint. Use from three to four ¼" (6.4 mm) tack welds.

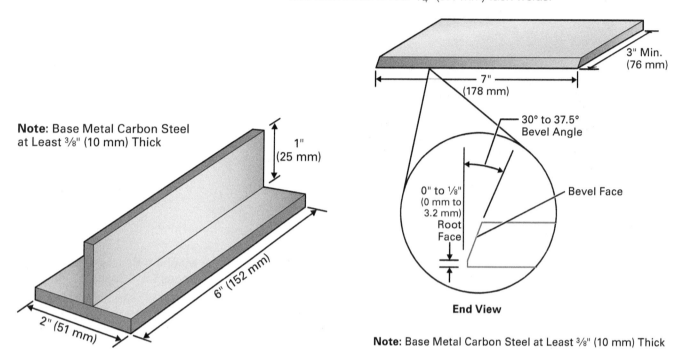

Note: Base Metal Carbon Steel at Least ⅜" (10 mm) Thick

Figure 5 Fillet weld coupons.

Note: Base Metal Carbon Steel at Least ⅜" (10 mm) Thick

Figure 6 Metal cut for V-groove weld coupons.

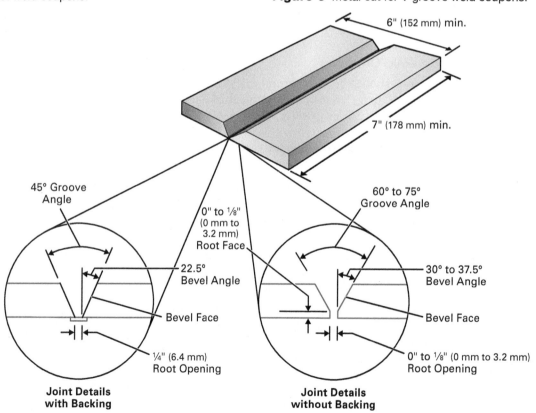

Note: Base Metal Carbon Steel, at Least ⅜" (10 mm) Thick

Figure 7 V-groove weld coupon.

1.3.3 The Welding Machine

Identify the proper welding machine for FCAW use and follow these steps to set it up for welding:

Step 1 Verify that the welding machine can be used for FCAW, with or without gas shielding as required.

Step 2 Verify the location of the primary disconnect.

Step 3 Configure the welding machine for FCAW-G or FCAW-S as directed by the instructor (*Figure 8*). Configure the gun polarity. Equip the gun with the correct liner material and contact tip for the diameter of the electrode being used and the correct nozzle for the application.

Step 4 In accordance with the manufacturer's instructions, load the wire feeder and gun with a flux-cored electrode of the proper diameter as directed by the instructor.

Step 5 If required, connect the proper shielding gas for the application as specified by the electrode manufacturer, Welding Procedure Specification (WPS), site quality standards, or instructor.

Step 6 Connect the clamp of the workpiece lead to the table or workpiece.

Step 7 Turn on the welding machine and, if necessary, purge the gun as directed by the gun manufacturer's instructions.

Step 8 Set the initial welding voltage and wire feed speed for the type and size of electrode as recommended by the manufacturer, directed by the instructor, or provided in the WPS.

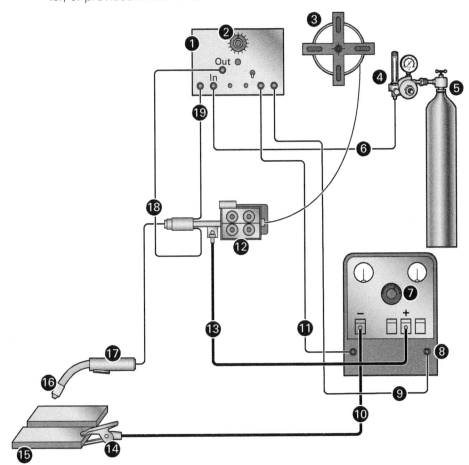

1. Control Unit
2. Wire Feed Speed Control
3. Electrode Wire Reel
4. Regulator/Flowmeter
5. Optional Shielding Gas Source
6. Gas-In Hose
7. Voltage Control
8. Power Source
9. Power to Control Unit
10. Workpiece Lead
11. Contactor Control Cable
12. Wire Feed Drive Assembly
13. Electrode Lead
14. Workpiece Clamp
15. Workpiece
16. Optional Gas Nozzle
17. FCAW Gun
18. Gas-Out Hose
19. Gun Control Cable

Note: The polarity of the gun and workpiece leads is determined by the type of filler metal and application.

Figure 8 Configuration diagram of a typical FCAW welding machine.

Submerged Arc Welding

No, this is not underwater welding. Submerged arc welding (SAW) was first patented in 1935. The process involves covering the electric arc with a bed of flux granules. SAW can use multiple electrodes at the same time to deposit weld metal. Like FCAW, the electrode is in wire form and may be tubular or solid. The arc zone and weld are protected from atmospheric contamination by being covered by a blanket of flux that becomes molten due to the arc's heat. Once liquefied, the flux becomes conductive, providing a current path between the electrode and the work. The flux also significantly reduces the UV radiation and spatter emitted from the weld site. SAW is normally done using automated equipment but can be done by hand. The process is usually limited to flat or horizontal welding positions. As shown in the image here on these quick demonstration beads, the cooled flux material snaps off easily in large pieces.

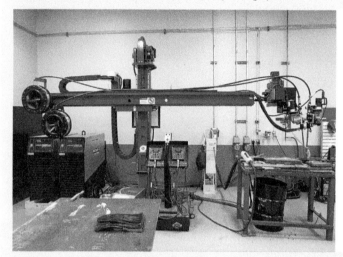 

1.0.0 Section Review

1. After a vessel has been thoroughly cleaned to prepare it for welding, it should be filled with water or purged with _____.
   a. oxygen
   b. compressed air
   c. sand or soil
   d. an inert gas

2. In the FCAW process, the shielding gas used to protect the weld zone typically comes from a(n) _____.
   a. flux-cored electrode
   b. electrode lead
   c. solid-cored wire
   d. gaseous filler metal

3. When FCAW equipment is being used for practice welding, the welding area should be enclosed with _____.
   a. shielding gas
   b. flash shields
   c. gas extinguishers
   d. gun liner material

2.0.0 Welding Procedures and Basic Beads

Objective

Describe equipment control and welding procedures for FCAW and explain how to produce basic weld beads.

a. Describe equipment control and welding techniques related to FCAW.

b. Explain how to produce basic FCAW weld beads.

Performance Tasks

1. Make multipass FCAW-G/GM (gas-shielded) and/or FCAW-S (self-shielded) fillet welds on carbon steel plate coupons in all four 1F through 4F positions.

2. Make multipass FCAW-G/GM (gas-shielded) and/or FCAW-S (self-shielded) V-groove welds on carbon steel plate coupons in all four 1G through 4G positions, with or without backing.

Effective welding requires the welder to properly control the equipment being used and employ various welding techniques to produce strong and acceptable weld beads. This section examines factors that affect FCAW and how these factors are controlled. The process for producing basic FCAW weld beads is also described.

2.1.0 Welding Procedures

Welding can be affected by factors such as the welding voltage, welding amperage, position and travel speed of the gun, and other aspects of the welding equipment and process. This section describes aspects of equipment control and welding techniques that pertain to FCAW.

2.1.1 Welding Voltage

Arc length is the distance from the electrode tip to the base metal or molten pool at the base metal. Arc length is determined by voltage, which is set at the power source. If the voltage is set too high, the arc will be too long, which could cause the electrode to fuse to the tip. An excessively high voltage also causes porosity and excessive spatter. Voltage and wire feed speed are correlated. Fine adjustments of either can be necessary to achieve the desired weld bead. Voltage must be increased or decreased as wire feed speed is increased or decreased. Set the voltage so that the arc length is just above the surface of the base metal (*Figure 9*).

Synergic welding machines control voltage, polarity, current, and wire feed moment by moment using a program customized for each welding situation. These machines will have a **trim** control that works similarly to the voltage control on other machines. Essentially, it overrides the program.

Synergic: Welding technology that controls voltage, current, and wire feed using a program optimized for each welding situation.

Trim: A control found on synergic welding machines similar to the voltage control on regular machines that acts as a kind of override to the program.

Electrode Manufacturer's Recommendations

Always obtain and follow the manufacturer's recommendations or WPS for shielding gas (if required) and for initially setting the welding voltage, current (if variable), and wire feed speed parameters. These balanced parameters are critical and are based on the welding position, size, and composition of the flux-cored electrode being used.

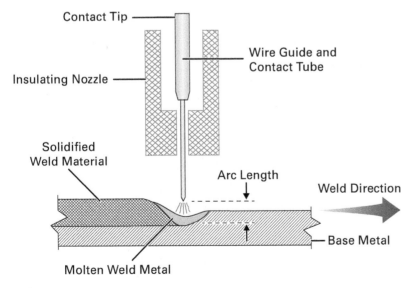

Figure 9 Arc length.

Arc Voltage

A minimum arc voltage is needed to maintain spray transfer. However, penetration is not directly related to voltage. Penetration will increase with voltage for a time, but beyond the optimum voltage, penetration will actually decrease as the voltage increases further.

Travel Speed and Wire Feed Speed

Beginner welders are tempted to turn down the wire feed speed if they experience difficulty controlling the weld puddle. Electrodes must be run at certain balanced parameters that cannot be changed individually. Voltage, wire feed speed, and travel speed are adjusted and balanced together to control the weld puddle. The voltage selected is usually determined by the thickness of the base metal or its heat absorption capacity.

2.1.2 Welding Amperage

The wire feed speed control on a standard power source manages the welding amperage after the welder has set the initial recommended setting. The welding power source provides the amperage necessary to melt the electrode while maintaining the selected welding voltage. Within limits, when the wire feed speed is increased, the welding amperage and deposition rate also increase. This results in higher welding heat, deeper penetration, and higher beads. When the wire feed speed is decreased, the welding amperage automatically decreases. With lower welding amperage and less heat, the deposition rate drops, and the weld beads are smaller and have less penetration.

Note that some constant-voltage (CV) welding machines used for FCAW provide varying degrees of current modification, such as slope or induction adjustments. Welding machines with a slope adjustment allow the amount of amperage change to vary in relation to the voltage range of the unit. This allows the correct short-circuiting current to be established for certain welding applications and electrode requirements.

Standard CV FCAW units have a current slope fixed by the manufacturer for general welding applications and conditions. Other units allow adjustment of the response time for short-circuit current by using an induction control. In an instantaneous short-circuit current draw mode of operation, the machine draws a high level of current as soon as the electrode touches the workpiece. In the adjustable induction mode of operation, the machine gradually increases the welding current after the initial contact of the electrode to the workpiece. This evens out the arc characteristics during short-circuiting transfer.

Pulse-transfer welding machines allow the adjustment of the peak current for the pulse and the background current between pulses to match specific welding applications and electrode requirements. In other cases, where a voltage-sensing wire feeder is used with a constant-current (CC) welding machine, the current is adjustable, and the wire feeder varies the wire feed speed to maintain the proper arc voltage at the electrode tip. Note that some codes prohibit the use of CC machines for FCAW.

2.1.3 Welding Travel Speed

Weld travel speed is the speed of the electrode tip across the base metal in the direction of the weld, measured in inches per minute (IPM) or a metric equivalent. Travel speed has a great effect on penetration and bead size. Slower travel speeds build higher beads with deeper penetration until the arc no longer contacts the base metal at the front of the puddle and can actually interfere with proper fusion. Faster travel speeds build smaller beads with less penetration. Ideally, the welding parameters should be adjusted so the electrode tip is positioned at the leading edge of the weld puddle during travel.

2.1.4 Gun Position

Gun position for FCAW affects weld penetration along with voltage and travel speed. The gun travel angle (*Figure 10*) is defined as an angle less than 90° between the electrode axis and a line perpendicular to the weld axis at the point of electrode contact in a plane determined by the electrode axis and weld axis. For pipe, the plane is determined by the electrode axis and a line tangent to the pipe surface at the same point. The following list describes different types of travel angles:

- *Neutral angle* — A neutral angle is the travel angle when the center line of the gun electrode is exactly perpendicular (90°) to the welding surface. The term is nonstandard for a 0° travel angle. Weld penetration is moderate when the gun electrode is used in this position.

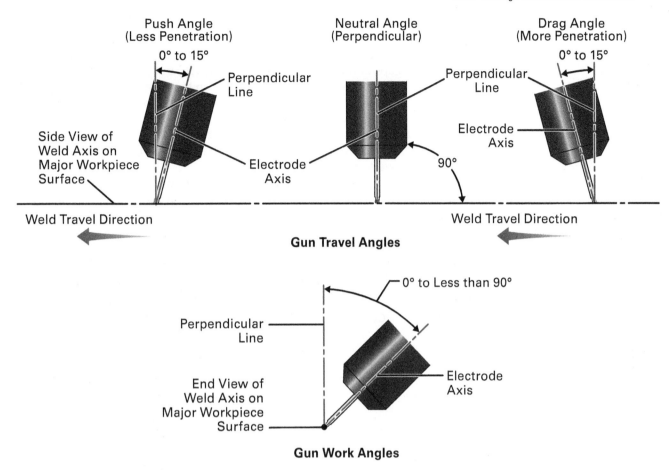

Figure 10 Gun travel and work angles.

- *Push angle* — A push angle is generally used for welding thin materials or when shallow penetration is required. A push angle is the travel angle created when the welding gun is tilted back so that the electrode tip precedes the gun in the direction of the weld. In this position, the electrode tip and shielding gas are being directed ahead of the weld bead. Welding with this angle produces a wider weld bead with less penetration. Push angles greater than 15° are not recommended for manual welding.

- *Drag angle* — A drag angle is the travel angle created when the gun is tilted forward so the electrode tip trails the gun in the direction of the weld. In this position, the electrode tip and shielding gas are being directed back toward the weld bead. Welding with this angle produces a higher and narrower weld bead with more penetration. Maximum penetration is achieved with a drag angle of approximately 25°.

The gun work angle, also shown in *Figure 10*, is an angle less than 90° between a line perpendicular to the major workpiece surface at the point of electrode contact and a plane determined by the electrode axis and weld axis. For a T-joint or corner joint, the line is 45° to the non-butting member. For pipe, the plane is determined by the electrode axis and a line tangent to the pipe surface at the same point.

2.1.5 Electrode Extension, Stickout, and Standoff Distance

Electrode extension is the length of the electrode that extends beyond the contact tip of the welding gun. Changing the extension influences the welding current for low-conductivity metal electrodes because it changes the preheating of the electrode caused by the resistance of the electrode to current flow. As the extension increases, preheating of the electrode increases.

Since the welding power source is self-regulating, it does not have to furnish as much welding current to melt the electrode, so the current output automatically decreases. This results in less penetration. Increasing extension is useful for bridging gaps and compensating for mismatch, but it can cause overlap or lack of fusion and a ropy bead appearance. When the extension is decreased, the power source is forced to increase its current output to burn off the electrode. Too little extension can cause the electrode to weld to the contact tip or develop porosity in the weld. For metal electrodes which are highly conductive, the preheating effect of electrode resistance is minimal, and electrode speed and voltage settings have a more direct effect.

FCAW-G extensions vary widely. Always check the manufacturer's recommendations. The extension for FCAW-S and an insulating nozzle is $5/8$" to 1" (16 mm to 25 mm), depending on the welding position and electrode size. *Figure 11* shows a typical electrode extension for an FCAW-S gun configuration.

Additional gun terminology is defined in *Figure 11*. Stickout is the distance from the insulating nozzle to the end of the electrode. Standoff distance is the distance from the insulating nozzle to the workpiece. Contact tip extension or setback is usually dependent on the transfer mode for the application.

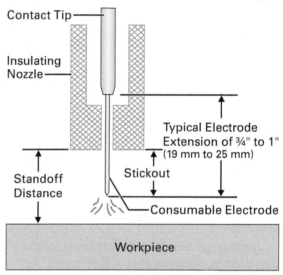

FCAW-S Gun with Insulating Nozzle

Figure 11 FCAW-S gun configuration.

2.1.6 Gas Nozzle Cleaning

As an FCAW-G welding machine is being used, weld spatter accumulates on the gas nozzle and contact tip. If the gas nozzle is not properly cleaned, it will restrict the shielding gas flow and cause porosity in the weld. When making out-of-position welds, accumulated spatter in the nozzle can result in a short circuit between the gas diffuser and the nozzle shell, causing the nozzle to stick to the workpiece.

Clean the gas nozzle with a reamer or round file or with the tang of a standard file. After cleaning, the nozzle can be sprayed or dipped in a special anti-spatter compound. The anti-spatter compound helps prevent the spatter from sticking to the nozzle. Ensure that any anti-spatter product used is compatible with the welding application and equipment in use.

2.2.0 FCAW Bead Types

The following sections explain how to create weld beads using FCAW equipment. The two basic bead types are stringer beads and weave beads.

CAUTION

Overuse of electrode extension is a common mistake for new welders. Too much electrode extension will reduce amperage and potentially risk weld quality and adherence to welding specifications. Paying attention to the amperage lost with every interval of electrode extension will help new welders gain a sense of the relationship.

CAUTION

When using anti-spatter, be sure not to let any of the compound come into contact with the welding wire. If contact does occur, trim off the contaminated wire. If the contaminated wire is not trimmed off, the compound on the wire could cause weld defects or discontinues.

2.2.1 Stringer Beads

Stringer beads are made with little or no side-to-side motion of the electrode. Practice running stringer beads in the flat position. Experiment with different voltages, travel angles, and stickouts. Change the wire feed speed and observe the bead size. Try different travel angles and observe the bead shape and size. If the practice coupon gets too hot between passes, it may have to be cooled in water.

WARNING!

Use pliers to handle the hot practice coupons. Wear gloves when placing the practice coupon in water. Steam will rise off the coupon and can burn or scald unprotected hands and arms.

Follow these steps to run stringer beads:

Step 1 Hold the gun at the desired travel angle (drag angle), with the electrode tip directly over the point where the weld is to begin and pull the gun trigger.

Step 2 Hold the arc in place until the weld puddle begins to form.

Step 3 Slowly advance the arc while maintaining the gun angle. Stay in the leading edge of the puddle to prevent incomplete fusion.

Step 4 Continue to weld until a bead about 2" to 6" (50 mm to 152 mm) long is formed.

Step 5 Stop, raise the gun to a neutral (0°) travel angle, pause until the crater is filled, and then release the trigger.

Step 6 Inspect the bead for the following:

- Straightness of the bead
- Uniform appearance of the bead face
- Smooth, flat transition with complete fusion at the toes of the weld
- No excessive porosity with no pores larger than $^3/_{32}$" (2.38 mm)
- No undercut greater than $^1/_{32}$" (0.8 mm) deep or 10% of the base metal thickness, whichever is less
- No overlap
- No lack of fusion
- Complete penetration
- No pinholes
- No inclusions
- Crater filled
- No cracks

Step 7 Continue practicing stringer beads until acceptable welds are made every time.

2.2.2 Weave Beads

Weave beads are made with wide, side-to-side motions of the electrode. The width of a weave bead is determined by the amount of side-to-side motion. *Figure 12* shows an example of a weave bead.

When making a weave bead, use care at the toes to be sure there is proper tie-in to the base metal. To ensure proper tie-in at the toes, slow down or pause slightly at the edges. The pause at the edges will also flatten out the weld, giving it the proper profile.

Practice running weave beads about $^5/_8$" (16 mm) wide in the flat position. Experiment with different weave motions, travel angles, and stickouts. Change the wire feed speed and observe the bead size. Try different gun angles and observe the bead shape and size.

CAUTION

The width of stringer beads is specified in the welding code or WPS being used at a site. Do not exceed the widths specified for the site.

CAUTION

Cooling with water (quenching) is done only on practice coupons. Never cool test coupons or on-the-job welds with water. Cooling with water can cause weld cracks and affect the mechanical properties of the base metal.

Gas Marks and Electrode Extension

The FCAW process sometimes results in gas marks, also known as worm tracks, on the weld surface. These marks are the result of excessive amounts of gases being trapped under the flux as it solidifies. If the flux solidifies before the gas escapes, pockets are left behind in the molten weld metal, causing the marks.

This surface discontinuity can be caused by an electrode extension that is too short. An extension that is too short allows dissolved hydrogen to remain trapped beneath the slag as it forms.

There are other potential causes as well. Gas marks are more likely with small-diameter electrodes operating at lower current levels than when using larger electrodes in the flat and horizontal position operating at higher currents. An excessive arc voltage, which increases the arc length, is also a common cause. Another possible cause is moisture trapped in the weld area, possibly entering from the surrounding air or from condensation in the shielding gas.

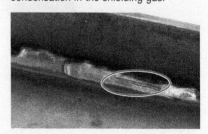

Source: Photo courtesy of The Lincoln Electric Company, Cleveland, OH, U.S.A.

Anti-Spatter Materials

Use only anti-spatter material specifically designed for FCAW gas nozzles. Water-based products may cause hydrogen embrittlement in hydrogen-sensitive steel. Always refer to the SDS/MSDS for specific safety considerations. Caution should be taken not to use too much anti-spatter, as using more anti-spatter than needed can cause porosity or contaminate the weld. Anti-spatter sprays and gels are usually water- or petrolatum-based. However, always check the SDS for the specific type of anti-spatter being used to ensure you know the materials you are working with. Additionally, some of these products are prohibited for use in code welding, so check the WPS before using them. For explosion and flammability reasons, be sure that all aerosol cans are a safe distance from the welding area.

Source: Photo courtesy of The Lincoln Electric Company, Cleveland, OH, U.S.A.

Figure 12 Close-up view of weave beads.
Source: Dan Sterry, TIC

Follow these steps to run weave beads:

Step 1 Hold the gun at the desired travel angle with the electrode tip directly over the point where the weld is to begin.

Step 2 Pull the gun trigger.

Step 3 Hold the arc in place until the weld puddle begins to form.

Step 4 Slowly advance the arc in a weaving motion while maintaining the gun angle (*Figure 13*).

Step 5 Continue to weld until a bead about 2" to 6" (50 mm to 152 mm) long is formed.

Step 6 Stop, raise the gun to a neutral (0°) travel angle, pause until the crater is filled, and then release the trigger.

Step 7 Inspect the bead for the following:

- Straightness of the bead
- Uniform appearance of the bead face
- Smooth, flat transition with complete fusion at the toes of the weld
- No excessive porosity with no pores larger than $3/_{32}$" (2.38 mm)
- No undercut greater than $1/_{32}$" (0.8 mm) deep or 10% of the base metal thickness, whichever is less
- No overlap
- No lack of fusion
- Complete penetration
- No pinholes
- No inclusions
- Crater filled
- No cracks

Step 8 Continue practicing weave beads until acceptable welds are made every time.

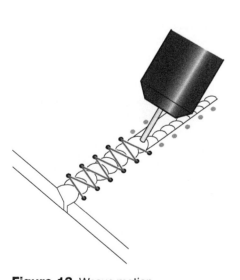

Figure 13 Weave motion.

2.2.3 Weld Restarts

A restart is the junction where a new weld connects to and continues the bead of a previous weld. Restarts are important because an improperly made restart creates a weld discontinuity. One advantage of the semiautomatic FCAW process is that restarts can be minimized. Plan the weld to eliminate restarts, if possible. If a restart has to be made, it must be made to blend smoothly with the previous weld and not stand out. The technique for making a restart is the same for both stringer and weave beads. Practice weld restarts using FCAW as directed by the instructor. Remember that the welding codes do not allow arc strikes outside the area of permanent welds.

Follow these steps to make a restart:

Step 1 With the weld clean, hold the gun at the proper travel angle while restarting the arc directly over the center of the crater. Remember that welding codes do not allow arc strikes outside the area of permanent welds.

Step 2 Pull the gun trigger.

Step 3 Move the electrode tip over the crater in a small circular motion to fill the crater with a molten puddle.

Step 4 Move to the leading edge of the puddle; as soon as the puddle fills the crater, continue the stringer or weave bead pattern being used.

Step 5 Inspect the restart. A properly made restart will blend into the bead, making it hard to detect. If the restart has undercut, not enough time was spent in the crater to fill it. If the undercut is on one side or the other, use more of a side-to-side motion while moving back into the crater. If the restart has a lump, it was overfilled; too much time was spent in the crater before resuming the forward motion.

Step 6 Continue to practice restarts until they are correct.

Bead Cleaning

There is no such thing as a self-cleaning electrode or shield gas. Welders must ensure that the surfaces to be welded are clean and free of contaminants. It is important that multipass welds are cleaned before each successive bead is laid down. Attempting to lay multipass beads without properly cleaning the previous bead will result in inclusions in the welded joint. Shown in the image here is a radiograph of a butt weld that has two slag lines easily visible in the weld root.

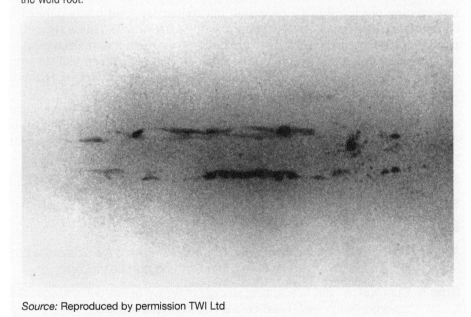

Source: Reproduced by permission TWI Ltd

2.2.4 Weld Terminations

A weld termination is made at the end of a weld. A termination normally leaves a crater. When making a termination, the welding codes require that the crater must be filled to the full cross section of the weld. This can be difficult since most terminations are at the edge of a plate, where welding heat tends to build up, making filling the crater more difficult. *Figure 14* shows the proper technique for terminating a weld.

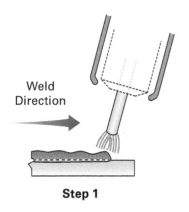
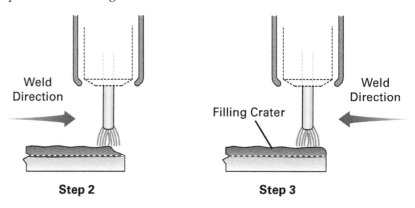

Figure 14 Terminating a weld.

Follow these steps to make a weld termination:

Step 1 Start to bring the gun up to a 0° travel angle and slow the forward travel as the end of the weld is approached.

Step 2 Stop the forward movement about $\frac{1}{8}$" (3.2 mm) from the end of the plate.

Step 3 With the gun at a 0° travel angle, stop and begin moving backward until the crater is filled. A $\frac{1}{8}$" to $\frac{1}{4}$" of backward movement is typical, depending upon the size of the crater.

Step 4 Release the trigger when the crater is completely filled.

Step 5 Inspect the termination. The crater should be filled to the full cross section of the weld.

Step 6 Continue to practice terminations until they are correct.

2.2.5 Overlapping Beads

Overlapping beads are made by depositing connective weld beads parallel to one another. The parallel beads overlap, forming a flat surface. This is also called padding. Overlapping beads are used to build up a surface and make multipass welds. Both stringer and weave beads can be overlapped. When viewed from the end, properly overlapped beads will have a relatively flat surface. *Figure 15* shows proper and improper overlapping string and weave beads.

Follow these steps to weld overlapping stringer or weave beads using a specific diameter flux-cored electrode and, if necessary, the proper shielding gas as directed by the instructor. Remember to chip and brush each bead before running the next bead.

Step 1 Mark out a square on a piece of steel.

Step 2 Weld a stringer or weave bead along one edge at a 0° work angle.

Step 3 Clean the weld.

Step 4 Position the gun at a work angle of 10° to 15° toward the side of the previous bead to get proper tie-in (*Figure 16*) and pull the trigger.

Step 5 Continue running overlapping stringer or weave beads until the square is covered. Clean each bead before running the new bead.

Step 6 Continue building layers of stringer or weave beads, one on top of the other, until perfecting the technique.

Build a pad using stringer beads (*Figure 17*). Repeat building the pad using weave beads.

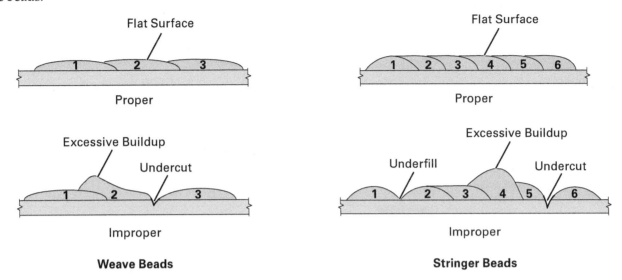

Figure 15 Proper and improper overlapping beads.

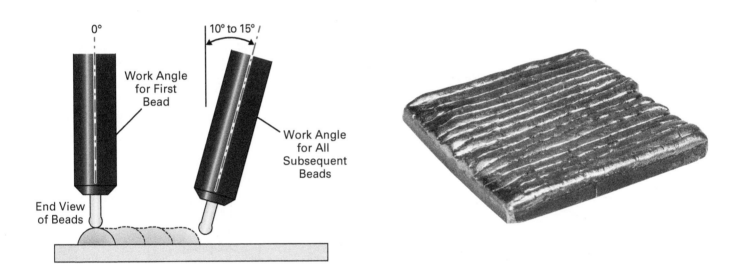

Figure 16 Work angles for overlapping beads.

Figure 17 Pad built with stringer beads.
Source: Photo courtesy of The Lincoln Electric Company, Cleveland, OH, U.S.A.

2.0.0 Section Review

1. In FCAW, the distance from the electrode tip to the base metal or the molten pool at the base metal is known as _____.
    a. slope reach
    b. strikeout distance
    c. arc length
    d. stickout length

2. The width of a weave bead is determined by the amount of _____.
    a. forward and backward gun movement
    b. stringer bead padding overlap
    c. filler material in the flux core
    d. side-to-side motion of the electrode

3.0.0 FCAW Fillet and V-Groove Welds

Performance Tasks

1. Make multipass FCAW-G/GM (gas-shielded) and/or FCAW-S (self-shielded) fillet welds on carbon steel plate coupons in all four 1F through 4F positions.

2. Make multipass FCAW-G/GM (gas-shielded) and/or FCAW-S (self-shielded) V-groove welds on carbon steel plate coupons in all four 1G through 4G positions, with or without backing.

Objective

Describe the welding procedures needed to produce proper fillet and V-groove welds using FCAW welding techniques.

a. Explain how to produce proper fillet welds using FCAW welding techniques.

b. Explain how to produce proper V-groove welds using FCAW welding techniques.

Fillet welds and V-groove welds are two common types of welds produced with FCAW. This section examines the procedures used to make each of these welds in four different positions: flat, horizontal, vertical, and overhead.

3.1.0 Fillet Welds

Fillet welds require little base metal preparation except for the cleaning of the weld area and the removal of any excess material from cut surfaces. Any dross from cutting will cause porosity in the weld. For this reason, the codes require that this material be entirely removed prior to welding.

The most common fillet welds are made in lap and T-joints. The weld position for plate is determined by the weld axis and the orientation of the workpiece. The positions for fillet welding on plate are flat (1F, where F stands for fillet), horizontal (2F), vertical (3F), and overhead (4F), as shown in *Figure 18*. In the 1F and 2F positions, the weld axis can be inclined up to 15°. Any weld axis inclination for the other positions varies with the rotational position of the weld face as specified in AWS standards.

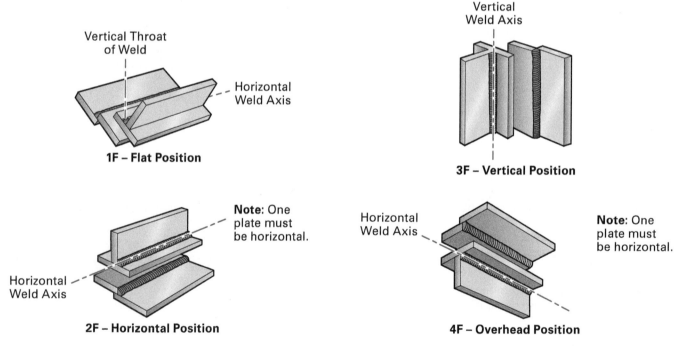

Figure 18 Fillet weld positions.

Fillet welds can be concave or convex, depending on the WPS or site quality standards. The welding codes require a fillet weld to have a uniform concave or convex face, although a slightly nonuniform face is acceptable. Convexity is the distance the weld extends above a line drawn between the toes of the weld. The convexity of a fillet weld or individual surface bead must not exceed that permitted by the applicable code or standard. In single-pass, weave-bead fillet welds where two workpieces are being joined at an angle (not lap joints), flat or slightly convex faces are usually preferred because weld stresses are more uniformly distributed through the fillet weld and workpieces. A fillet weld must be repaired if the profile has defects.

A fillet weld is unacceptable and must be repaired if the profile has insufficient throat, excessive convexity, undercut, overlap, insufficient leg, or incomplete fusion. *Figure 19* shows ideal, acceptable, and unacceptable fillet weld profiles.

3.1.1 Practicing Flat (1F) Position Fillet Welds

Practice flat (1F) position fillet welds by welding multipass (six-pass) convex fillet welds in a T-joint. Use an instructor-specified size of flux-cored electrode wire with appropriate shielding gas as required. When making flat fillet welds, pay close attention to the work angle and travel speed. For the first bead, the work angle is vertical (45° to both plate surfaces). The angle is adjusted for all subsequent beads. Increase or decrease the travel speed to control the amount of weld metal buildup. *Figure 20* shows the work angles for 1F position fillet welds.

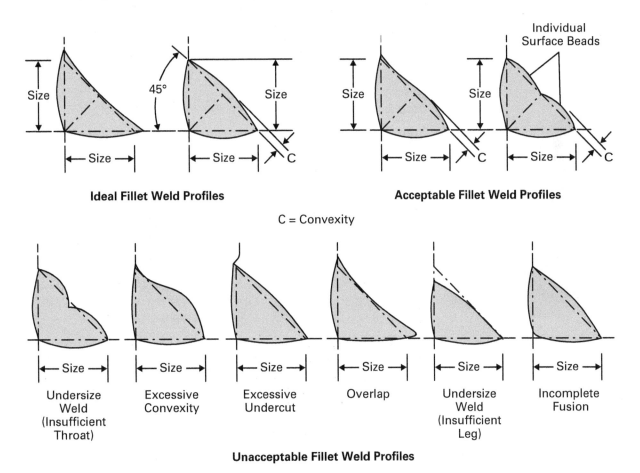

Figure 19 Ideal, acceptable, and unacceptable fillet weld profiles.

T-Joint Heat Dissipation

In T-joints, the welding heat dissipates more rapidly in the thicker or non-butting member. On various bead passes, the arc may have to be concentrated slightly more on the thicker or non-butting member to compensate for the heat loss.

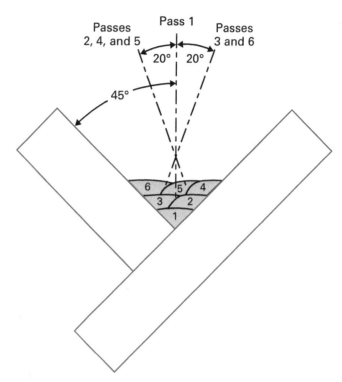

Figure 20 Multipass 1F bead sequence and work angles.

Follow these steps to make a flat fillet weld. Remember to clean all weld beads before beginning the next bead.

Step 1 Tack two plates together to form a T-joint for the fillet weld coupon (*Figure 21*). Clean the tack welds.

Step 2 Clamp or tack-weld the coupon in the flat position.

Step 3 Run the first bead along the root of the joint, using a work angle of 45° and a 5° to 10° drag angle. Use a slight side-to-side motion (oscillation).

Step 4 Run the second bead along a toe of the first weld, overlapping about 75% of the first bead. Alter the work angle as shown in *Figure 20* and use a 5° to 10° drag angle with a slight oscillation.

Step 5 Run the third bead along the other toe of the first weld, filling the groove created when the second bead was run. Use the work angle shown in *Figure 20* and a 5° to 10° drag angle with a slight oscillation.

Step 6 Run the fourth bead along the outside toe of the second weld, overlapping about half the second bead. Use the work angle shown in *Figure 20* and a 5° to 10° drag angle with a slight oscillation.

Step 7 Run the fifth bead along the inside toe of the fourth weld, overlapping about half the fourth bead that was run. Use the work angle shown in *Figure 20* and a 5° to 10° drag angle with a slight oscillation.

Step 8 Run the sixth bead along the toe of the fifth weld, filling the groove created when the fifth bead was run. Use the work angle shown in *Figure 20* and a 5° to 10° drag angle with a slight oscillation.

Figure 21 Fillet weld coupon.
Source: Holley Thomas, SME

Step 9 Inspect the weld. The weld is acceptable if it has the following features:

- Uniform rippled appearance of the bead face
- Craters and restarts filled to the full cross section of the weld
- Uniform weld size $\pm^1\!/_{16}$" (1.6 mm)
- Acceptable weld profile in accordance with the applicable code or standard
- Smooth, flat transition with complete fusion at the toes of the weld
- No excessive porosity with no pores larger than $^3\!/_{32}$" (2.38 mm)
- No undercut greater than $^1\!/_{32}$" (0.8 mm) deep or 10% of the base metal thickness, whichever is less
- No overlap
- No inclusions
- No cracks

3.1.2 Practicing Horizontal (2F) Position Fillet Welds

Practice horizontal (2F) fillet welding by placing multipass fillet welds in a T-joint using flux-cored electrodes and, if required, appropriate shielding gas as directed by the instructor. When making horizontal fillet welds, pay close attention to the gun angles and travel speed. For the first bead, the electrode work angle is 45°. The work angle is adjusted for all other welds. Increase or decrease the travel speed to control the amount of weld metal buildup.

Follow these steps to make a horizontal fillet weld:

Step 1 Tack two plates together to form a T-joint for the fillet weld coupon. Clean the tack welds.

Step 2 Clamp or tack-weld the coupon in the horizontal position.

Step 3 Run the first bead along the root of the joint using a work angle of approximately 45° with a 5° to 10° drag angle and a slight oscillation (*Figure 22*).

Step 4 Clean the weld.

Step 5 Run the remaining passes at the appropriate work angles using a 5° to 10° drag angle and a slight oscillation at the appropriate work angles. Overlap each previous pass and clean the weld after each pass.

Step 6 Have the instructor inspect the weld. The weld is acceptable if it has the following features:

- Uniform rippled appearance of the bead face
- Craters and restarts filled to the full cross section of the weld
- Uniform weld size $\pm^1\!/_{16}$" (1.6 mm)
- Acceptable weld profile in accordance with the applicable code or standard
- Smooth, flat transition with complete fusion at the toes of the weld
- No excessive porosity with no pores larger than $^3\!/_{32}$" (2.38 mm)
- No undercut greater than $^1\!/_{32}$" (0.8 mm) deep or 10% of the base metal thickness, whichever is less
- No overlap
- No inclusions
- No cracks

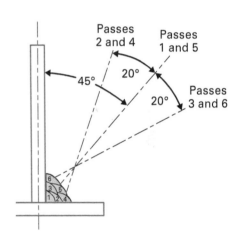

> **NOTE**
>
> While the goal is to produce a weld with no porosity or undercut, in some instances, welding students and instructors may use the absence of excessive porosity or undercut as their criteria for weld evaluation. In those instances, parameters for evaluation of porosity and undercut would be as follows: no single porosity would exceed $^3\!/_{32}$" (2.4 mm) diameter, and no undercut would be more than 10% of the thickness of the weld coupon or $^1\!/_{32}$" (0.8 mm).

Figure 22 Multipass 2F bead sequence and work angles.

Shielding Gas Blender

The wall-mounted Harris US-S10 gas blender shown in the image is designed to blend two gases from different cylinders. The mixture ratio is easily adjusted on the panel. Gases can be mixed and routed to a manifold, providing the same mixture to multiple welding stations. Another model is available that allows the blending of three different gases.

3.1.3 Practicing Vertical (3F) Position Fillet Welds

Practice vertical (3F) fillet welding by placing multipass fillet welds in a T-joint using flux-cored electrodes and, if required, appropriate shielding gas as directed by the instructor. Normally, vertical welds are made welding uphill (also called *vertical welding*) from the bottom to the top, using a gun push angle (up-angle). However, downhill welding can also be accomplished using a slight drag angle. When vertical welding, either stringer or weave beads can usually be used. However, weave beads are avoided on stainless steel or in downhill welding. On the job, the site WPS or quality standard will specify which technique to use.

When making vertical fillet welds, pay close attention to the gun angles and travel speed. Increase or decrease the travel speed to control the amount of weld metal buildup.

Follow these steps to make an uphill fillet weld with weave beads:

Step 1 Tack two plates together to form a T-joint for the fillet weld coupon.

Step 2 Clamp or tack-weld the coupon in the vertical position.

Step 3 Starting at the bottom, run the first bead along the root of the joint using a work angle of approximately 45° and a 10° to 15° push angle. Pause in the weld puddle to fill the crater. Ensure it's filled to the full cross section at the end of the pass.

Step 4 Clean the weld.

Step 5 Run the remaining passes using a 5° to 10° push angle and a side-to-side weave technique with a 45° work angle (*Figure 23*). Use a slow motion across the face of the weld, pausing at each toe for penetration and to fill the crater. Adjust the travel speed across the face of the weld to control the buildup. Overlap each previous pass as required and clean the weld after each pass.

Step 6 Have the instructor inspect the weld. The weld is acceptable if it has the following features:

- Uniform rippled appearance of the bead face
- Craters and restarts filled to the full cross section of the weld
- Uniform weld size $\pm^1/_{16}$" (1.6 mm)
- Acceptable weld profile in accordance with the applicable code or standard
- Smooth, flat transition with complete fusion at the toes of the weld
- No excessive porosity with no pores larger than $^3/_{32}$" (2.38 mm)
- No undercut greater than $^1/_{32}$" (0.8 mm) deep or 10% of the base metal thickness, whichever is less
- No overlap
- No inclusions
- No cracks

Repeat vertical fillet (3F) welding using stringer beads. Use a 0° to 10° push angle and the work angles shown in *Figure 23* as required. Depending on the electrode type, note that a drag angle may be required. Follow the manufacturer's guidance for the electrode type.

3.1.4 Practicing Overhead (4F) Position Fillet Welds

Practice overhead (4F) fillet welding by welding multipass fillet welds in a T-joint using flux-cored electrodes and, if required, appropriate shielding gas as directed by the instructor. When making overhead fillet welds, pay close attention to the gun angles and travel speed. For the first bead, the work angle is approximately 45°. The work angle is adjusted for all other welds. Increase or decrease the travel speed to control the amount of weld metal buildup.

Follow these steps to make an overhead fillet weld:

Step 1 Tack two plates together to form a T-joint for the fillet weld coupon.

Step 2 Clamp or tack-weld the coupon so it is in the overhead position.

Step 3 Run the first bead along the root of the joint, using a work angle of approximately 45° with a 5° to 10° drag angle. Use a slight oscillation to tie in the weld at the toes.

Step 4 Clean the weld.

Step 5 Using a slight oscillation, run the remaining passes using a 5° to 10° drag angle and the work angles shown in *Figure 24*. Overlap each previous pass and clean the weld after each pass.

Step 6 Have the instructor inspect the weld. The weld is acceptable if it has the following features:

- Uniform rippled appearance of the bead face
- Craters and restarts filled to the full cross section of the weld
- Uniform weld size $\pm^1/_{16}$" (1.6 mm)
- Acceptable weld profile in accordance with the applicable code or standard
- Smooth, flat transition with complete fusion at the toes of the weld
- No excessive porosity with no pores larger than $^3/_{32}$" (2.38 mm)
- No undercut greater than $^1/_{32}$" (0.8 mm) deep or 10% of the base metal thickness, whichever is less
- No overlap
- No inclusions
- No cracks

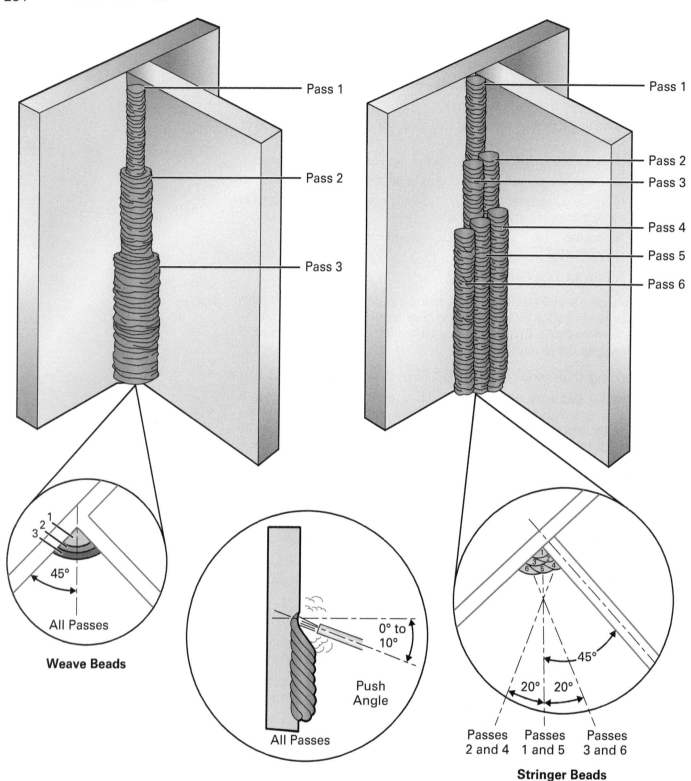

Pass 1
Pass 2
Pass 3

Pass 1
Pass 2
Pass 3
Pass 4
Pass 5
Pass 6

45°

All Passes

Weave Beads

0° to 10°

Push Angle

All Passes

45°

20° 20°

Passes 2 and 4   Passes 1 and 5   Passes 3 and 6

Stringer Beads

Figure 23 Multipass 3F bead sequences and work angles for weave and stringer beads.

3.2.0 Open V-Groove Welds

The open V-groove weld is a common groove weld normally made on plate and pipe. Practicing the open V-groove weld on plate will help with making the more difficult pipe welds covered in *Welding Level Three*.

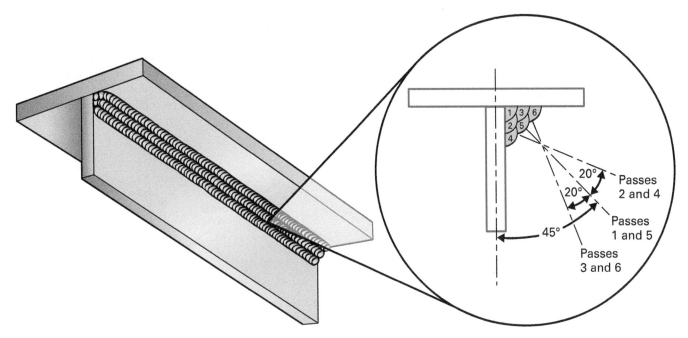

Figure 24 Multipass 4F bead sequence and work angles.

3.2.1 Root Pass

The most difficult part of making an open V-groove weld is the root pass. The root pass is made from the V-groove side of the joint and must have complete penetration but not an excessive amount of root reinforcement. The penetration is controlled with a side-to-side motion or oscillation.

After the root pass has been run, it should be cleaned and inspected. Inspect the face of the weld for excess buildup or undercut (*Figure 25*), which could trap slag if an FCAW root pass is run. Remove excess buildup or undercut with a hand grinder by grinding the face of the root pass with the edge of a grinding disk. Take care not to grind through the root pass.

NOTE

It's very important for you to know that open-root passes are not typically done with FCAW. Short-circuit GMAW, GTAW, or SMAW processes are typically used for root passes on open V-groove welds. For this module, you can use other processes for the root pass of an open V-groove weld or the FCAW process with a backing bar.

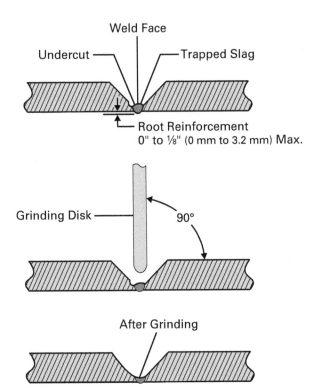

Figure 25 Grinding a root pass.

3.2.2 Groove Weld Positions

As shown in *Figure 26*, groove welds can be made in all positions. The weld position for plate is determined by the axis of the weld and the orientation of the workpiece. Groove weld positions for plate are flat (1G, where G stands for groove), horizontal (2G), vertical (3G), and overhead (4G).

In the 1G and 2G positions, the weld axis can be inclined up to 15°. Any weld axis inclination for the other positions varies with the rotational position of the weld face as specified in the WPS.

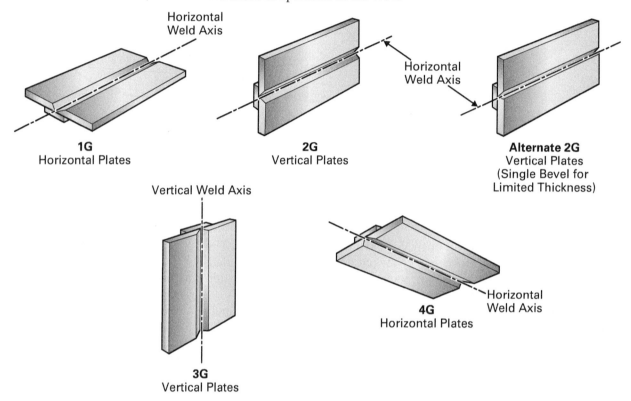

Figure 26 Groove weld positions for plate.

3.2.3 Acceptable and Unacceptable Groove Weld Profiles

Groove welds should be made with slight reinforcement and a gradual transition to the base metal at each toe. As shown in *Figure 27*, groove welds must not have excessive reinforcement, undercut, underfill, or overlap. If a groove weld has any of these defects, it must be repaired.

3.2.4 Practicing Flat (1G) Position V-Groove Welds

Practice flat (1G) V-groove welds, as shown in *Figure 28*. Use flux-cored electrodes and, if required, appropriate shielding gas as directed by the instructor. Use a stringer bead for the root pass, keeping the gun angle at 90° to the plate surface (0° work angle) with a 5° to 10° drag angle. For fill and cover passes, stringer or weave beads may be used. When using stringer beads, use a 5° to 10° drag angle, but adjust the work angle to tie in and fill the bead. For weave beads, also use a 5° to 10° drag angle, but keep the gun angle at 90° to the plate surface (0° work angle). Pay special attention at the termination of the weld to fill the crater.

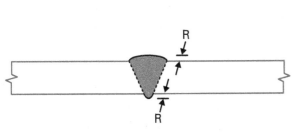

R = Face and Root Reinforcement
Not to Exceed ⅛" (3.2 mm), per Code

Acceptable Weld Profile

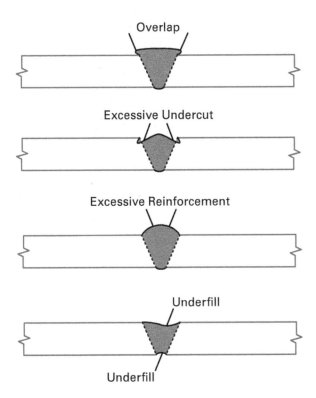

Overlap

Excessive Undercut

Excessive Reinforcement

Underfill

Underfill

Unacceptable Weld Profiles

Figure 27 Acceptable and unacceptable groove weld profiles.

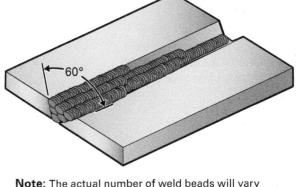

Note: The actual number of weld beads will vary depending on the plate thickness.

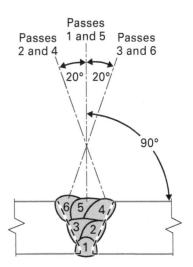

Passes
Passes 1 and 5 Passes
2 and 4 3 and 6
20° 20°
90°

Stringer Bead Sequence

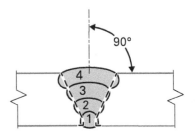

90°

Weave Bead Sequence

Figure 28 Multipass 1G bead sequences and work angles.

Follow these steps to practice open V-groove welds in the flat position:

Step 1 Tack-weld the practice coupon together, as explained earlier.

Step 2 Clamp or tack-weld the weld coupon in the flat position above the welding table surface.

Step 3 If using FCAW and a backing bar, run the root pass using a 5° to 10° drag angle and a slight side-to-side motion to control penetration. For an open-root V-groove weld, run the root pass using GTAW or SMAW before proceeding.

Step 4 Clean the root pass and grind if necessary (and if grinding is allowed).

Step 5 Run the remaining passes using a 5° to 10° drag angle and applicable work angles. Overlap each previous pass and clean the weld after each pass.

NOTE

Avoid using weave beads for stainless steel.

Root Pass Welding

In practice, open V-groove root passes are usually made on various metals using GTAW, SMAW, or GMAW methods. The fill and cover passes are then made using FCAW or GMAW.

Step 6 Inspect the weld. The weld is acceptable if it has the following:

- Uniform rippled appearance of the bead face
- Craters and restarts filled to the full cross section of the weld
- Uniform weld width $\pm\frac{1}{16}$" (1.6 mm)
- Acceptable weld profile in accordance with the applicable code or standard
- Smooth, flat transition with complete fusion at the toes of the weld
- Complete uniform root and face reinforcement at least flush with the base metal to a maximum buildup of $\frac{1}{8}$" (3.2 mm)
- No excessive porosity with no pores larger than $\frac{3}{32}$" (2.38 mm)
- No undercut greater than $\frac{1}{32}$" (0.8 mm) deep or 10% of the base metal thickness, whichever is less
- No inclusions
- No cracks

3.2.5 Practicing Horizontal (2G) Position V-Groove Welds

A horizontal (2G) V-groove weld is shown in *Figure 29*. Practice horizontal open V-groove welds using flux-cored electrodes and, if required, appropriate shielding gas as directed by the instructor. For the gun angles, use a 5° to 10° drag angle and adjust the work angle as required. Pay particular attention to filling the crater at the termination of the weld.

Follow these steps to practice welding open V-groove welds in the horizontal position:

Step 1 Tack-weld the practice coupon together, as explained earlier. Use the standard or alternate horizontal weld coupon as directed by the instructor.

Step 2 Tack-weld the flat plate weld coupon into the horizontal position.

Step 3 If using FCAW and a backing bar, run the root pass using a 5° to 10° drag angle and a slight side-to-side motion to control penetration. For an open-root V-groove weld, run the root pass using GTAW or SMAW before proceeding.

Step 4 Clean the weld.

Step 5 Run the remaining passes using a 5° to 10° drag angle and appropriate work angles. Overlap each previous pass and clean the weld after each pass.

Step 6 Inspect the weld. The weld is acceptable if it has the following:

- Uniform rippled appearance of the bead face
- Craters and restarts filled to the full cross section of the weld
- Uniform weld width $\pm\frac{1}{16}$" (1.6 mm)
- Acceptable weld profile in accordance with the applicable code or standard
- Smooth, flat transition with complete fusion at the toes of the weld
- Complete uniform root and face reinforcement at least flush with the base metal to a maximum buildup of $\frac{1}{8}$" (3.2 mm)
- No excessive porosity with no pores larger than $\frac{3}{32}$" (2.38 mm)
- No undercut greater than $\frac{1}{32}$" (0.8 mm) deep or 10% of the base metal thickness, whichever is less
- No inclusions
- No cracks

Note: The actual number of weld beads will vary depending on the plate thickness.

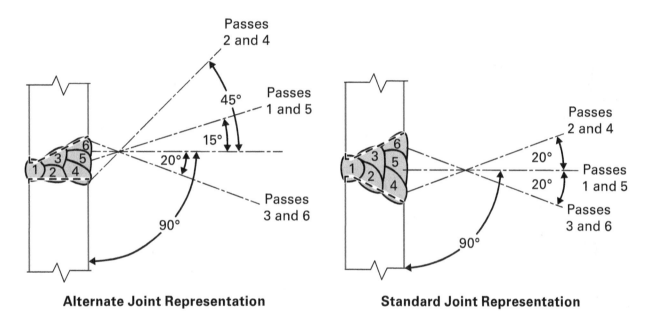

Alternate Joint Representation   **Standard Joint Representation**

Figure 29 Multipass 2G bead sequences and work angles.
Source: Terry Lowe (photo)

3.2.6 Practicing Vertical (3G) Position V-Groove Welds

Practice vertical (3G) multipass groove welds using stringer beads or weave beads. Set the welding parameters at the lower ends of the suggested ranges. Out-of-position welds are made at lower welding voltages and currents than flat or horizontal welds to prevent the molten weld metal from sagging or running.

When welding vertically, either stringer or weave beads can be used. However, weave beads are not normally used on stainless steel. On the job, the WPS or site quality standards will specify which technique to use. Check with the instructor to see if stringer beads, weave beads, or both techniques should be practiced. The root, fill, and cover beads can be run either uphill or downhill. However, downhill weave beads are not normally used. Use the direction preferred at the site or as specified in the WPS or site quality standards.

Run the root bead using a work angle of 0° (nozzle centerline bisects the V-groove). The work angle is adjusted for all other welds.

Practice 3G open V-groove welds using FCAW as directed by the instructor. Also, use flux-cored electrodes and appropriate shielding gas as directed by the instructor.

Note: The actual number of weld beads will vary with the metal thickness.

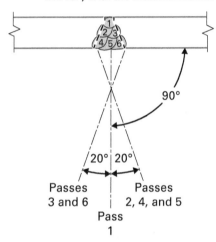

Figure 30 Multipass 3G bead sequence and work angles for stringer beads.
Source: Terry Lowe (photo)

Follow these steps to make a vertical groove weld with stringer beads:

Step 1 Tack-weld the practice coupon together, as explained earlier.

Step 2 Clamp or tack-weld the coupon in the vertical position.

Step 3 If using FCAW and a backing bar, run the root pass uphill with a work angle of 0° (nozzle centerline bisects the V-groove) and a 5° to 15° push angle. For an open-root V-groove weld, run the root pass using GTAW or SMAW before proceeding.

Step 4 Clean the weld.

Step 5 Run the remaining passes at the work angles shown in *Figure 30*, using a 5° to 15° push angle. Overlap each previous pass and use a slight side-to-side oscillation to tie in the welds at the toes. Clean the weld after each pass.

Step 6 Inspect the weld. The weld is acceptable if it has the following features:

- Uniform rippled appearance of the bead face
- Craters and restarts filled to the full cross section of the weld
- Uniform weld width $\pm^1/_{16}$" (1.6 mm)
- Acceptable weld profile in accordance with the applicable code or standard
- Smooth, flat transition with complete fusion at the toes of the weld
- Complete uniform root and face reinforcement at least flush with the base metal to a maximum buildup of $^1/_8$" (3.2 mm)
- No excessive porosity with no pores larger than $^3/_{32}$" (2.38 mm)
- No undercut greater than $^1/_{32}$" (0.8 mm) deep or 10% of the base metal thickness, whichever is less
- No inclusions
- No cracks

Follow these steps to make a vertical groove weld with weave beads:

Step 1 Tack-weld the practice coupon together, as explained earlier.

Step 2 Clamp or tack-weld the coupon in the vertical position.

Step 3 Run the root pass uphill with a work angle of 0° (90° from the plate) and a 5° to 15° push angle. Use a slight oscillation to control the penetration.

Step 4 Clean the weld.

Step 5 Run the remaining passes as uphill weave beads across the root bead at a 0° work angle. Clean the weld after each pass. Use a zigzag weave pattern across the root bead and pause at each toe for good fusion. Increase or decrease the travel speed to control the amount of metal buildup. *Figure 31* shows the zigzag pattern and bead sequence for weave beads.

Step 6 Inspect the weld. The weld is acceptable if it has the following features:

- Uniform rippled appearance of the bead face
- Craters and restarts filled to the full cross section of the weld
- Uniform weld width $\pm \frac{1}{8}$" (3.2 mm)
- Acceptable weld profile in accordance with the applicable code or standard
- Smooth, flat transition with complete fusion at the toes of the weld
- Complete uniform root and face reinforcement at least flush with the base metal to a maximum buildup of $\frac{1}{8}$" (3.2 mm)
- No excessive porosity with no pores larger than $\frac{3}{32}$" (2.38 mm)
- No undercut greater than $\frac{1}{32}$" (0.8 mm) deep or 10% of the base metal thickness, whichever is less
- No inclusions
- No cracks

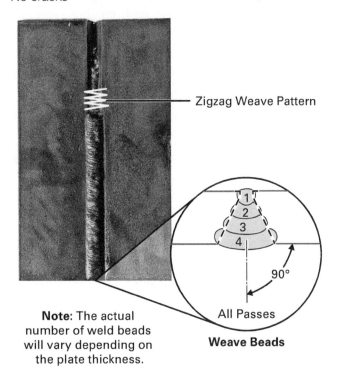

Note: The actual number of weld beads will vary depending on the plate thickness.

Figure 31 Multipass 3G bead sequence and work angle for weave beads.
Source: Terry Lowe (photo)

3.2.7 Practicing Overhead (4G) Position V-Groove Welds

Practice overhead (4G) V-groove welds using FCAW as directed by the instructor. Use flux-cored electrodes and, if required, appropriate shielding gas as directed by the instructor. To reduce heat and prevent the molten weld metal from sagging or running, FCAW overhead welds are made with lower welding voltages and currents than flat or horizontal welds. Use a work angle of 0° (90° to plate) and a 5° to 15° drag angle for the root bead.

Follow these steps to make an overhead groove weld with stringer beads:

Step 1 Tack-weld the practice coupon together, as explained earlier.

Step 2 Clamp or tack-weld the coupon in the overhead position.

Step 3 Run the root pass, using a work angle of 0° and a 5° to 15° drag angle. Use a slight oscillation to control the penetration.

Step 4 Clean the weld.

Step 5 Run the remaining passes at the work angles shown in *Figure 32* using a 5° to 10° drag angle. Overlap each previous pass and use a slight side-to-side motion to tie in the welds at the toes. Each bead should be cleaned before running the next bead.

Step 6 Inspect the weld. The weld is acceptable if it has the following features:

- Uniform rippled appearance of the bead face
- Craters and restarts filled to the full cross section of the weld
- Uniform weld width $\pm^1\!/_{16}$" (1.6 mm)
- Acceptable weld profile in accordance with the applicable code or standard
- Smooth, flat transition with complete fusion at the toes of the weld
- Complete uniform root and face reinforcement at least flush with the base metal to a maximum buildup of $^1\!/_8$" (3.2 mm)
- No excessive porosity with no pores larger than $^3\!/_{32}$" (2.38 mm)
- No undercut greater than $^1\!/_{32}$" (0.8 mm) deep or 10% of the base metal thickness, whichever is less
- No inclusions
- No cracks

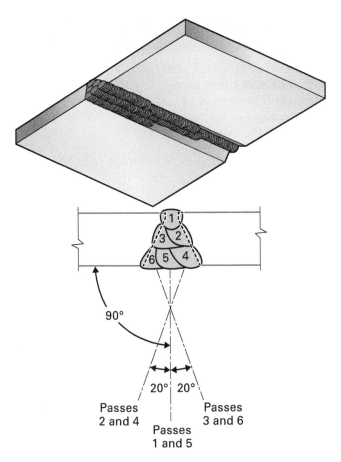

Stringer Bead Sequence

Figure 32 Multipass 4G bead sequence and work angles.

3.0.0 Section Review

1. When flat (1F) fillet welds are being made, the travel speed can be increased or decreased to control the amount of _____.
   a. base metal fusion
   b. weld metal buildup
   c. root penetration
   d. flux-core shielding

2. A 4G weld is a(n) _____.
   a. vertical plate weld
   b. overhead groove weld
   c. flat groove weld
   d. horizontal plate weld

Module 29210   Review Questions

1. The filler metal in a FCAW process is provided by _____.
   a. individual rods, or fillets, of steel
   b. a continuous flux-cored electrode
   c. the expansion of molten base metal
   d. the material in the flux-cored wiring

2. The lenses used in safety glasses, face shields, or helmets worn during FCAW procedures *must* be _____.
   a. made of Plexiglas
   b. properly tinted
   c. spatter resistant
   d. opaque and blued

3. A gas that is commonly used to purge areas or containers prior to FCAW operations is _____.
   a. nitrogen
   b. propane
   c. oxygen
   d. acetylene

4. If possible, practice weld coupons for FCAW should be _____.
   a. $\frac{1}{4}$" (6.4 mm) thick stainless steel
   b. $\frac{3}{8}$" (10 mm) thick carbon steel
   c. $\frac{3}{4}$" (19.1 mm) thick low-alloy plate
   d. $\frac{7}{8}$" (22.2 mm) thick composite steel

5. Setting up a FCAW-G machine requires connecting the proper shielding gas for the application as well as _____.
   a. locking out and tagging out the primary disconnect
   b. removing any liner material from the feeder and gun
   c. attaching the ground lead to the gun handle
   d. loading the wire feeder and gun with the proper electrode

6. After a welder has made the initial recommended setting, the welding amperage of an FCAW system is typically controlled by the _____.
   a. wire feed speed control
   b. arc length and porosity
   c. shielding gas flow regulator
   d. gun position and travel

7. When weave beads are made, the motion of the electrode is _____.
   a. quickly forward and backward
   b. slightly side-to-side
   c. widely side-to-side
   d. slowly circular and forward

8. An improperly made FCAW restart will create a _____.
   a. bead overlap
   b. weave bead
   c. stringer bead
   d. weld discontinuity

9. Unless the proper technique is used, a weld termination normally leaves _____.
   a. a crack
   b. porosity
   c. spatter
   d. a crater

10. Weld beads used to build up a surface and make multipass welds are known as _____.
    a. fillet beads
    b. layer beads
    c. crater beads
    d. overlapping beads

11. The most common fillet welds are made in _____.
    a. horizontal piping
    b. root passes
    c. lap and T-joints
    d. vertical piping

12. What work angle should be used for the first bead in a horizontal (2F) fillet weld?
    a. 25°
    b. 45°
    c. 60°
    d. 90°

13. With FCAW, groove welds can be made in _____.
    a. only the 1G and 2G positions
    b. only the 2G and 3G positions
    c. only the 1G, 2G, and 3G positions
    d. all positions

14. A flat (1G), open V-groove weld is acceptable if it has complete fusion at the toes of the weld and a smooth, flat _____.
    a. cross section
    b. transition
    c. overlap
    d. crater

15. When making a 3G open V-groove weld with weave beads, use a zigzag pattern across the root bead and pause at each toe _____.
    a. for complete penetration
    b. to prevent excessive spatter
    c. for good fusion
    d. to fill in the crater

Answers to odd-numbered questions are found in the Review Question Answer Keys at the back of this book.

Answers to Section Review Questions

Answer	Section Reference	Objective
Section 1.0.0		
1. d	1.1.2	1a
2. a	1.2.1	1b
3. b	1.3.1	1c
Section 2.0.0		
1. c	2.1.1	2a
2. d	2.2.2	2b
Section 3.0.0		
1. b	3.1.1	3a
2. b	3.2.2	3b

MODULE 29207

GTAW – Equipment and Filler Metals

Source: Photo courtesy of The Lincoln Electric Company, Cleveland, OH, U.S.A.

Objectives

Successful completion of this module prepares you to do the following:

1. Identify GTAW-related safety practices and describe the electrical characteristics that affect GTAW.
   a. Identify GTAW-related safety practices.
   b. Describe the electrical characteristics that affect GTAW.
2. Identify and describe GTAW equipment and consumables.
   a. Identify and describe GTAW welding machines.
   b. Identify and describe GTAW torches.
   c. Identify and describe GTAW torch nozzles and electrodes.
   d. Identify and describe GTAW shielding gases.
   e. Identify and describe GTAW filler metals.
3. Explain how to set up for GTAW welding.
   a. Explain how to select and position the welding machine.
   b. Explain how to connect and set up the shielding gas flow rate.
   c. Explain how to select and prepare the tungsten electrode.
   d. Explain how to select and install the nozzle along with the tungsten electrode.

Performance Tasks

Under supervision, you should be able to do the following:

1. Select a GTAW shielding gas.
2. Select a GTAW filler metal.
3. Connect the shielding gas and set the flow rate.
4. Select and prepare the tungsten electrode.
5. Break down and reassemble a GTAW torch.

Overview

To produce consistent, high-quality welds, welders must select, install, and use the equipment and materials involved in the gas tungsten arc welding (GTAW) process properly. This module provides an overview of the equipment and safety concerns that are associated with GTAW. It describes GTAW equipment and consumables, including power sources, torches, nozzles, electrodes, shielding gases, and filler metals. It also explains how to select, prepare, and set up GTAW equipment safely and efficiently.

Digital Resources for Welding

Scan this code using the camera on your phone or mobile device to view the digital resources related to this craft.

SCAN ME

1.0.0 GTAW

Performance Tasks

There are no Performance Tasks in this section.

Objective

Identify GTAW-related safety practices and describe the electrical characteristics that affect GTAW.

a. Identify GTAW-related safety practices.

b. Describe the electrical characteristics that affect GTAW.

Tungsten: A nonstandard common name for the tungsten electrode used in GTAW; also, the principal metal from which the electrodes are constructed.

Gas **tungsten** arc welding (GTAW) is an arc welding process that uses a welding power source to produce an electric arc between a tungsten electrode and the base metal. The arc melts and fuses the base metal to form the weld. Because tungsten is a nonconsumable electrode, compared to shielded metal arc welding (SMAW) and gas metal arc welding (GMAW) / flux-cored arc welding (FCAW), a filler metal or rod is typically added. Welding processes where a filler metal is not added are referred to as *autogenous welding*.

In manual welding (*Figure 1*), the filler metal is usually a handheld metal rod or wire that is fed into the leading edge of the weld metal pool. An inert gas or gas mixture is used to shield the electrode and molten weld metals to prevent oxidation and contamination from the atmosphere. Also, since no filler metal is transported within the arc, the process produces little or no spatter.

Torch

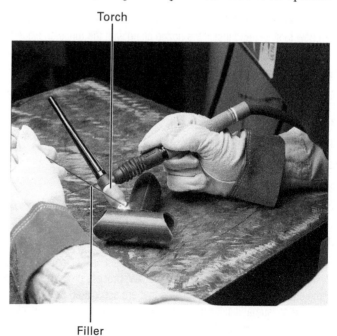

Filler
Metal

Figure 1 Performing manual GTAW.
Source: Photo courtesy of The Lincoln Electric Company, Cleveland, OH, U.S.A.

GTAW is used to weld a variety of metals. The process is slower than GMAW but produces high-quality welds. GTAW has been commonly used for root pass welds on pipe where the subsequent fill and cover passes are performed with SMAW, GMAW, or FCAW. In the past, GTAW was called *tungsten inert gas (TIG) welding*. Today, the nonstandard term *TIG* is still used, including in the control nomenclature marked on some welding machines.

1.1.0 Safety Summary

The following sections summarize the safety procedures and practices that welders must observe when cutting or welding. Complete safety coverage is provided in NCCER Module 29101, *Welding Safety*, which should be completed before continuing with this module. The safety procedures and practices are in place to prevent potentially life-threatening injuries from welding accidents. Above all, be sure to wear appropriate protective clothing and equipment when welding or cutting.

1.1.1 Protective Clothing and Equipment

Welding activities can cause injuries unless you wear all the protective clothing and equipment designed specifically for the welding industry. The following safety guidelines about protective clothing and equipment should be followed to prevent injuries:

- Wear a face shield over snug-fitting cutting goggles or safety glasses for gas welding or cutting. The welding hood should be equipped with an approved shade for the application. A welding hood equipped with a properly tinted lens is best for all forms of welding.

- Wear proper protective leather and/or flame-retardant clothing along with welding gloves that protect the welder from flying sparks, molten metal, and heat.

- Wear high-top safety shoes or boots. Make sure the tongue and lace area of the footwear will be covered by a pant leg. If the tongue and lace area is exposed or the footwear must be protected from burn marks, wear leather spats under the pants or chaps over the front and top of the footwear.

- Wear a 100% cotton cap with no mesh material included in its construction. The bill of the cap points to the rear. If a hard hat is required for the environment, use one that allows the attachment of rear deflector material and a face shield. A hard hat with a rear deflector is generally preferred when working overhead and may be required by some employers and jobsites.

- Wear earmuffs, or at least earplugs, to protect your ear canals from sparks.

WARNING!

Using proper personal protective equipment (PPE) for the hands and eyes is particularly important. The most common injuries that welders experience during GTAW operations are injuries to the fingers and eyes. When they are holding and grinding the tungsten electrode by hand, welders who fail to use proper gloves, safety glasses, and face shields can cut or burn their fingers, have splintered tungsten electrode lodge in their hands or fingers, or get small slivers of the electrode stuck in their eyes.

1.1.2 Fire/Explosion Prevention

Welding activities usually involve the use of fire or extreme heat to melt metal. Whenever fire is used in a weld, the fire must be controlled and contained. Welding or cutting activities are often performed on vessels that may have once contained flammable or explosive materials. Residues from those materials can catch fire or explode when a welder begins work on such a vessel. The following fire and explosion prevention guidelines associated with welding contribute to a safe work zone:

- Never carry matches or gas-filled lighters in your pockets. Sparks can cause the matches to ignite or the lighter to explode, causing serious injury.

- Never use oxygen to blow dust or dirt from clothing. The oxygen can remain trapped in the fabric for a time. If a spark hits clothing during this time, the clothing can burn rapidly and violently out of control.

- Make sure any flammable material in the work area is moved or shielded by a fire-resistant covering. Approved fire extinguishers must be available before attempting any heating, welding, or cutting operations. If a hot work permit and a fire watch are required, be sure those items are in place before beginning and that all site requirements have been observed.

- Never release a large amount of oxygen or use oxygen as compressed air. The presence of oxygen around flammable materials or sparks can cause rapid and uncontrolled combustion. Keep oxygen away from oil, grease, and other petroleum products.

- Never release a large amount of fuel gas, especially acetylene. Propane tends to concentrate in and along low areas and can ignite at a considerable distance from the release point. Acetylene is lighter than air but is even more dangerous than propane. When mixed with air or oxygen, acetylene will explode at much lower concentrations than any other fuel.

- To prevent fires, maintain a neat and clean work area, and make sure that any metal scrap or slag is cold before disposing of it.

- Before cutting containers such as tanks or barrels, check to see if they have contained any explosive, hazardous, or flammable materials, including petroleum products, citrus products, or chemicals that decompose into toxic fumes when heated. As a standard practice, always clean and then fill any tanks or barrels with water or purge them with a flow of inert gas such as nitrogen to displace any oxygen. Containers must be cleaned by steam cleaning, flushing with water, or washing with detergent until all traces of the material have been removed.

WARNING!

Welding or cutting must never be performed on drums, barrels, tanks, vessels, or other containers until they have been emptied and cleaned thoroughly, eliminating all flammable materials and all substances (such as detergents, solvents, greases, tars, or acids) that might produce flammable, toxic, or explosive vapors when heated. Clean containers only in well-ventilated areas, as vapors can accumulate during cleaning, causing explosions or injury.

Proper procedures for cutting or welding hazardous containers are described in the *American Welding Society (AWS) F4.1, Safe Practices for the Preparation of Containers and Piping for Welding, Cutting, and Allied Processes.*

1.1.3 Work Area Ventilation

Vapors and fumes tend to rise in the air from their sources. Welders must often work above the welding area where the fumes are being created. Welding fumes can cause personal injuries including long-term respiratory harm. Good work area ventilation helps to remove the vapors and protect the welder. The following is a list of work area ventilation guidelines to consider before and during welding activities:

- Follow confined space procedures before conducting any welding or cutting in a confined space.
- Never use oxygen for ventilation in confined spaces.
- Always perform cutting or welding operations in a well-ventilated area. Cutting or welding operations involving zinc or cadmium materials or coatings result in toxic fumes. For long-term cutting or welding of such materials, always wear an approved full-face, supplied-air respirator that uses breathing air supplied from outside of the work area. For occasional, very short-term exposure, a high-efficiency particulate arresting-rated (also called HEPA-rated) or metal fume filter may be used on a standard respirator.
- Make sure confined spaces are properly ventilated for cutting or welding purposes. Use powered extraction systems when available.

> **WARNING!**
>
> Backup flux may contain silica, fluorides, or other toxic materials. Alcohol and acetone used for the mixture are flammable. Weld only in well-ventilated spaces. Failure to provide proper ventilation could result in personal injury or death.

1.1.4 GTAW-Specific Safety

GTAW, also referred to as *Heliarc* or TIG welding, is a different process than those used in SMAW, GMAW, and FCAW. In other forms of welding, the electrode or wire is consumed in the welding process. However, in GTAW, a separate rod of filler metal is consumed rather than the electrode.

Vapors from the melted base metal, filler metal, and shielding gases used in the GTAW process can create respiratory issues. As always, welders performing any kind of work must be aware of the vapors generated from the welding processes. Argon gas is often used for shielding in the GTAW process. A combination of the arc and gas generates substantial levels of ultraviolet (UV) light, which reacts with the oxygen near the weld. UV radiation can ionize oxygen in the atmosphere near the weld, which can lead to a rise in ozone and nitrogen dioxide levels. If the work is being performed in well-ventilated areas, the risks are minimal. If work must be performed in confined spaces, consider using additional ventilation or respirators. Follow your employer's safety policies in these situations.

Because GTAW work is often performed on delicate welds, the welder is frequently positioned very close to the work. The risk of UV radiation being emitted from the electrical arc is higher because the shielding gas used with GTAW is more transparent than gases used with other forms of welding. This transparency is due to the minimal smoke produced from the weld. Because of the brightness of the GTAW process, safety representatives recommend that the welder wear darker clothing to minimize UV reflections under the welding helmet. They also recommend that the walls of the work area be painted with a paint that reduces reflections from the UV light generated by the GTAW process. Because aluminum and stainless steel reflect more light than common carbon steels, welders need to take extra precautions to protect their vision when performing GTAW work on these metals.

The GTAW process requires the base metal and filler metal to be extremely clean. Cleaning solvents are used to prepare the metals. Welders must read all Safety Data Sheet (SDS) information about any cleaning agent before using it. Some cleaning agents are flammable, while others are nonflammable. Make sure you know which kind you are using. Pay special attention to any storage requirements for such cleaning agents.

> **WARNING!**

Chlorinated cleaning solvents can produce toxic gases. For this reason, they should never be used for cleaning metals in a welding setting.

The tungsten electrodes used for GTAW become contaminated over time and with use. When electrodes lose their original shapes, welders must grind the tungsten tips to reshape them and make them suitable for continued use. The tungsten dust and flying particles from these grinding activities can damage the welder's vision and present a respiratory hazard. Read and understand the SDS that comes with each batch of electrodes to protect your health in the workplace.

> **WARNING!**

Observe proper respiratory, hazard, and environmental contamination procedures when using a grinder. Dust from tungsten electrode grinding can be hazardous. Refer to the manufacturer's SDS/MSDS to determine any specific hazards. Safety glasses must be worn when using a grinder.

1.2.0 Welding Current

The welding arc required for GTAW is produced by electrical current from a constant-current (CC) welding machine. Electrical current is the flow of electrons along a conductor. The current flows through two welding cables, one connected to the workpiece and the other connected to the torch, which holds a tungsten electrode.

An arc is established in the gap between the end of the tungsten electrode and the work. The gap has a high electrical resistance. When the electrical current meets the resistance, heat is generated. When the arc is struck across the gap, the resistance generates intense heat of 6,000°F to 10,000°F (3,300°C to 5,500°C), melting the base metal and forming a weld. The area surrounding the arc is flooded with a shielding gas to prevent oxidation of the tungsten electrode, filler rod, and base metal surface during the welding process. *Figure 2* shows the GTAW welding process.

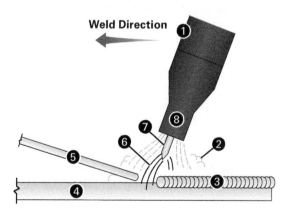

Figure 2 Diagram of the GTAW welding process.

1.2.1 Characteristics of Welding Current

The current produced by a welding machine to perform welding has different characteristics than the current flowing through utility power lines. Welding current has low voltage and high amperage, while the power line current has high voltage and low amperage.

Voltage is the measure of the electromotive force or pressure that causes current to flow in a circuit. There are two voltage levels associated with welding current: *open-circuit voltage* and *operating voltage*. Open-circuit voltage is the voltage that is present when the machine is on but no welding is taking place. For GTAW, there are usually ranges of open-circuit voltages that can be selected, up to about 80 V. Operating voltage, or *arc voltage*, is the voltage after the arc has been struck. With GTAW, arc voltage is generally much lower than the open-circuit voltage. The arc voltage drops rapidly with increasing current flow.

Amperage is the electrical current flow in a circuit. The unit of measurement for amperage is the ampere, or amp. The number of amps produced by the welding machine determines the intensity of the arc and the amount of heat generated. The higher the amperage, the hotter the arc. Usually, welders use a finger control or a foot pedal to control the GTAW welding machine current.

1.2.2 DC Welding Current

Direct current (DC) is electrical current that flows in one direction only. The direction in which the current is flowing is called the *polarity*.

The direction in which the current flows in a circuit determines polarity. Direct current electrode negative (DCEN), also referred to by the nonstandard term *straight polarity*, is almost always used for performing GTAW on ferrous and nonferrous metals.

When the welding circuit is set up as DCEN, the tungsten electrode of the GTAW torch is connected to the negative terminal of the welding machine and the workpiece lead is connected to the positive terminal. DCEN produces the greatest amount of heat in the base metal, compared to other types of welding current. About two-thirds of the arc heat goes to the base metal and one-third goes to the tungsten electrode. Welders usually use a tungsten electrode with a pointed end that is slightly **truncated**, or blunted, for DCEN. The weld bead produced is narrow but with deep penetration. DCEN is used to weld steels, stainless steels, nickel, and titanium. Because DCEN provides limited cleaning action, the base metal surfaces must first be thoroughly cleaned of any contaminants such as rust, mill scale, or oxides.

Direct current electrode positive (DCEP), also referred to by the nonstandard term *reverse polarity*, produces the least amount of heat in the base metal. About one-third of the arc heat goes to the base metal and two-thirds goes to the tungsten electrode. Because of this, DCEP usually requires a large tungsten electrode with a rounded end. Unlike DCEN, the weld bead produced is wide but with shallow penetration. This current type also has a strong cleaning action on the surface of oxidized base metals. DCEP is only used for special applications in which shallow penetration is required. *Figure 3* shows typical DCEN and DCEP welding circuits.

Truncated: Having the pointed top (vertex) of a conical shape cut off, leaving only a plane section parallel to the base of the cone.

Polarity Switches

Many welding machines have a polarity switch. Instead of having to physically disconnect the welding leads and reconnect them to change polarity, the welder changes polarity by turning a switch.

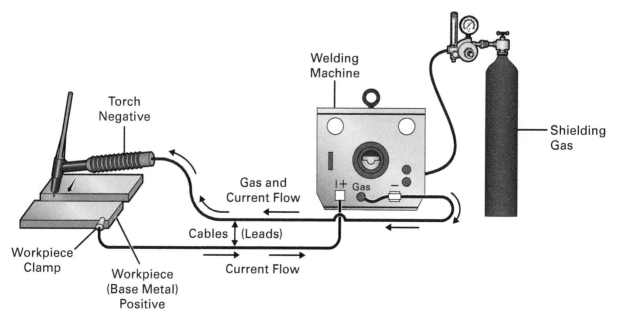

Direct Current Electrode Negative (DCEN)
(Recommended)

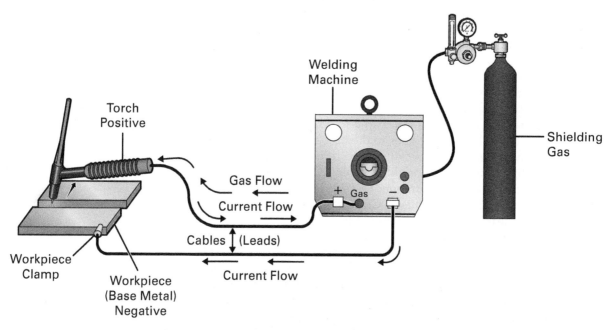

Direct Current Electrode Positive (DCEP)

Figure 3 Representative DCEN and DCEP welding circuits.

1.2.3 AC Welding Current

Alternating current (AC) is electrical current that alternates between positive and negative values. In the positive half of the cycle, the current flows in one direction; during the negative half, the current reverses itself. The number of cycles completed in one second is called the *frequency*. In the United States, AC is almost always 60 cycles per second, or 60 Hertz (Hz). In other countries, the frequency is often 50 Hz. *Figure 4* shows one cycle of AC.

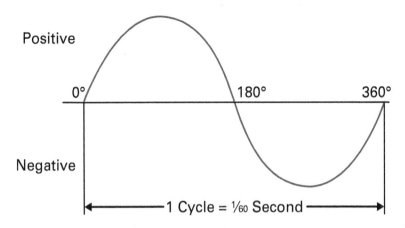

Figure 4 AC cycle.

AC is DCEN for half the time and DCEP for the remainder of the time. For this reason, AC characteristics fall in between those of DCEN and DCEP. About half of the heat goes to the base metal and the other half to the tungsten electrode. The weld bead size and penetration are midway between the beads produced by DCEN and DCEP. Some cleaning action does occur with AC. The AC tungsten electrode has a small, rounded (balled) tip. Newer welding machines with variable AC frequency controls do not require a balled tungsten tip. AC is used with GTAW to weld aluminum, magnesium, and their alloys.

1.2.4 High Frequency

High frequency is an electrical current with a very high voltage (3,000V to 5,000V) and very low amperage. It is called high frequency because it cycles at millions of times per second compared to standard current, which has a frequency of 60 Hz. This high-voltage, high-frequency current creates a very stable arc that can jump a gap of about $\frac{1}{2}$" (13 mm). Because high-frequency current has such low amperage, it creates very little heat.

High frequency has two purposes: to stabilize AC, and to allow a DC arc to be started without touching the tungsten electrode to the work. The high voltage allows it to jump the gap between the two.

When AC is used with GTAW, each time the arc cycles, it goes out and must be reestablished in the opposite direction. Even though this happens very fast (60 times per second), it can cause the arc to extinguish on the reverse polarity part of the AC cycle when the electron flow is weakest. If this happens, oxides are absorbed into the weld, causing porosity and weakening the weld. To prevent the AC arc from extinguishing, a high-frequency current is added to the alternating welding current. The high-frequency current, with its high voltage and low amperage, establishes a path on which the alternating welding current can travel, preventing the arc from extinguishing as it cycles. When non-square-wave AC is being used, the high frequency must be on continuously.

When using DC for GTAW without high frequency, welders must physically touch the tungsten to the work to establish the arc. Except for newer GTAW machines with **lift-arc technology**, this may not be desirable since it can cause contamination of the electrode and defects in the weld. To prevent this, high-frequency current can be added to the direct welding current. When the tungsten electrode is brought close to the work, the high-voltage, high-frequency arc jumps the gap, establishing a path for the direct welding current to travel without touching the tungsten to the work. Typically, a timer will automatically shut off the high frequency about two to three seconds after the welding arc has been established. Since DC flows in only one direction, it is very stable. The high voltage and frequency is only needed for starting the arc.

Lift-arc technology: A type of GTAW technology in which the tungsten electrode may be touched to the work, but the arc will not strike until the tip is lifted away to an appropriate distance.

Welding power sources specifically designed for GTAW have a high-frequency generator built into them. If conventional power sources are used for GTAW, a separate auxiliary high-frequency generator can be attached if required. The generators require 120 volts of alternating current (VAC) to operate. Some auxiliary high-frequency generators only have high-frequency controls; others may have a combination of high-frequency controls and gas and cooling water controls. Most will have a torch switch that plugs into the console. The switch attaches to the torch and can be used to turn the power, shielding gas, and cooling water (if used) on or off. The auxiliary high-frequency generator is connected to the power source by the electrode cable. The GTAW torch is connected to the auxiliary high-frequency generator (*Figure 5*).

Starting DC GTAW without High Frequency

DC GTAW is often performed without a high-frequency start for noncritical welds. The arc is touch-started in the weld zone or on a piece of copper. Copper is less likely to contaminate the tungsten. Some machines have a built-in selectable no high-frequency start mode.

**Remote-Controlled Arc Start
Switch on GTAW Torch**

**GTAW Welding
Machine**

**GTAW Machine on Top of
High-Frequency Generator**

Figure 5 Auxiliary high-frequency generator with a GTAW torch.
Source: Photo courtesy of The Lincoln Electric Company, Cleveland, OH, U.S.A.

High-Frequency Operation Precaution

When operating most transformer and transformer-rectifier machines in SMAW mode, welders must turn off the high-frequency generators.

1.0.0 Section Review

1. When performing long-term cutting or welding of zinc or cadmium materials or coatings, *always* _____.
   a. ventilate the area with a continuous oxygen supply
   b. connect the welding nozzle to a nitrogen purge
   c. wear an approved full-face, supplied-air respirator
   d. use a water-cooled, multi-flame (rosebud) torch

2. DCEN produces the *greatest* amount of heat in the _____.
   a. base metal
   b. filler metal
   c. electrode
   d. torch

2.0.0 GTAW Equipment

Objective

Identify and describe GTAW equipment and consumables.

   a. Identify and describe GTAW welding machines.
   b. Identify and describe GTAW torches.
   c. Identify and describe GTAW torch nozzles and electrodes.
   d. Identify and describe GTAW shielding gases.
   e. Identify and describe GTAW filler metals.

Performance Tasks

1. Select a GTAW shielding gas.
2. Select a GTAW filler metal.
5. Break down and reassemble a GTAW torch.

GTAW operations involve the use of multiple different types of equipment and consumables. Examples include GTAW welding machines, torches, nozzles, electrodes, shielding gases, and filler metals.

2.1.0 Welding Power Source Types

Welders use CC welding machines to perform GTAW. The type of welding current (DCEN, DCEP) depends on the type and thickness of the base metal.

Many types of welding machines may supply welding current for GTAW. Usually, any welding machine that produces constant current, such as those used for SMAW, can be used. In addition, there are welding machines specially designed for GTAW. These special machines have built-in high frequency and controls for shielding gas and cooling water. Types of welding machines that can be used for GTAW include the following:

- Transformer welding machines
- Transformer-rectifier welding machines
- Inverter welding machines
- Engine-driven generator and alternator welding machines

DC welding machines are usually designed to produce either constant-current or constant-voltage direct welding current. Constant current means that welding current is produced over a wide voltage range regardless of arc length. This is typical of a SMAW or GTAW welding machine. A constant-voltage machine maintains a constant voltage as the output current varies. This is typical of a GMAW or FCAW welding machine. The output voltages and output currents of a welding machine can be plotted on a graph to form a curve. These curves show how the output voltage relates to the output current as either of these factors changes (*Figure 6*).

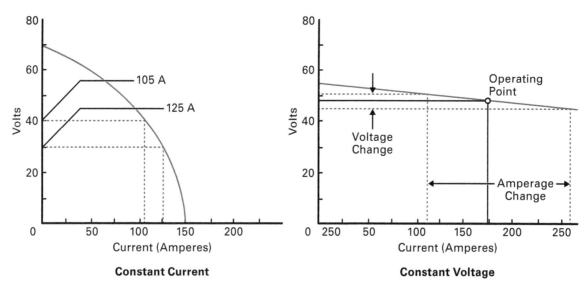

Figure 6 Constant-current and constant-voltage output curves.

2.1.1 Characteristics of Welding Current

The current produced by a welding machine has different characteristics from the current flowing through utility power lines. Welding current has low voltage and high amperage, while the power line current has high voltage and low amperage.

Voltage is the measure of the electromotive force or pressure that causes current to flow in a circuit. The specific unit of measure is the volt (V). There are two types of voltage associated with welding current: open-circuit voltage and operating voltage. Open-circuit voltage is the voltage present when the machine is on, but no welding is being done. There are usually ranges of open-circuit voltages that can be selected up to about 40 V. Operating voltage, or arc voltage, is the voltage after the arc has been struck.

Arc voltage is generally slightly lower than the open-circuit voltage. The arc voltage is typically 2 V to 3 V lower than the open-circuit voltage for every 100 amperes (A) of current, but it depends on the range selected.

Amperage is the measure of electric current flow in a circuit. The specific unit of measure for amperage is the ampere (A), commonly abbreviated to amp.

In welding, the current flows in a closed loop through two welding cables; the workpiece lead connects the power source to the base metal, and the other cable connects the power source to the wire electrode. During welding, an arc is established between the end of the wire electrode and the workpiece. The arc generates intense heat at 6,000°F to 10,000°F (3,316°C to 5,538°C), melting the base metal and the wire electrode to form the weld. The amount of amperage produced by the welding machine determines the intensity of the arc and the amount of heat available to melt the workpiece and the electrode.

2.1.2 Transformer Welding Machines

Transformer welding machines produce alternating welding current only. They use a voltage step-down transformer, which converts high-voltage, low-amperage current from commercial power lines to low-voltage, high-amperage welding current. The primary electrical power input required for a transformer welder can be 120 VAC, 208/230 VAC single or three phase, or 460 VAC three phase. Special light-duty transformer welding machines used for sheet metal work are designed to be plugged into a 120 VAC outlet. However, most light-duty transformer welding machines require 230 VAC primary current. Heavy-duty industrial transformer welders require 460 VAC three-phase primary current.

Transformer welders are not as common as other types of welding machines on the jobsite, but they are used for special jobs. A transformer welder has an On/Off switch, amperage control, and terminals for connecting the electrode lead and the workpiece lead. Transformer welding machines used for GTAW must have high-frequency capabilities built in or they must use an auxiliary high-frequency generator.

2.1.3 Transformer-Rectifier Welding Machines

The transformer-rectifier welding machine uses a transformer to convert the primary current to welding current and a rectifier to change the current from AC to DC. Transformer-rectifier welding machines can be designed to produce either AC and DC or DC only. Transformer-rectifiers that produce both AC and DC are usually lighter duty than those that produce DC only. Transformer-rectifier welding machines that produce DC only are sometimes called rectifier power sources or just rectifiers. Depending on the size, transformer-rectifier welding machines may require 120/240 VAC single-phase input power, 208/230/240 VAC three-phase input power, or 480 VAC three-phase input power.

The welding cables (electrode and workpiece) are connected to terminals marked Electrode and Ground, or Positive (+) and Negative (−). Welding machines often have selector switches to select direct current electrode negative (DCEN) or direct current electrode positive (DCEP). If there is no selector switch, the cables must be manually changed on the machine terminals to select the type of current desired. Most high-amperage machines designed specifically for GTAW are the transformer-rectifier type.

Transformer rectifiers designed specifically for GTAW generally have the following:

- High-frequency generator
- High-frequency selector switch with Start (for DC), Continuous (for AC), or Off (no high frequency)
- Shielding gas preflow timer
- Control to automatically start cooling water flow when an arc is struck
- Shielding gas postflow timer control to prevent weld contamination (keeps shielding gas flowing for a set time after the arc has been terminated)
- Hand- or foot-operated remote current control
- Remote/local selector switch for remote control
- Start amperage control

Welders often use GTAW for critical welds on difficult-to-weld materials. For this reason, many special advanced features and controls have been developed and are available on machines designed for GTAW. These advanced features and controls include the following:

- High-frequency intensity control for better arc starting
- High-frequency stabilizer control to control tungsten spitting
- Crater fill timer to automatically taper the welding current from the selected setting to a minimum current at the end of a weld
- Balanced wave adjustment for penetration/cleaning control (sets percentage of time AC remains on the straight or reverse side of the cycle)

Regardless of the machine used, welders should always consult the manufacturer's documentation for specific operating information. *Figure 7* shows a typical industrial, high-amperage, heavy-duty transformer-rectifier welding machine.

Key Controls

A. Minimum Output Preset Control & Display Switch
B. Setup Menu
C. Digital Amperage Meter
D. Local/Remote Current Switch
E. Maximum Output Preset Control
F. Postflow Time
G. Thermal Shutdown Light
H. Downslope Time
I. Pulse Background Current Control
J. Pulse % On Time Control
K. Spot Time
L. Pulse Frequency Control
M. Polarity Switch
N. Power Switch
O. Pulse/Spot Time Mode Switch
P. Trigger Switch
Q. AC Balance Control
R. Mode Switch
S. Remote Receptacle (Not Shown)

Setup Menu

- DC TIG Start Modes: High-Frequency, Scratch, and Touch Start TIG®
- Adjustable Preflow Time
- Adjustable Start Pulse for Soft or Forceful Starts
- Adjustable TIG Hot Start
- Adjustable Upslope Time
- Adjustable Stick Hot Start
- Adjustable Stick Arc Force

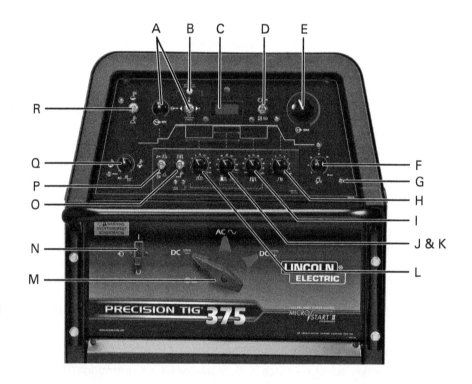
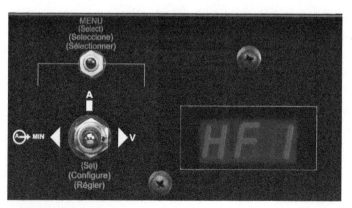

Figure 7 Transformer-rectifier welding machine designed for SMAW, GTAW, and pulsing.
Source: Photo courtesy of The Lincoln Electric Company, Cleveland, OH, U.S.A.

2.1.4 Inverter Welding Machines

Inverter welding machines (*Figure 8*) increase the frequency of the incoming primary power. This provides a smaller, lighter power source with a faster response time and much more waveform control for pulse welding. Welders generally use inverter welding machines in applications where space is limited and portability is important. The controls on these machines vary according to size and application.

The welding machine shown in *Figure 8* weighs less than 30 lb (14 kg) and is a full-function inverter intended for alloy fabrication or critical maintenance work. The welding machine produces up to 90 A with 100% duty cycle for GTAW, and 60 A with 100% duty cycle for SMAW on an input power supply of 120 VAC. However, when the input power is 208/230 VAC, the welding machine can produce up to 130 A for both SMAW and GTAW.

Many production inverter welding machines used today are referred to as *advanced inverter welding machines*. They usually include all the controls listed for transformer-rectifier machines and basic inverter machines plus the following:

- Pulse arc control that allows the welding current to be pulsed between a high-current setting and a low-current setting for puddle control
- Hot-start current control to enable a controlled surge of welding current to establish a puddle quickly
- Slope-up control for direct hot-start current, which controls how quickly the current rises to the hot-start setting
- Optional spot arc timer control for GTAW spot welding
- Variable AC frequency control
- AC balance-wave control

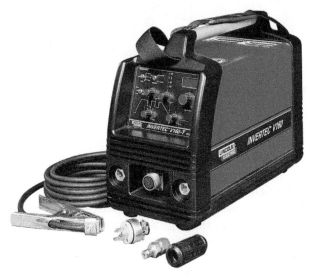
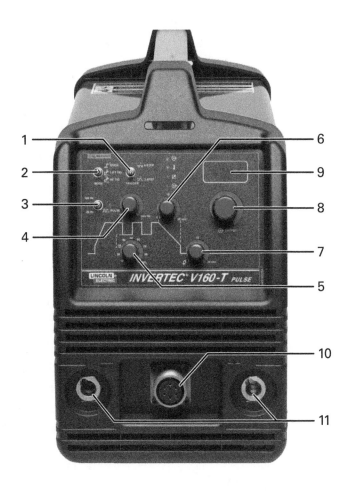

1. High-Frequency, Scratch, and Touch Start TIG®
2. 2-Step/4-Step Trigger Interlock
3. Dual Range Pulsing
4. Adjustable Pulse Frequency
5. Adjustable Pulse Background Current
6. Adjustable Down Slope
7. Adjustable Postflow
8. Output Control
9. Digital Meter (Preset/Actual)
10. MS-Type Remote Connector
11. Twist Mate™ Torch/Gas & Work Connections

Figure 8 Basic GTAW inverter welding machine.
Source: Photo courtesy of The Lincoln Electric Company, Cleveland, OH, U.S.A.

Advanced inverter welding machines are available in capacities of up to 500 A that weigh approximately 100 lb (45 kg).

2.1.5 Engine-Driven Generator Welding Machines

Welding machines can also be powered by gasoline or diesel engines. The engine is connected to a generator (*Figure 9*) that produces the power needed for the welding machine. The size and type of engine used depends on the requirements of the welding machine.

Figure 9 Engine-driven generator welding machine.
Source: © Miller Electric Mfg. LLC

To produce sufficient electrical power, the engine generator must turn at a required number of revolutions per minute (rpm). The engines of power generators have governors to control the engine speed. Most governors have a welding speed switch. The switch can be set to idle the engine when no welding is taking place. When the electrode is touched to the base metal, the governor automatically increases the engine's speed to the required welding speed. If no welding takes place for about 15 seconds, the engine will automatically return to idle. The switch can also be set for the engine to run continuously at the welding speed.

Engine-driven generators often have an auxiliary power unit that produces 120 VAC current for lighting, power tools, and other common equipment. When 120 VAC current is required, the engine-driven generator must run continuously at the welding speed.

Engine-driven generators have engine controls and welding current controls. The engine controls vary with the type and size of the unit but normally include the following:

- Starter
- Voltage gauge
- Temperature gauge
- Fuel gauge
- Hour meter
- Oil pressure indicator

The following are the common welding current controls:

- Amperage control
- Current range switch
- Amperage and voltage gauge
- Polarity switch

The advantage of engine-driven generators is that they are portable and can be used in the field where electricity is not available to power other types of welding machines. The disadvantage is that engine-driven generators are more costly to purchase, operate, and maintain.

2.1.6 Welding Machine Ratings

The rating of a welding machine is determined by the amperage output of the machine at a given duty cycle. The duty cycle of a welding machine is based on a 10-minute period. It is the percentage of 10 minutes during which the machine can continuously produce its rated amperage without overheating. For example, a machine with a rated output of 300 A with a 60% duty cycle can deliver 300 A welding current for 6 minutes out of every 10 without overheating.

The duty cycle of a welding machine will generally be 10%, 20%, 30%, 40%, 60%, or 100%. A welding machine having a duty cycle of 10% to 40% is considered a light- to medium-duty machine; welding can only be maintained for 2 to 4 minutes out of each 10-minute period. Most industrial, heavy-duty machines for manual welding have a 60% or 100% duty cycle rating. Machines designed for automatic welding operations have a 100% duty cycle rating.

With the exception of the 100% duty cycle machines, the maximum amperage that a welding machine can produce is always higher than its rated capacity. A welding machine rated for 300 A with a 60% duty cycle generally puts out a maximum of 375 A to 400 A. However, since the duty cycle is a function of its rated capacity, the duty cycle will decrease as the amperage is raised over 300 A. Welding at 375 A with a welding machine rated for 300 A with a 60% duty cycle will lower the duty cycle to about 30%. If welding continues for more than 3 out of 10 minutes under these conditions, the machine will overheat. Note that most welding machines have a heat-activated circuit breaker that will shut off the machine automatically when it overheats. Leave the machine plugged in and turned on to allow the cooling fans to continue running when this occurs. The machine cannot be turned back on until it has cooled below a preset temperature.

Conversely, if the amperage is set below the rated amperage, the duty cycle increases. Setting the amperage at 200 A for a welding machine rated at 300 A with a 60% duty cycle will increase the duty cycle to 100%. *Figure 10* is a graph that shows the relationship between amperage and duty cycle.

2.1.7 Welding Cable

Cables used to carry welding current are designed for maximum strength and flexibility. The conductors inside the cable are made of fine strands of copper wire. The copper strands are covered with layers of rubber reinforced with nylon or Dacron® cord. *Figure 11* shows the construction of a welding cable.

The size of a welding cable is based on the number of copper strands it contains. Large-diameter cable has more copper strands and can carry more welding current. Typically, the smallest cable size is number 4 and the largest is number 3/0 (spoken as "3 aught"), as shown in *Table 1*.

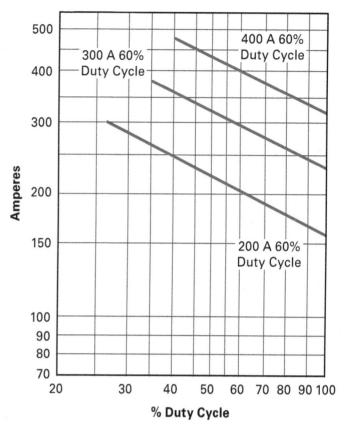

Figure 10 Relationship between amperage and duty cycle.

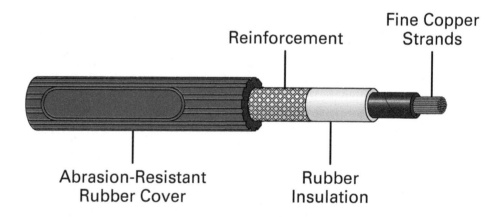

Figure 11 Construction of a welding cable.

When selecting a welding cable size, consider the amperage load as well as the distance the current will travel. The longer distance the current has to travel, the larger the cable must be to reduce voltage drop and heating caused by the electrical resistance in the welding cable. When selecting welding cable, use the rated capacity of the welding machine for the cable amperage requirement. For the distance, measure both the electrode and workpiece leads and add the two lengths together. To identify the welding cable size required, refer to a table of recommended welding cable sizes. These tables are furnished by most welding cable manufacturers.

TABLE 1 Welding Cable Sizes

Machine Size (A)	Duty Cycle (%)	Copper Cable Sizes for Combined Lengths of Electrodes Plus Workpiece Lead				
		Up to 50' (15 m)	50' to 100' (15 m to 30 m)	100' to 150' (30 m to 46 m)	150' to 200' (46 m to 61 m)	200' to 250' (61 m to 76 m)
100	20	#8	#4	#3	#2	#1
180	20	#5	#4	#3	#2	#1
180	30	#4	#4	#3	#2	#1
200	50	#3	#3	#2	#1	#1/0
200	60	#2	#2	#2	#1	#1/0
225	20	#4	#3	#2	#1	#1/0
250	30	#3	#3	#2	#1	#1/0
300	60	#1/0	#1/0	#1/0	#2/0	#3/0
400	60	#2/0	#2/0	#2/0	#3/0	#4/0
500	60	#2/0	#2/0	#3/0	#3/0	#4/0
600	60	#3/0	#3/0	#3/0	#4/0	**
650	60	#3/0	#3/0	#4/0	*	**

*Use double strand of #2/0
**Use double strand of #3/0

2.1.8 Remote Current Control

With GTAW, control of the welding current is often performed with remote controls such as a foot- or hand-operated switch and a **potentiometer**. Remote controls are generally designed for use with a specific welding machine, so they cannot be used universally on all machines. Remote controls with only a switch can control one or more of the following:

- High frequency
- Shielding gas
- Cooling water
- Welding current (on and off only)

Remote controls equipped with a potentiometer can be used to continuously vary the welding current as needed, from zero to the maximum current set on the power source.

The most useful remote control has both switches and a potentiometer, and it controls the following:

- High frequency
- Shielding gas
- Cooling water
- Welding current (on and off)
- Welding current up or down as needed (zero to maximum setting on the power source)

Figure 12 shows various remote control units.

Potentiometer: A three-terminal resistor with an adjustable center connection; used for variable voltage control.

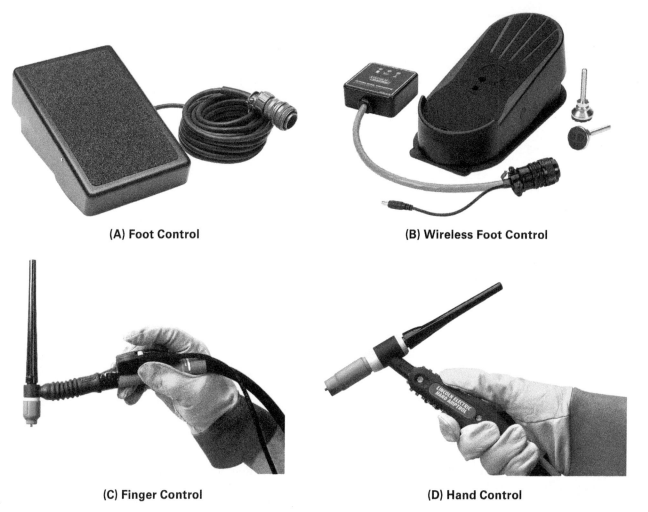

(A) Foot Control

(B) Wireless Foot Control

(C) Finger Control

(D) Hand Control

Figure 12 Typical foot-, finger-, and hand-operated remote controls.
Source: Photo courtesy of The Lincoln Electric Company, Cleveland, OH, U.S.A.

2.2.0 GTAW Torches

In addition to a CC welding machine (and a high-frequency generator for AC welding), GTAW requires the following:

- GTAW torch
- Shielding gas
- Optional torch cooling system

In addition, a remote amperage control is often used with GTAW. *Figure 13* shows a typical GTAW welding system.

The GTAW torch provides an electrical path for the welding current between the welding current lead and the tungsten electrode. There are many different manufacturers of torches, but all torches are basically the same even though their parts are not often interchangeable. All torch bodies have a built-in or replaceable collet body and electrode collet. The electrode collet and collet body come in a variety of sizes to match the size of the electrode being used. The collet holds the tungsten electrode in place and is secured by a back cap. The back cap, which is threaded onto the torch, is loosened to change the electrode or to adjust its stickout. The reserve length of electrode that extends from the back of the collet is insulated and protected by the back cap. Back caps are made in several lengths to cover tungstens of various lengths. The shorter back caps are used with short tungstens in confined spaces where torch clearance is inadequate.

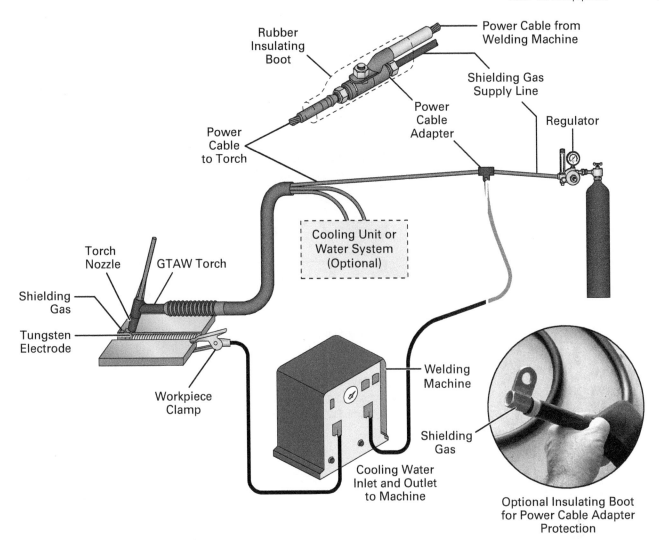

Figure 13 Basic GTAW welding system.

The torch also contains passages for the shielding gas or cooling water, if needed. The shielding gas exits the torch through an insulated gas nozzle, sometimes called a *cup*, which surrounds the tungsten. *Figure 14* provides some GTAW torch terminology.

Heat is removed from GTAW torches either by air cooling with a shielding gas or by water cooling. *Figure 15* shows examples of both air-cooled and water-cooled GTAW torches. Two styles of air-cooled torches are common. One style uses a single plastic hose that has a power cable inside it, as shown in *Figure 15*. The shielding gas flows around the power cable, cooling it and the torch before being discharged to shield the electrode and weld. A special power cable adapter attaches to the power terminal on the power source. This connector needs to be covered by a rubber insulating boot to prevent shorts. The single hose connects to one side of the adapter. This connection also makes the electrical connection. The shielding gas supply hose connects to the other side of the adapter.

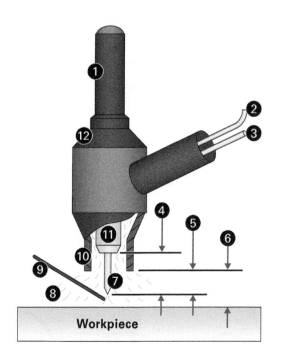

1	Back Cap
2	Electrode Lead
3	Shielding Gas Inlet
4	Electrode Extension
5	Electrode Stickout
6	Standoff Distance
7	Tungsten Electrode
8	Shielding Gas Plume
9	Welding Wire or Welding Rod
10	Gas Nozzle
11	Collet Body
12	Torch Body

Figure 14 GTAW torch terminology.

Source: AWS A3.0M/A3.0:2020, Figure 36, adapted with permission from the American Welding Society (AWS), Miami, FL, USA

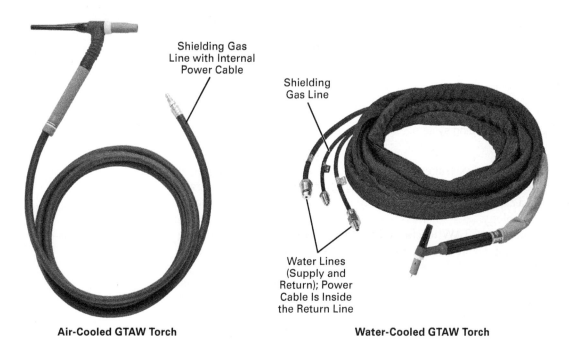

Shielding Gas Line with Internal Power Cable

Shielding Gas Line

Water Lines (Supply and Return); Power Cable Is Inside the Return Line

Air-Cooled GTAW Torch   **Water-Cooled GTAW Torch**

Figure 15 GTAW torches.

Source: Photo courtesy of The Lincoln Electric Company, Cleveland, OH, U.S.A.

The second type of air-cooled torch has a separate power cable and shielding gas hose. Either the single- or double-hose torch can be purchased with a gas control valve in the torch to manually control the shielding gas flow. This is necessary when the power source being used does not have a solenoid valve to automatically start and stop the shielding gas flow. *Figure 16* shows the various parts of an air-cooled GTAW torch.

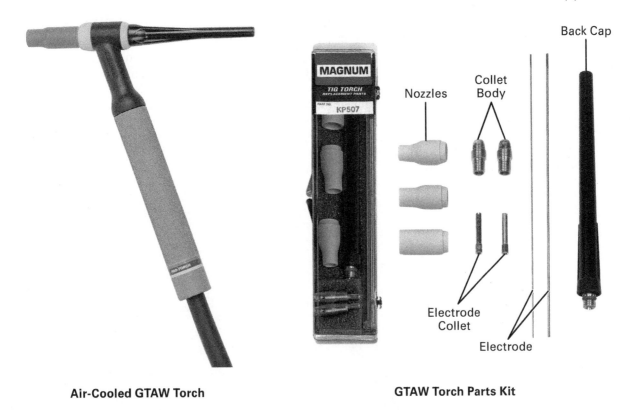

Air-Cooled GTAW Torch   GTAW Torch Parts Kit

Figure 16 Air-cooled GTAW torch.
Source: Photo courtesy of The Lincoln Electric Company, Cleveland, OH, U.S.A.

Water-cooled torches have three hoses, each with their own function. One hose supplies shielding gas, one delivers cooling water to the torch, and one contains the power cable and the cooling water return or discharge. The cooling water may be supplied from the domestic water system or from a closed-loop cooling system. Regardless of the system used, the system should include a fusible link in the torch cooling line. A fusible link is a temperature-based safety device that will open and stop the welding current flow if the coolant flow to the torch is interrupted. *Figure 17* shows the parts of a water-cooled GTAW torch.

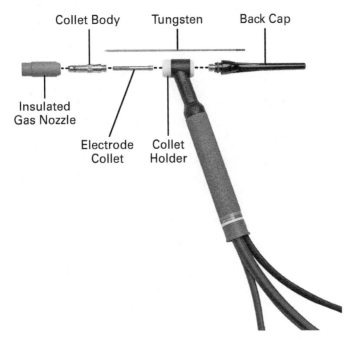

Figure 17 Water-cooled GTAW torch.
Source: Photo courtesy of The Lincoln Electric Company, Cleveland, OH, U.S.A.

Closed-loop cooling systems recirculate demineralized water or special cooling fluids that will not corrode internal torch surfaces or plug the cooling passages with mineral scale. After cooling the torch, the water is typically returned to the system through a hose that also contains the torch power cable. A typical closed-loop cooling system usually consists of a reservoir, circulating pump, heat transfer equipment to reject any collected heat to the atmosphere, and the connecting lines. The reservoir may also serve as the portable unit's base and be equipped with wheels for easy transport. *Figure 18* shows a closed-loop cooling unit. This unit is designed to stack with welding machines that have a matching chassis.

Figure 18 Typical closed-loop cooling unit.
Source: © Miller Electric Mfg. LLC

Cable Covers

To prevent damage from hot or molten metal to GTAW torch cables containing hoses for gas or water, leather or fire-resistant cable covers can be added as protection. These covers zip or snap around the torch cable for 10" to 20" (25 cm to 50 cm) behind the torch handle.

2.3.0 Nozzles, Collets, and Electrodes

GTAW collets and electrodes are designed to work together, with the tungsten electrode fitting precisely into the collet. For this reason, welders must take both devices into consideration when selecting either the proper collet or electrode for a particular job. Note, the gas lens and collets range in sizes which are determined by the tungsten electrode diameter.

2.3.1 Gas Nozzles

The GTAW gas nozzle shapes and directs the shielding gas flow as it exits the torch. Nozzles, sometimes called *cups*, are usually threaded onto the torch to form a gastight joint. They are made of ceramic material, chrome-plated steel, Pyrex®, or glass. Ceramic nozzles can be used up to about 300 A. Above that amperage, water-cooled, metal-coated ceramic nozzles or water-cooled ceramic nozzles must be used.

Gas nozzles are available in different lengths and diameters (refer to *Figure 16* and *Figure 17*). Nozzle length is determined by the job requirements, such as limited clearance or deep grooves. Nozzle size is the exit orifice inside diameter depending on the torch type and size and the electrode diameter. Nozzle size may be specified in fractions of an inch ($\frac{1}{4}$", $\frac{3}{8}$", $\frac{7}{16}$", $\frac{1}{2}$", $\frac{3}{4}$") or in millimeters (6 mm, 10 mm, 11 mm, 13 mm, 19 mm). Metric size nozzles are not, however, designed to be equivalent in size to imperial versions, and metric size and imperial size nozzles cannot be substituted for each other. Sometimes nozzle diameter is given as a size number, such as 4, 6, 7, 8, or 12. These numbers are the nozzle opening diameters in multiples of $\frac{1}{16}$". For example, a Number 6 nozzle is 6 × $\frac{1}{16}$", or $\frac{3}{8}$" in diameter.

A GTAW torch may also be equipped with a gas lens collet body. A gas lens collet body contains an assembly of fine screens that straightens the gas flow from the nozzle and eliminates turbulence, which ensures the gas to flow smoothly past the tungsten electrode. This reduces the chance that the turbulence will pull atmosphere into the weld zone and cause contamination and weld defects. The use of a gas lens allows longer tungsten stickout. This is useful for welding in tight quarters. Increasing tungsten stickout decreases the current-carrying capacity of the tungsten. *Figure 19* shows a typical gas lens collet body and torch with nozzles, collets, and electrodes.

Gas Nozzles

Some manufacturers recommend that the inside diameter of the nozzle should be a minimum of three times the electrode diameter.

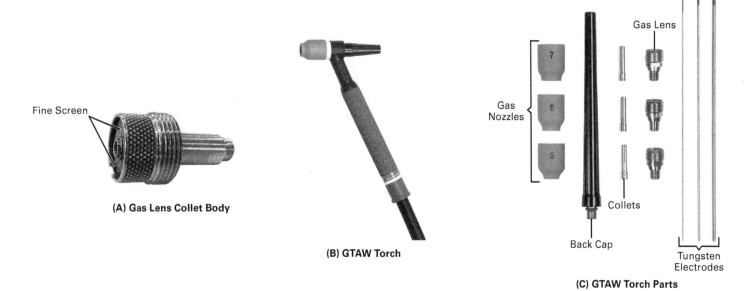

(A) Gas Lens Collet Body

(B) GTAW Torch

(C) GTAW Torch Parts

Figure 19 Typical gas lens collet body and torch with nozzles.
Source: Zachry Group

Think About It

Nozzle Size

What size nozzle would a welder select if a $\frac{1}{2}$" diameter opening was needed?

2.3.2 Tungsten Electrodes

Tungsten electrodes used for GTAW are manufactured in different formulations and sizes to meet the requirements of *AWS Specification A5.12*. Electrode diameters range from 0.01" (0.3 mm) up to $\frac{1}{4}$" (6.4 mm) for current ratings of 5 A to 1,000 A. Electrodes are made in lengths of 3" (76 mm), 6" (152 mm), 7" (175 mm), 12" (305 mm), 18" (457 mm), and 24" (610 mm). Manual welding is usually done with 7" (175 mm) or shorter electrodes.

Electrodes are manufactured with two different finishes—chemically cleaned or ground. The ground finish is the more expensive option.

Multiple different tungsten electrodes are manufactured. They include the following:

- Pure tungsten
- Zirconiated tungsten
- Thoriated tungsten
- Ceriated tungsten
- Lanthanated tungsten
- Rare earth tungsten
- Cryogenically treated thoriated tungsten

Pure tungsten electrodes, often simply referred to as *tungstens*, are used for AC welding. They provide good arc stability and good resistance against contamination. When used for DC welding, they are easily contaminated. Although relatively inexpensive per unit, pure tungsten is more difficult to start, less able to maintain a stable arc, and has a shorter service life than other types of tungsten electrodes.

Zirconiated tungstens are also used for AC welding. They have a small percentage of zirconium added to the tungsten. They are used for AC welding where tungsten inclusions cannot be tolerated. The zirconium also gives these electrodes easy arc-starting characteristics.

Thoriated tungsten electrodes are used for DC welding. They have a small percentage of thorium added to do the following:

- Make the arc easier to start
- Increase the current range
- Help prevent tip melt
- Reduce the tendency to stick or freeze to the work
- Increase resistance to contamination, when properly used

Thoriated tungsten electrodes are manufactured most commonly in these two concentrations of thorium dioxide (thoria): EWTh-1 (approximately 1% thoria) and EWTh-2 (approximately 2% thoria).

WARNING!

Thoriated tungsten electrodes contain low-level radioactive thorium. The grinding dust from these electrodes is considered a hazard. *AWS Safety and Health Fact Sheet 27* provides further information about this subject. If possible, always use tungsten electrodes containing cerium or lanthanum instead of thorium.

Ceriated and lanthanated tungsten electrodes have the advantage of not being radioactive. Ceriated tungsten electrodes can be balled and used for AC welding.

EWG electrodes, which combine three nonradioactive rare earth materials into one electrode, are increasingly used as replacements for 2% thoriated electrodes. Besides their advantage of not being radioactive, these mixed tungsten electrodes start and reignite well and have very good service life for applications in which welding cycles of at least 15 minutes are used.

Thoriated electrodes that have been treated with a multistep cryogenic process are another good replacement for 2% thoriated electrodes. The cryogenic treatment maximizes grain structure and improves electron flow. The cryogenically treated electrodes are easier to ignite and have longer tip life, longer overall electrode life, and higher amperage tolerance than nontreated 2% thoriated tungsten.

Other tungsten electrodes continue to be developed. For example, tungsten electrodes with less common oxides, such as yttrium and magnesium, are also becoming available.

Tungsten electrodes are identified by a color band at one end. *Table 2* shows GTAW electrode color codes and AWS classifications.

TABLE 2 GTAW Electrode Color Codes and AWS Classifications

Electrode	Color Band	Electrode AWS Classification
Pure Tungsten	Green	EWP (Minimum 99.5% Tungsten)
1% Thoriated	Yellow	EWTh-1 (1% Thoria)
2% Thoriated	Red	EWTh-2 (2% Thoria)
Zirconiated	Brown	EWZr-1 (0.15% to 0.40% Zirconium Oxide)
Zirconiated	White	EWZr-8 (0.8% Zirconium Oxide)
Ceriated	Gray	EWCe-2 (2% Ceria)
Lanthanated	Black	EWLa-1.0 (1% Lanthana)
Lanthanated	Gold	EWLa-1.5 (1.5% Lanthana)
Lanthanated	Blue	EWLa-2.0 (2% Lanthana)
EWG	No standard*	Undefined (e.g., various combinations of rare earth alloys, yttrium oxide, magnesium oxide, etc.)

*Manufacturers may select any color not already in use.

Around the World

Tungsten Standards

The United States, Canada, Europe, and Japan each have their own published standards for tungsten, as shown in the table below. These standards specify the dimensions, packaging, and manufacturing requirements that tungsten manufacturers must meet for the given market area.

Market Area	Standard Name
United States	ANSI/AWS A5.12
Canada	ASME/SFA
Europe	ISO 6848
Japan	JIS

Tungsten Electrode Packaging

Most tungsten electrodes are packaged as shown to prevent contamination or damage to the electrodes. An inner divider containing the electrodes slides out of the package. Note the orange color band (may appear reddish due to printing color variations) on the end of each electrode, identifying the type in agreement with *Table 2*. All color bands typically discolor after use due to the heat produced. The color band end is not used when sharpening an electrode.

2.4.0   Shielding Gases

The GTAW process always uses inert shielding gases to displace the atmosphere from the weld zone and prevent oxidation and contamination of the tungsten electrode, weld puddle, and filler metal. The two principal shielding gases used for GTAW are argon (Ar) and helium (He). The use of helium was what prompted the naming of GTAW initially as *heliarc welding*. These gases may be used alone or as mixtures. Occasionally, small percentages of other gases are added for special conditions.

Each shielding gas and shielding gas mixture has distinct performance characteristics. Each affects the arc differently and produces different weld characteristics. Mixtures of gases often have the best features of the individual gases within them. The following sections explain shielding gas characteristics and their principal uses.

Gas flow rate from the gun tip is important because it affects the quality and the cost of a weld. Too low a flow rate will not shield the weld zone adequately, which will result in a poor-quality weld. Excessive gas flow wastes expensive gases and can cause turbulence at the weld. The turbulence can pull atmosphere into the weld zone, causing porosity in the weld. Welding specifications indicate appropriate shielding gas flow rates.

Always Use the Correct Gas
Make sure to always use the correct gas or gas mixture. Using an incorrect gas mixture, like a common one used for GMAW, which is 75% argon / 25% CO_2, will immediately cause the tungsten electrode to be consumed and deposited in the weld as a contaminant.

2.4.1 Argon

Argon is the most common shielding gas used with GTAW. Argon provides a smooth, quiet arc that requires a lower arc voltage than other shielding gases for a given arc length. This means that it gives the welder the greatest tolerance for arc gap variation. Argon also works well with AC, and it provides better base metal cleaning than helium. Argon is ten times heavier than helium, so it forms a better gas shield than helium, which tends to rise at the same flow rates. Argon is used both alone and in combination with other shielding gases.

When compared with helium, argon has the following advantages:

- A smoother, quieter arc
- Lower arc voltage for a given arc length
- Easier arc starting
- Better cathode cleaning on aluminum and magnesium with AC
- Lower shielding flow rate
- Better shielding in cross drafts
- Least expensive shielding gas

Small amounts of hydrogen are sometimes mixed with argon for deeper penetration in welding stainless steel. Hydrogen cannot be used with aluminum or carbon steels because it produces porosity and underbead cracking. The most common argon-hydrogen mixture is 85% argon and 15% hydrogen.

Nitrogen is sometimes added to argon to stabilize austenitic stainless steel and to increase penetration in welding copper.

2.4.2 Helium

Helium is used when deeper penetration and higher travel speed are required. Arc stability is not as good as it is with argon. However, helium can deliver more heat on the base metal than is possible with argon, although it does require a higher voltage. This makes helium better for welding thick sections of highly heat-conductive metals such as aluminum and copper.

When compared with argon, helium has the following advantages:

- Deeper penetration into the weld joint
- Increased welding speed
- Welding of highly heat-conductive metals

Helium is often mixed with argon. A common mixture is 75% helium and 25% argon. This mixture gives good cathodic cleaning action and deep weld penetration, which are good characteristics of both gases.

2.4.3 Gas Mixtures

The principal shielding gases are often mixed to improve their overall welding characteristics. For steel, small amounts of oxygen added to argon stabilize the arc, increase filler metal deposition rate, and improve wetting and bead shape. Helium is added to argon to increase the arc voltage and heat for welding nonferrous metals. Small amounts of nitrogen and oxygen are added to argon for welding stainless steel when the austenitic characteristics should be preserved. Many other typical gas mixtures are available.

2.4.4 Cylinder Safety

Shielding gases may be supplied in bulk liquid tanks, in liquid cylinders, or in high-pressure cylinders of various sizes. The most common container is the high-pressure cylinder, which is portable and can be easily moved as needed.

When transporting and handling cylinders, always observe the following rules:

- Always install the safety cap over the valve, except when the cylinder is connected for use.

WARNING!

Do not remove the protective valve cap unless the cylinder is secured. If a cylinder falls over and the valve breaks off, the cylinder may rocket away uncontrollably, causing severe injury or death to anyone in its way.

- Secure the cylinder to prevent it from falling when it is in use. Chain or clamp it to the welding machine or to a post, beam, or pipe.
- Always use a cylinder cart to transport cylinders.
- Never hoist cylinders with a sling or magnets. Cylinders can slip out of the sling or fall from the magnets. Always use a hoisting basket or similar device.
- Always open the cylinder valve slowly. Once pressure has been applied to the system, open the valve completely to prevent leakage around the valve stem.

2.4.5 Gas Regulators / Flowmeters

A gas pressure regulator and a flowmeter are required to supply the shielding gas to the torch at the proper pressure and flow rate (*Figure 20*). A typical regulator/flowmeter has a preset pressure regulator with a cylinder valve spud, a flow-metering needle valve, and a flow rate gauge. The pressure regulator is usually equipped with a cylinder pressure gauge to indicate the cylinder gas pressure. The metering valve is used to adjust the gas flow to the torch nozzle. The flow gauge indicates the gas flow rate in cubic feet per hour (cfh) or liters per minute (L/min) as shown by a ball-type flow indicator. Although the regulator and flowmeter are shown separately in *Figure 20*, their functions may also be combined into a single device.

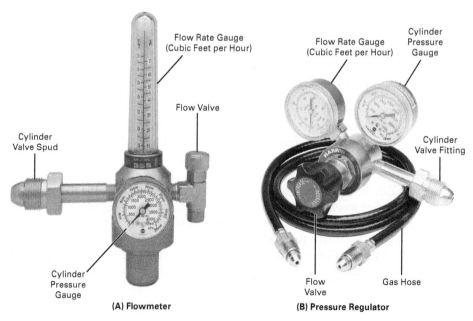

Figure 20 Shielding gas flowmeter and gas regulator.
Source: Photo courtesy of The Lincoln Electric Company, Cleveland, OH, U.S.A.

Gas flow to the torch is started and stopped either by a manually operated valve or by an electric solenoid valve. The manual valve may be on the torch or in the gas line to the torch. An electric solenoid valve is controlled automatically by the welding current flow or by a manual switch.

2.5.0 GTAW Filler Metals

Filler metal for manual GTAW is generally supplied in 36" (914.4 mm) lengths in diameters from $\frac{1}{16}$" (1.6 mm) to $\frac{1}{4}$" (6.4 mm). There are some automatic and manually operated wire feeders used with manual GTAW, but they are usually found only in high-production facilities. Filler metals for GTAW are drawn from high-grade pure alloys compounded for specific applications. Generally, the rods are not coated. They are bare, without a flux component added, except for a corrosion-resistant copper electroplating on some carbon steel rods. However, some special-purpose rods may have flux cores or coatings.

Several industry organizations and the US government publish specification standards for filler metals. The most common standards are those published by the American Welding Society. The purpose of the AWS specifications is to set standards that all manufacturers must follow when manufacturing welding consumables. This ensures consistency for the user regardless of who manufactured the product. The specifications set standards for the following:

- Classification system, identification, and marking
- Chemical composition of the deposited weld metal
- Mechanical properties of the deposited weld metal

Examples of AWS specifications that apply to GTAW filler metals are the following:

- *AWS A5.7, Specification for Copper and Copper-Alloy Bare Welding Rods and Electrodes*
- *AWS A5.9/A5.9M, Specification for Bare Stainless Steel Welding Electrodes and Rods*
- *AWS A5.10/A5.10M, Specification for Bare Aluminum and Aluminum-Alloy Welding Electrodes and Rods*
- *AWS A5.14/A5.14M, Specification for Nickel and Nickel-Alloy Bare Welding Electrodes and Rods*
- *AWS A5.16/A5.16M, Specification for Titanium and Titanium-Alloy Welding Electrodes and Rods*
- *AWS A5.18/A5.18M, Specification for Carbon Steel Electrodes and Rods for Gas Shielded Arc Welding*
- *AWS A5.19, Specification for Magnesium Alloy Welding Electrodes and Rods*
- *AWS A5.22/A5.22M, Specification for Stainless Steel Flux Cored and Metal Cored Welding Electrodes and Rods*
- *AWS A5.28/A5.28M, Specification for Low-Alloy Steel Electrodes and Rods for Gas Shielded Arc Welding*

NOTE

An AWS specification number is generally followed by a four-digit number that indicates the year the specification was last revised. When referring to an AWS specification on the job, always make sure to use the version called out by design and drawing documents rather than defaulting to the most current version.

Industry designations for electrodes are specified in *AWS A5.01M/5.01, Welding and Brazing Consumables — Procurement of Filler Metals and Fluxes*. AWS's specific testing schedules of electrodes are designated with letters F through K:

- *Schedule F* — Standard testing schedule of each manufacturer.
- *Schedule G* — Tests from production runs of the electrode lot. Tests to be performed within 12 months of purchase order.
- *Schedule H* — Chemical analysis of each electrode lot.
- *Schedule I* — Required tests for each electrode lot.
- *Schedule J* — Tests which the classification called for in the relevant AWS, ISO, or other welding consumable standard for each electrode lot.
- *Schedule K* — Tests specified by the purchaser for each electrode lot.

Factors that affect the selection of a GTAW filler metal include the following:

- Base metal chemical composition
- Base metal mechanical properties
- Weld joint design
- Service or specification requirements
- Shielding gas used

The following are commonly used filler metal types:

- Carbon steel
- Low-alloy steel
- Stainless steel
- Aluminum and aluminum alloy
- Copper and copper alloy
- Nickel and nickel alloy
- Magnesium and magnesium alloy
- Titanium and titanium alloy

Table 3 summarizes the GTAW filler metal specifications and the major AWS class covered by each specification.

The AWS classification of metal filler rod is identified in a durable way on its container. Acceptable methods of identification, following the AWS classification, include stamping, coining, embossing, imprinting, flag tagging, or color coding. In the case of stamping, each rod is usually marked by the AWS classification. In the case of flag tagging, a paper or plastic tab is wrapped around the rod. In the case of color coding, it's important to note that the purchaser and the supplier will cooperatively agree on a color and ensure the color appears on packaging. In some cases, classification may also be printed on the rod with ink. *Figure 21* shows two common types of identification methods.

Given the importance of metal filler rod identification in selecting the right material for the job, pay attention to how filler metal rods are handled. Consume filler metal rods so that the ID marker is not destroyed. Markers are located at each end. Cutting a rod in half allows you to retain the marker on the used stub. Retaining the stubs with the markings until the weld is complete provides evidence that the correct filler metal was used.

TABLE 3 GTAW Filler Metal Specifications and Classification System

Material	Filler Metal		X Designator Description	Example
	AWS Spec.	AWS Class.		
Carbon steel	A5.18	ERXXS-Y	Tensile strength × 1,000 (psi)	ER70S-3
		EXXC-Y		E70C-3
Low-alloy steel	A5.28	ERXXS-Y	Tensile strength × 1,000 (psi)	ER80S-B2
		EXXC-Y		E80C-B2
Stainless steel	A5.9	ERXXXY	Stainless alloy (308, 410, etc.)	ER308L
				EC308L
Stainless steel (flux-cored)	A5.22	RXXXT1-5	Stainless alloy (308, 309, etc.)	R309LT1-5
Aluminum	A5.10	ERXXXX-Y	Aluminum alloy (4043, 5083, etc.)	ER4043
Nickel	A5.14	ERNiXX-Y	Major alloying elements (Cr, Fe, Mo, etc.)	ERNiCr-3
Copper	A5.7	ERCuXX-Y	Major alloying elements (Al, Ni, Si, etc.)	ERCuAl-A2
Magnesium	A5.19	ERXXYYY	Major alloying elements (Al, Zn, etc.)	ERAZ92A
Titanium	A5.16	ERTi-Y	—	ERTi-5

Notes:
E — Filler metal may be used as an electrode.
R — Filler metal may be used as a rod.
S — Solid filler metal.
C — Composite or stranded filler metal.
Y — Designator (or combination of designators) that describes specific alloy, shielding gas to be used, diffusible hydrogen limit, etc. Refer to the appropriate AWS filler metal specification shown above for explanation.

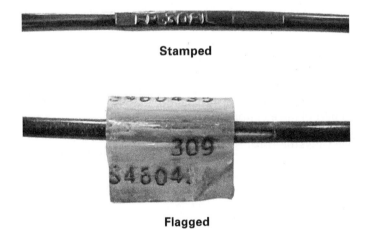

Stamped

Flagged

Figure 21 Filler rod markings.
Source: Zachry Group

2.5.1 Carbon Steel and Low-Alloy Steel Filler Metals

Carbon steel filler metals are identified by *AWS A5.18*. Low-alloy steel filler metals are identified by *AWS A5.28*. The rod classification number is found on a label on the packaging. *Figure 22* and *Figure 23* show the AWS classifications for carbon steel and low-alloy steel filler metals.

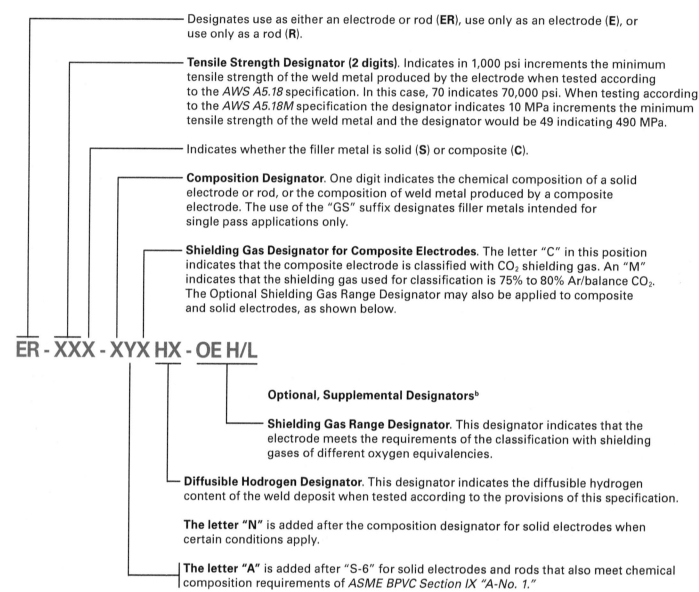

Mandatory Classification Designators[a]

Designates use as either an electrode or rod (**ER**), use only as an electrode (**E**), or use only as a rod (**R**).

Tensile Strength Designator (2 digits). Indicates in 1,000 psi increments the minimum tensile strength of the weld metal produced by the electrode when tested according to the *AWS A5.18* specification. In this case, 70 indicates 70,000 psi. When testing according to the *AWS A5.18M* specification the designator indicates 10 MPa increments the minimum tensile strength of the weld metal and the designator would be 49 indicating 490 MPa.

Indicates whether the filler metal is solid (**S**) or composite (**C**).

Composition Designator. One digit indicates the chemical composition of a solid electrode or rod, or the composition of weld metal produced by a composite electrode. The use of the "GS" suffix designates filler metals intended for single pass applications only.

Shielding Gas Designator for Composite Electrodes. The letter "C" in this position indicates that the composite electrode is classified with CO_2 shielding gas. An "M" indicates that the shielding gas used for classification is 75% to 80% Ar/balance CO_2. The Optional Shielding Gas Range Designator may also be applied to composite and solid electrodes, as shown below.

ER - XXX - XYX HX - OE H/L

Optional, Supplemental Designators[b]

Shielding Gas Range Designator. This designator indicates that the electrode meets the requirements of the classification with shielding gases of different oxygen equivalencies.

Diffusible Hodrogen Designator. This designator indicates the diffusible hydrogen content of the weld deposit when tested according to the provisions of this specification.

The letter "N" is added after the composition designator for solid electrodes when certain conditions apply.

The letter "A" is added after "S-6" for solid electrodes and rods that also meet chemical composition requirements of *ASME BPVC Section IX "A-No. 1."*

[a] The combination of these designators constitutes the electrode (or rod) classification.
[b] The designators are optional and do not constitute a part of the electrode (or rod) classification.

A5.18/A5.18M Classification System

Figure 22 AWS classification for carbon steel filler metals.

Source: AWS A5.18/A5.18M:2005, Figure A.1, Reproduced with permission from the American Welding Society (AWS), Miami, FL, USA

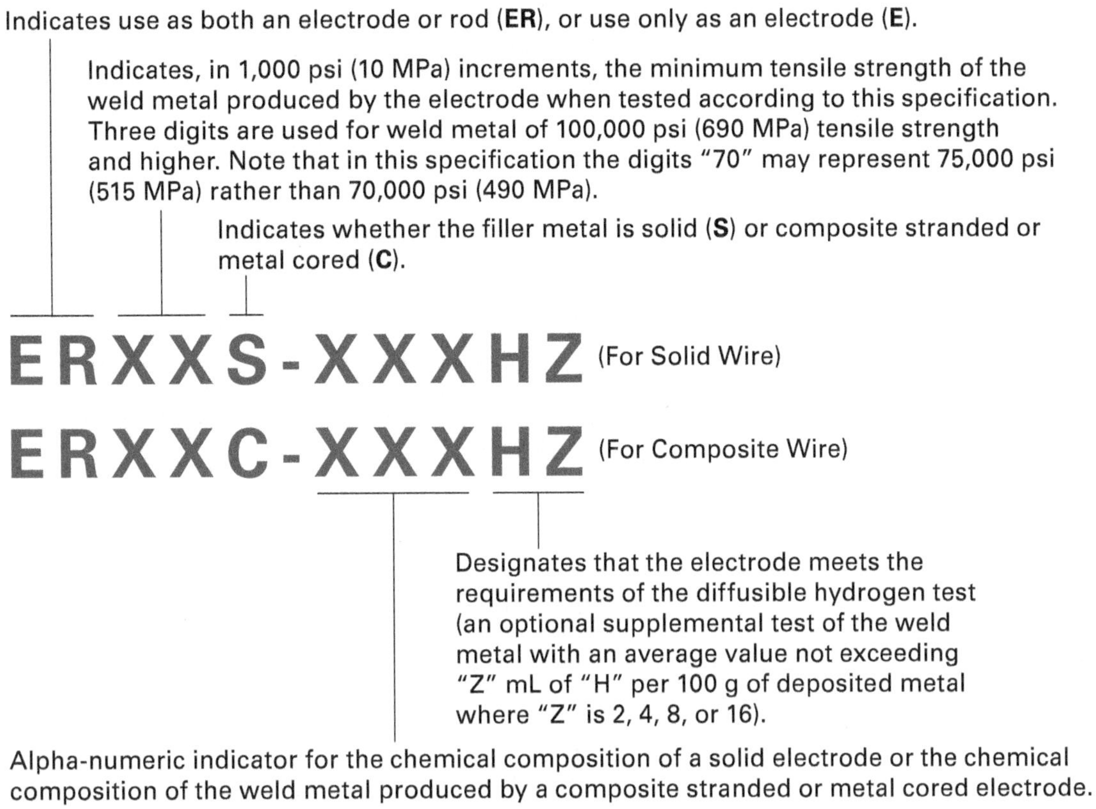

Indicates use as both an electrode or rod (**ER**), or use only as an electrode (**E**).

Indicates, in 1,000 psi (10 MPa) increments, the minimum tensile strength of the weld metal produced by the electrode when tested according to this specification. Three digits are used for weld metal of 100,000 psi (690 MPa) tensile strength and higher. Note that in this specification the digits "70" may represent 75,000 psi (515 MPa) rather than 70,000 psi (490 MPa).

Indicates whether the filler metal is solid (**S**) or composite stranded or metal cored (**C**).

ERXXS-XXXHZ (For Solid Wire)

ERXXC-XXXHZ (For Composite Wire)

Designates that the electrode meets the requirements of the diffusible hydrogen test (an optional supplemental test of the weld metal with an average value not exceeding "Z" mL of "H" per 100 g of deposited metal where "Z" is 2, 4, 8, or 16).

Alpha-numeric indicator for the chemical composition of a solid electrode or the chemical composition of the weld metal produced by a composite stranded or metal cored electrode.

Figure 23 AWS classification for low-alloy steel filler metals.

All steel filler metals contain alloys such as silicon, manganese, aluminum, and carbon. Other alloys such as nickel, chromium, and molybdenum are also often added. The purpose of the alloys is as follows:

- *Silicon (Si)* — Concentrations of 0.40% to 1.00% are used to deoxidize the puddle and to strengthen the weld. Silicon above 1% may make the welds crack sensitive.
- *Manganese (Mn)* — Concentrations of 1% to 2% are also used as a deoxidizer and to strengthen the weld. Manganese also decreases hot crack sensitivity.
- *Aluminum (Al), titanium (Ti), and zirconium (Zr)* — One or more of these elements may be added in very small amounts for deoxidizing. These elements may also increase strength.
- *Carbon (C)* — Concentrations of 0.05% to 0.12% are used to add strength without adversely affecting ductility, porosity, or toughness.
- *Nickel (Ni), chromium (Cr), and molybdenum (Mo)* — These elements may be added in small amounts to improve corrosion resistance, strength, and toughness.

2.5.2 Stainless Steel Filler Metals

Stainless steel filler metals are identified by *AWS A5.9*. Designations for a typical AWS stainless steel electrode rod classification are shown in *Figure 24*.

Select stainless steel filler metal to closely match the alloy composition of the base metal. Stainless steel filler metals also require specific shielding gases or gas mixtures.

Rod and Electrode Designations

The letter R that follows the E in an AWS classification number indicates that the classification is used as a metal filler rod as well as an electrode. Electrode wire is used in GMAW. Since a tungsten-based electrode is used in GTAW, filler metal classifications with an ER must be used.

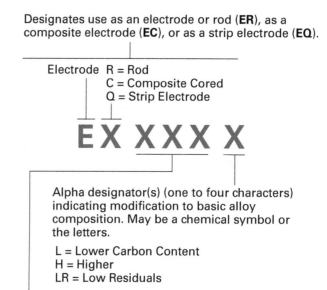

Figure 24 Typical AWS stainless steel rod classification.

2.5.3 Aluminum and Aluminum Alloy Filler Metals

Aluminum electrode rods are covered by *AWS A5.10* as shown in *Figure 25*. Aluminum filler metals usually contain magnesium, manganese, zinc, silicon, or copper for increased strength. Corrosion resistance and ease of welding are also considerations. Aluminum filler metals are designed to weld specific types of aluminum and should be selected for compatibility. The most widely used aluminum rods are ER4043 (contains silicon) and ER5356 (contains magnesium).

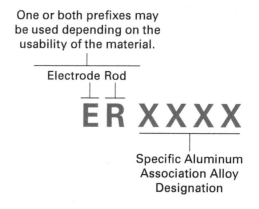

Figure 25 Typical AWS aluminum filler metal classification.

2.5.4 Copper and Copper Alloy Filler Metals

Copper and copper alloy filler metals (*Figure 26*) are covered by *AWS A5.7*. Most copper filler metals contain other elements to increase strength, deoxidize the weld metal, and match the base metal composition.

2.5.5 Nickel and Nickel Alloy Filler Metals

Nickel-based filler metals are covered by *AWS A5.14*. These filler metals contain other elements to match base metal applications and to increase the strength and quality of the weld metal. For GTAW, DCEN is used with high-purity argon, helium, or both argon and helium used as a shielding gas. *Figure 27* shows the AWS classification for nickel filler metals.

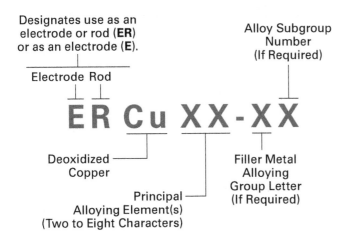

Figure 26 Typical AWS copper filler metal classification.

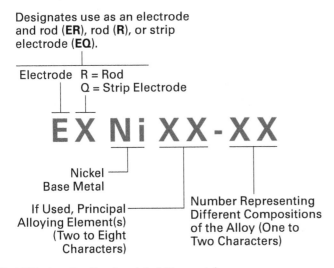

Figure 27 AWS classification for nickel filler metals.

2.5.6 Magnesium Alloy Filler Metals

Magnesium alloy filler metals (*Figure 28*) are covered by *AWS A5.19*. These filler metals are usually used with GTAW, GMAW, and plasma arc welding (PAW) processes. For GTAW welding, the techniques and equipment are similar to those for aluminum. Argon, helium, or both are used for shielding. Alternating current is preferred for arc cleaning and penetration. Direct current is also employed with DCEP, used for thin materials; and DCEN, used for mechanized welding with helium, for deep penetration. GTAW is recommended for the defect repair of clean magnesium castings. Magnesium fines are flammable, so don't allow them to accumulate.

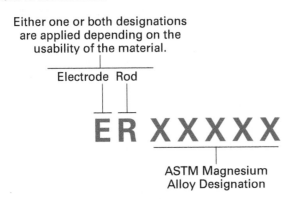

Figure 28 AWS classification for magnesium filler metals.

Magnesium Fires

While ignition of magnesium is a very remote possibility when welding, the fire will cease when the heat source is removed. Ignition of the weld pool is prevented by the gas shielding used in GTAW, GMAW, and PAW processes. Most magnesium fires occur when fines from grinding or filing, or chips from machining, are allowed to accumulate. Take care to prevent the accumulation of fines on clothing. Graphite-based or salt-based powders recommended for extinguishing magnesium fires should be stored in the work area. If large amounts of fines are produced, collect them in a water-wash dust collector designed for use with magnesium. Follow special precautions for the handling of wet magnesium fines.

2.5.7 Titanium and Titanium Alloy Filler Metals

Titanium and titanium alloy filler metals (*Figure 29*) are covered by *AWS A5.16*. These filler metals are generally used with GTAW, GMAW, submerged arc welding (SAW), and PAW processes. Titanium is sensitive to embrittlement by oxygen, nitrogen, and hydrogen at temperatures above 500°F (260°C). Like aluminum, titanium requires weld cleaning, high-purity gas shielding, and an especially adequate postflow shielding to avoid embrittlement. Titanium can be successfully fusion welded to zirconium, tantalum, niobium, and vanadium. Titanium should not be welded to copper, iron, nickel, or aluminum. Like magnesium, titanium fines are flammable.

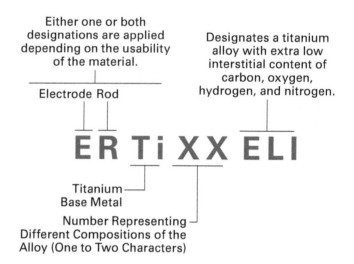

Figure 29 AWS classification for titanium filler metals.

2.5.8 Stainless Steel Flux-Cored Electrodes and Rods

Figure 30 shows the *AWS A5.22* specification format for stainless steel flux-cored electrodes and rods.

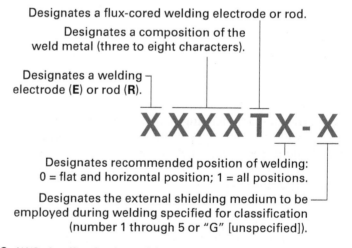

Figure 30 AWS classification for stainless steel flux-cored electrodes and rods.

2.0.0 Section Review

1. A type of welding machine that is designed to produce either direct welding current only or both alternating and direct welding currents is a(n) _____.
   a. transformer welding machine
   b. transformer-rectifier welding machine
   c. alternator welding machine
   d. inverter power source

2. How many hoses does a water-cooled GTAW torch have?
   a. One
   b. Two
   c. Three
   d. Four

3. The device that contains an assembly of fine screens that eliminate turbulence by straightening the gas flow from a GTAW gas nozzle is called a _____.
   a. laminar flow gas lens
   b. shielding gas supply hose
   c. closed-loop gas rectifier
   d. gas lens collet body

4. In GTAW, small amounts of hydrogen are sometimes mixed with argon for deeper penetration when welding _____.
   a. titanium
   b. aluminum
   c. carbon steel
   d. stainless steel

5. To increase strength, deoxidize the weld metal, and match the base metal composition, most copper filler rods contain _____.
   a. other elements
   b. gas lens filters
   c. magnesium castings
   d. anodized quenchers

3.0.0 Welding Equipment Setup

Objective

Explain how to set up for GTAW welding.

a. Explain how to select and position the welding machine.

b. Explain how to connect and set up the shielding gas flow rate.

c. Explain how to select and prepare the tungsten electrode.

d. Explain how to select and install the nozzle along with the tungsten electrode.

Performance Tasks

3. Connect the shielding gas and set the flow rate.

4. Select and prepare the tungsten electrode.

5. Break down and reassemble a GTAW torch.

To weld safely and efficiently, all the necessary welding equipment must be properly set up. The following sections explain how to properly set up GTAW equipment.

3.1.0 The GTAW Welding Machine

Select the welding machine and current type for the base metal to be welded, considering the following factors:

- GTAW requires a constant-current power source. SMAW power sources and welding machines are commonly used for GTAW.
- The welding current type required, either AC or DC. Generally, carbon steels, stainless steels, and alloy steels are welded with DC, while aluminum and magnesium are welded with AC at high frequency.
- The maximum amperage required.
- Whether there is an electrical power source and appropriate receptacles in which to plug a welding machine. Otherwise, an engine-driven welding machine is needed.

Because of the limited length of GTAW torch cables, the shielding gas supply must be located reasonably close to the welding site. The welding machine can be some distance away, but it will not be convenient to operate unless remote controls are installed.

Select a site where the GTAW equipment will not be in the way but will be protected from welding, cutting, or grinding sparks. There should be good air circulation to keep the welding machine cool. The environment should be free from explosive or corrosive fumes and as free as possible from dust and dirt. Welding machines have internal cooling fans that will pull these particles into the machines if they are present. The site should also be free of standing water or water leaks. If an engine-driven generator is used, position it so that it can be easily refueled and serviced.

There should be easy access to the site so that the equipment can be started, stopped, or adjusted as needed. If the machine will be plugged into an outlet, be sure that the outlet has been properly installed and properly grounded by a licensed electrician. Also, be sure to identify the location of the electrical power disconnect switch or circuit breaker before plugging the welding machine into the outlet.

3.1.1 Moving Welding Machines

Large engine-driven generators are mounted on a trailer frame and can easily be moved by a pickup truck or tractor with a trailer hitch. Other types of welding machines may have a skid base or be mounted on steel or rubber wheels. Be careful when moving wheel-mounted welding machines by hand. Some machines are top-heavy and could fall over in a tight turn or if the floor or ground is uneven or soft.

> **WARNING!**
>
> Secure or remove the wire feeder and gas cylinder before attempting to move a welding machine. If a welding machine starts to fall over, do not attempt to hold it. Welding machines are very heavy, and severe crushing injuries can occur if a welding machine falls on you.

Most welding machines have a lifting eye used to move machines mounted on skids or lift any machine. Before lifting a welding machine, check the equipment specifications for the weight. Be sure the lifting device and tackle can handle the machine's weight. Always use a sling and/or a shackle. Never attempt to lift a machine by placing the lifting hook directly in the machine's lifting eye because the safety latch on the hook cannot be closed. Also, before lifting or moving a welding machine, always inspect the rigging equipment before each use and be sure the welding cables are secure.

3.1.2 Placing the Workpiece Clamp

The workpiece clamp must be properly placed to prevent damage to surrounding equipment. If the electrical welding current travels through a bearing, seal, valve, or contacting surface, it could cause severe damage from heat and arcing, requiring these items be replaced. Carefully check the area to be welded and position the workpiece clamp so the welding current will not pass through any contacting surface. When in doubt, ask your supervisor for assistance before proceeding.

> **WARNING!**
>
> If welding is to be done near a battery, the battery must be removed. Batteries produce hydrogen gas, which is extremely explosive. A welding spark could cause the battery to explode, showering the area with battery acid.

CAUTION

Welding current passing through electrical or electronic equipment will cause severe damage to the affected equipment. Also, before welding on any type of equipment with a battery, the ground lead at the battery must be disconnected to protect the electrical system.

Workpiece clamps must never be connected to pipes carrying flammable or corrosive materials. The welding current could cause overheating or sparks, resulting in an explosion or fire.

The workpiece clamp must make good electrical contact when it is connected. Dirt and paint will inhibit the connection and cause arcing, resulting in overheating of the workpiece clamp. Dirt and paint also make the welding current unstable and can cause defects in the weld. Clean the surface before connecting the workpiece clamp. If the workpiece clamp is damaged or does not close securely onto the surface, replace it.

3.1.3 Energizing the Power Source

Electrically powered welding machines are energized by plugging them into an electrical outlet. The electrical requirements will be located on the equipment specification tag prominently displayed on the machine. Most machines requiring single-phase, 240 VAC power have a three-pronged plug. Machines requiring three-phase, 208 VAC, or 460 VAC power have a four-pronged plug.

If a welding machine does not have a power plug, an electrician must connect it. The electrician will add a plug or hard-wire the machine directly into an electrical box.

WARNING!

Never use a welding machine until you identify the location of the electrical disconnect switch for that specific machine or receptacle. In an emergency, you must be able to turn off the power quickly to the welding machine at the disconnect switch.

3.1.4 Starting Engine-Driven Generators

Before welding can take place with an engine-driven generator, the engine must be checked and started. As with a car engine, the engine powering the generator must also have routine maintenance performed.

Many sites have prestart checklists that must be completed and signed prior to starting or operating an engine-driven generator. Check with your supervisor. If your site has such a checklist, complete and sign it. If your site does not have a prestart checklist, perform the following checks before starting the engine:

- Check the oil using the engine oil dipstick. If the oil is low, add the appropriate grade of oil for the conditions according to the manufacturer's maintenance instructions.
- Check the coolant level in the radiator if the engine is liquid-cooled. If the coolant level is low, add the proper coolant.
- Check the fuel. The unit may have a fuel gauge or a dipstick. If the fuel is low, add the correct fuel (diesel or gasoline) to the fuel tank. The type of fuel required should be marked on the fuel tank. If it is not marked, contact your supervisor to verify the fuel required and have the tank marked.
- Check the battery water level unless the battery is sealed. Add demineralized water if the battery water level is low.
- Check the electrode holder to be sure it is not grounded. If the electrode holder is grounded, it will arc and overheat the welding system when the welding machine is started. This may cause damage to the equipment.
- Open the fuel shutoff valve if the equipment has one. If there is a fuel shutoff valve, it should be located in the fuel line between the fuel tank and carburetor.
- Record the hours from the hour meter. An hour meter records the total number of hours that the engine runs. This information is used to determine when the engine needs to be serviced. The hours will be displayed on a gauge similar to an odometer.

CAUTION

Do not add plain water to radiators that contain antifreeze. Antifreeze not only protects radiators from freezing in cold weather, but it also has rust inhibitors and additives to protect heat-transfer surfaces. If the antifreeze is diluted, it will not provide the proper protection.

CAUTION

Adding gasoline to a diesel engine or diesel to a gasoline engine will cause severe engine problems and may create a fire hazard. Always be sure to add the correct fuel to the fuel tank.

- Clean the unit. Use a compressed air hose to blow off the engine and generator or alternator. Use a rag to remove heavier deposits that cannot be removed with the compressed air.

> **WARNING!**
>
> Always wear eye protection when using compressed air to blow dirt and debris from surfaces. Never point the compressed air nozzle at yourself or anyone else.

Most engines have an On/Off ignition switch and a starter. They may be combined into a key switch similar to the ignition on a car. To start the engine, turn on the ignition switch and press the starter. Release the starter when the engine starts. The engine speed is controlled by the governor. If the governor switch is set for idle, the engine will slow to idle after a few seconds. If the governor is set to welding speed, the engine will continue to run at the welding speed.

Small engine-driven generators may have an On/Off switch and a pull cord. These are started by turning on the ignition switch and pulling the cord, similar to how a lawn mower is started. Engine-driven generators should be started about 5 minutes to 10 minutes before they are needed for welding. This will allow the engine to warm up before a welding load is applied.

If no welding is required for 30 minutes or more, stop the engine by turning off the ignition switch. If you are finished with the welding machine for the day, close the fuel valve if there is one.

CAUTION

To prevent damage to the equipment, perform preventive maintenance as recommended by the site procedures or by the manufacturer's maintenance schedule in the equipment manual.

Engine-driven generators require regular preventive maintenance to keep the equipment operating properly. Most sites have a preventive maintenance schedule based on the hours that the engine operates. In severe conditions, such as in very dusty environments or cold weather, maintenance may have to be performed more frequently.

The responsibility for performing preventive maintenance varies by site. Check with your supervisor to determine who is responsible for performing preventive maintenance. Even though the welder may not be responsible for conducting preventive or periodic maintenance, all welders share an obligation to monitor and protect the equipment and to report problems. As is true for most crafts, a welder cannot produce welds without functional and reliable equipment.

Dust Removal

Blow out the welding machine and drive unit with compressed air occasionally to remove dust. A significant buildup of dust inside the machine can provide an alternate path for the electricity and could result in some of the components shorting out.

When performing preventive maintenance, follow the manufacturer's guidelines in the equipment manual. Typical procedures to be performed as a part of preventive maintenance include the following:

- Changing the oil
- Changing the gas filter
- Changing the air filter
- Checking/changing the antifreeze
- Greasing the undercarriage
- Repacking the wheel bearings

3.2.0 The Shielding Gas

The hose from the shielding gas regulator and flowmeter connects to the welding machine gas solenoid if the welding machine is designed for GTAW. If a standard welding machine is used, the shielding gas hose connects to the torch cable. To connect the shielding gas, perform the following steps:

Step 1 Identify the shielding gas required by referring to the WPS or site quality standard.

Step 2 Locate a cylinder of the correct gas or mixture and secure it nearby. Be sure to secure the cylinder so that it cannot fall over.

> **WARNING!**
>
> Even though they are available, do not use an adapter to connect a regulator/flowmeter equipped with one type of Compressed Gas Association (CGA) connection to a gas cylinder with a different CGA connection. The CGA connections are specific to the types of gas and cylinder pressures permitted for the connection.

Step 3 Remove the cylinder's protective cap, momentarily crack open the cylinder valve to blow out any dirt, and then close it again.

Step 4 Using a regulator or combination regulator/flowmeter with the correct CGA connection for the cylinder, mount the regulator on the cylinder.

Step 5 Connect the gas hose to the flowmeter and to the gas solenoid on the welding machine or to the end of the torch cable.

Step 6 Check to be sure the flowmeter adjusting valve is closed.

Step 7 Very slowly open the cylinder valve slightly. After the pressure gauge indicates the cylinder pressure and that it is stable, open the valve completely.

> **WARNING!**
>
> Do not open the valve quickly, or the flowmeter gauges may break. If a gauge breaks, the glass may shatter, and the gas may escape, potentially injuring the operator.

> **NOTE**
>
> If the gas hose is connected to the torch cable and the torch does not contain a gas shutoff valve, install a valve in the line between the torch cable and the regulator/flowmeter.

3.2.1 Setting the Shielding Gas Flow Rate

Gas flow rate is measured in cubic feet per hour (ft^3/hr or cfh) or liters per minute (L/min), which represents the volume of gas flowing from the torch nozzle. Flow rate is important because it affects the quality and the cost of a weld. A flow rate that is too low will not shield the weld zone adequately and will result in a poor-quality weld. An excessively high gas flow rate wastes expensive gas and can generate turbulence in the gas column above the weld. This turbulence can pull air into the weld zone and cause oxidation and weld contamination. The nozzle must be large enough to gently flood the weld pool with inert gas. Larger nozzles generally require higher flow rates. Welding specifications contain nozzle sizes and shielding gas flow rates.

Factors that may affect the shielding gas flow rate include the following:

- *Drafts* — The flow rate must be increased in a drafty location to maintain the gas shield around the weld zone.
- *Specific gas used* — For example, helium usually requires a higher flow rate than argon because helium rises much faster due to its low density.
- *Welding current* — High welding currents require higher flow rates.
- *Nozzle size (exit opening)* — Larger nozzles require higher flow rates.
- *Weld joint type* — Welds on flat surfaces require higher flow rates than welds in deep grooves or fillets.
- *Welding speed* — Fast advance speeds require higher flow rates than slower advance speeds.
- *Weld position* — Vertical- and horizontal-position welds require higher flow rates than flat or overhead welds.

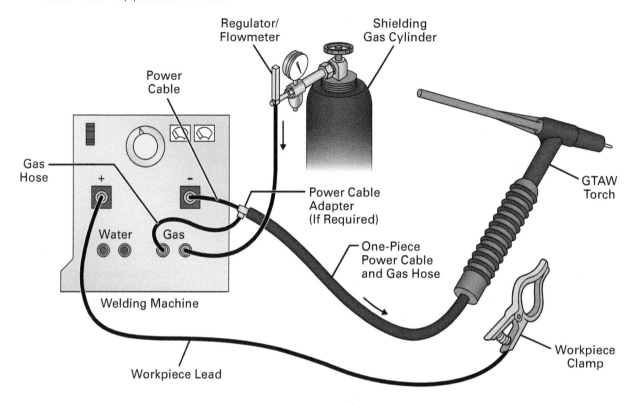

Warning: If a power cable adapter is used, it must be covered by an insulating rubber boot.

Single-Hose, Gas-Cooled Torch Setup

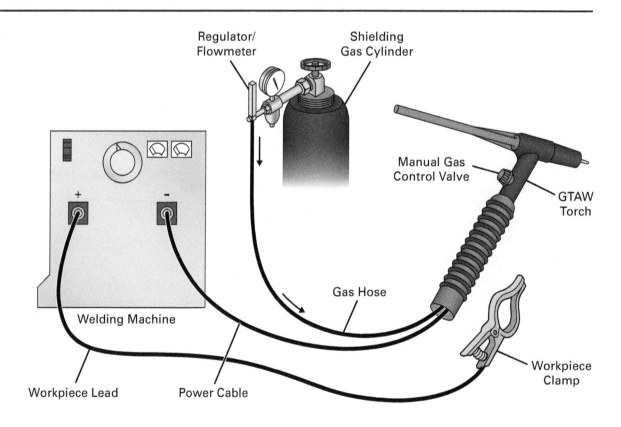

Double-Hose, Gas-Cooled Torch Setup

Figure 31 Connecting the shielding gas.

Adjust the flow for the type and thickness of the base metal being welded, or per applicable WPS. To set the shielding gas flow rate, perform the following steps:

Step 1   Start the gas flow by depressing the remote control foot pedal or hand-operated switch to open the gas solenoid on the welder, or manually open the hand valve on the torch.

Step 2   Set the gas flow by turning the flow-metering adjustment knob out until the ball inside the glass tube rises to the desired flow rate (*Figure 32*). The flow rate is usually read from the top of the floating ball. However, always check the manufacturer's instructions to determine the correct adjustment method.

Step 3   When the proper gas flow has been achieved, release the remote foot pedal or hand switch, or manually close the valve on the torch to stop the gas flow.

Step 4   Check for leaks before starting to weld.

The following sections provide general recommendations for flow rates for various types of base metals. Note that these flow rates can vary; always check the WPS for the appropriate flow rate.

3.2.2 Gas Flow Rates for Carbon Steel

Argon flow rates for GTAW on low- and medium-carbon steels and low-alloy steels (*Table 4*) typically vary from 15 cfh (7 L/min) for light welding at 60 A, to 20 cfh (9 L/min) for heavier welding at 200 A or more.

> **CAUTION**
>
> Some flowmeters are preset for a specific gas. These flowmeters do not require any adjustment, but they can only be used with the specific shielding gases indicated on the flowmeter. Due to density differences in the various shielding gases, using this type of flowmeter with the wrong gas will result in incorrect flow rate readings.

> **NOTE**
>
> Some flowmeters are equipped with several scales of different calibrations around the same sight tube for monitoring the flows of different types (densities) of gases. Be sure to rotate the scales or read the correct side for the gas type being used.

Figure 32 Adjusting the flowmeter.
Source: Terry Lowe

3.2.3 Gas Flow Rates for Stainless Steel

Argon flow rates for GTAW on stainless steel (*Table 5*) typically vary from 10 cfh (5 L/min) for light welding at 80 A, to 15 cfh (7 L/min) for heavy welding at 350 A or more.

3.2.4 Gas Flow Rates for Aluminum

Argon flow rates for GTAW on aluminum (*Table 6*) typically vary from 15 cfh (7 L/min) for light welding at 60 A, to 30 cfh (14 L/min) for heavy welding at 350 A or more.

TABLE 4 Carbon Steel—DCEN

Metal Thickness	Joint Type	Amperage	Electrode Diameter	Filler Rod Diameter	Shielding Gas	Cubic Feet per Hour
$\frac{1}{16}$" (1.6 mm)	Butt Lap Corner Fillet	60 A to 70 A 70 A to 90 A 60 A to 70 A 70 A to 90 A	$\frac{1}{16}$" (1.6 mm)	$\frac{1}{16}$" (1.6 mm)	Argon	15
$\frac{1}{8}$" (3.2 mm)	Butt Lap Corner Fillet	80 A to 100 A 90 A to 115 A 80 A to 100 A 90 A to 115 A	$\frac{1}{16}$" to $\frac{3}{32}$" (1.6 mm to 2.4 mm)	$\frac{3}{32}$" (2.4 mm)	Argon	15
$\frac{3}{16}$" (5 mm)	Butt Lap Corner Fillet	115 A to 135 A 140 A to 165 A 115 A to 135 A 140 A to 170 A	$\frac{3}{32}$" (2.4 mm)	$\frac{1}{8}$" (3.2 mm)	Argon	20
$\frac{1}{4}$" (6 mm)	Butt Lap Corner Fillet	160 A to 175 A 170 A to 200 A 160 A to 175 A 175 A to 210 A	$\frac{1}{8}$" (3.2 mm)	$\frac{5}{32}$" (4 mm)	Argon	20

TABLE 5 Stainless Steel—DCEN

Metal Thickness	Joint Type	Amperage (Flat Position)	Electrode Diameter	Filler Rod Diameter	Shielding Gas	Cubic Feet per Hour
$\frac{1}{16}$" (1.6 mm)	Butt Lap Corner Fillet	80 A to 100 A 100 A to 120 A 80 A to 100 A 90 A to 110 A	$\frac{1}{16}$" (1.6 mm)	$\frac{1}{16}$" (1.6 mm)	Argon	10
$\frac{1}{8}$" (3.2 mm)	Butt Lap Corner Fillet	120 A to 140 A 130 A to 150 A 120 A to 140 A 130 A to 150 A	$\frac{1}{16}$" (1.6 mm)	$\frac{3}{32}$" (2.4 mm)	Argon	10
$\frac{3}{16}$" (5 mm)	Butt Lap Corner Fillet	200 A to 250 A 225 A to 275 A 200 A to 250 A 225 A to 275 A	$\frac{3}{32}$" or $\frac{1}{8}$" (2.4 mm or 3.2 mm)	$\frac{1}{8}$" (3.2 mm)	Argon	15
$\frac{1}{4}$" (6 mm)	Butt Lap Corner Fillet	275 A to 350 A 300 A to 375 A 275 A to 350 A 300 A to 375 A	$\frac{1}{8}$" (3.2 mm)	$\frac{5}{32}$" (4 mm)	Argon	15
$\frac{1}{2}$" (12.5 mm)	Butt Lap Fillet	350 A to 450 A 375 A to 475 A 375 A to 475 A	$\frac{1}{8}$" or $\frac{3}{16}$" (3.2 mm or 4.8 mm)	$\frac{3}{16}$" (4.8 mm)	Argon	15

3.2.5 Gas Flow Rates for Copper

A typical argon flow rate for GTAW on deoxidized copper up to $\frac{1}{8}$" (3.2 mm) thick is 15 cfh (7 L/min) from 110 A to 250 A. With copper over $\frac{1}{8}$" (3.2 mm) thick, use helium at 30 cfh to 40 cfh (14 L/min to 19 L/min) from 190 A to 525 A.

TABLE 6 Aluminum—AC with High-Frequency Stabilization

Metal Thickness	Joint Type	Amperage (Flat Position)	Electrode Diameter	Filler Rod Diameter	Shield Gas	Cubic Feet per Hour
$\frac{1}{16}$" (1.6 mm)	Butt Lap Corner Fillet	60 A to 85 A 70 A to 90 A 60 A to 85 A 75 A to 100 A	$\frac{1}{16}$" (1.6 mm)	$\frac{1}{16}$" (1.6 mm)	Argon	15
$\frac{1}{8}$" (3.2 mm)	Butt Lap Corner Fillet	125 A to 150 A 130 A to 160 A 120 A to 140 A 130 A to 160 A	$\frac{3}{32}$" (2.4 mm)	$\frac{3}{32}$" or $\frac{1}{8}$" (2.4 mm or 3.2 mm)	Argon	20
$\frac{3}{16}$" (5 mm)	Butt Lap Corner Fillet	180 A to 225 A 190 A to 240 A 180 A to 225 A 190 A to 240 A	$\frac{1}{8}$" (3.2 mm)	$\frac{1}{8}$" (3.2 mm)	Argon	20
$\frac{1}{4}$" (6 mm)	Butt Lap Corner Fillet	240 A to 280 A 250 A to 320 A 240 A to 280 A 250 A to 320 A	$\frac{3}{16}$" (4.8 mm)	$\frac{1}{8}$" or $\frac{3}{16}$" (3.2 mm or 4 mm)	Argon	25
$\frac{1}{2}$" (12.5 mm)	Butt Lap Corner Fillet	400 A to 450 A 400 A to 450 A 400 A to 450 A 420 A to 470 A	$\frac{3}{16}$" or $\frac{1}{4}$" (4.8 mm or 6.4 mm)	$\frac{3}{16}$" or $\frac{1}{4}$" (4.8 mm or 6.4 mm)	Argon	30

Reading Torch Flow Rate

Welders must maintain accurate and consistent torch flow rates to create consistent welds and adhere to a WPS. The device shown here can be used to quickly measure shielding gas flow directly at the torch tip. This simple peashooter style is durable and convenient. Digital models offer a bit more sophistication. They can also be mounted on the gas line, providing a constant digital display of flow rate.

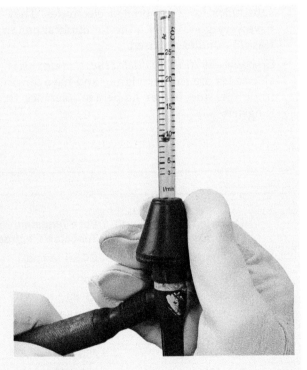

Source: Terry Lowe

3.3.0 The Electrode

The GTAW tungsten-based electrode must be selected and the end properly prepared before it can be installed in the torch. Correct preparation of the electrode is absolutely essential. If the tip is improperly shaped, it will not produce the required arc shape and characteristics.

When selecting a GTAW electrode, choose the type recommended for the welding current required by the base metal. The type of electrode will be specified in the WPS or site quality standards. General recommendations are as follows:

- *Pure tungsten (EWP)* — These electrodes are used for welding aluminum and magnesium with AC and high frequency.
- *Zirconiated tungsten (EWZr)* — Used for welding aluminum and magnesium with AC and high frequency for welds where tungsten inclusions are not tolerated, and higher current capacity is desired.
- *Thoriated tungsten (EWTh-1 or EWTh-2)* — These electrodes are used for mild steels, alloy steels, and stainless steels with DCEN current.

> **WARNING!**
>
> Thoriated tungsten electrodes contain low levels of radioactivity and may pose health risks.

- *Ceriated electrodes* — These electrodes also operate successfully with AC current. Ceriated and lanthanated electrodes are used as nonradioactive substitutes for thoriated electrodes.
- *Rare earth electrodes (EWG)* — These electrodes are used as nonradioactive substitutes for 2% thoriated electrodes. They start and reignite well and have very good service life for applications in which welding cycles of at least 15 minutes are used.
- *Cryogenically treated thoriated tungsten electrodes* — The cryogenically treated electrodes are easier to ignite and have longer tip life, longer overall electrode life, and higher amperage tolerance than nontreated 2% thoriated tungsten.

There Is No Universal Electrode Size

Contrary to some popular beliefs, the $3/32$" (2.4 mm) electrode is not a universal size that can be used for everything. Always use the electrode size recommended for the amperage required. Using an electrode that is too big will result in arc starting and stability problems. Using an electrode that is too small can result in tungsten spitting (tungsten melting and migration into the weld).

TABLE 7 Electrode Current Ranges

Electrode Diameter	GTAW Modes			
	Pure Tungsten	Thoriated or Rare Earth Electrodes*	Pure Tungsten, Rare Earth, or Thoriated Tungsten Electrodes	
	AC (amps)**	AC (amps)**	DCEN (amps)	DCEP (amps)
0.020" (0.5 mm)	5–15	8–20	8–20	—
0.040" (1.0 mm)	10–60	15–80	15–80	—
$1/16$" (1.6 mm)	50–100	70–150	70–150	10–20
$3/32$" (2.4 mm)	100–160	140–235	150–250	15–30
$1/8$" (3.2 mm)	150–210	225–325	250–400	25–40
$5/32$" (4.0 mm)	200–275	300–425	400–500	40–55
$3/16$" (4.8 mm)	250–350	400–525	500–800	55–80
$1/4$" (6.4 mm)	325–475	500–700	800–1,000	80–125

*Balled electrode tip ends can best be sustained at these current levels.
**Maximum values shown have been determined using an unbalanced AC power source. If a balanced AC power source is used, either reduce these values by about 30% or use the next size electrode. This is necessary because of the higher heat input to the electrode in a balanced AC setup.

Select the size of electrode rated for the amperage to be used. *Table 7* lists examples of electrode current ranges by electrode size and type when used with argon shielding gas under various GTAW modes.

3.3.1 Preparing the Electrode

The different types of GTAW welding current require differently shaped electrode tips. For manual DCEN welding, a taper-ground and truncated point is the best shape. For manual DCEP welding, a rounded end is the best shape. For manual AC welding, a tapered and balled end is the best shape. *Figure 33* shows GTAW electrode end shapes.

Electrode tips must be formed on all new electrodes and on used electrodes that have damaged or contaminated tips. The contaminated section must be reshaped. Do not prepare the end of the electrode that contains the color code band, or else you will not be able to identify the electrode's type in the future.

The degree of taper on an electrode tip affects weld penetration and bead width. A long, narrow taper (small angle, more pointed) produces a wide bead with shallow penetration. A short, wide taper (large angle, less pointed) produces a narrow bead and deep penetration. During any welding operation, maintain the same taper on the electrode during sharpening or replacement so that the bead characteristics remain the same.

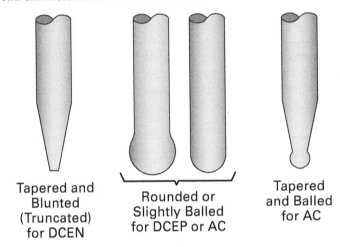

Tapered and
Blunted
(Truncated)
for DCEN

Rounded or
Slightly Balled
for DCEP or AC

Tapered
and Balled
for AC

Figure 33 GTAW electrode end shapes for different current types.

3.3.2 Pointing the Electrode Tip

For DCEN welding, electrodes can be made pointed by dipping the red hot end into a special chemical powder, or they can be tapered by grinding or sanding.

Perform the following steps to use a chemical powder to sharpen a tungsten electrode:

Step 1   Place the tungsten electrode in the torch so that it extends about 1" (25 mm) from the gas nozzle.

Step 2   Heat the tungsten electrode by shorting it on a copper plate or by striking an arc. Be sure shielding gas is flowing to protect the tungsten electrode.

Step 3   When the tip of the tungsten electrode is cherry red for a length about four times the electrode diameter, place the end of the electrode in the chemical powder. For example, if the electrode diameter is $\frac{1}{8}$" (3.2 mm), about $\frac{1}{2}$" (13 mm) of the length should be cherry red.

Step 4   Hold the end of the tungsten electrode in the powder for the time recommended on the container for the diameter of the electrode being pointed. The chemical powder will dissolve the tungsten electrode and form a sharp point at the proper angle. For an untapered (new or broken off) tungsten, the procedure may have to be repeated several times to obtain the desired shape.

Figure 34 shows the chemical powder used for sharpening.

Figure 34 Tungsten sharpening powder.

More often, tungsten electrodes are sharpened by grinding or sanding. Since tungsten electrodes are very hard, they are best ground with diamond-impregnated silicon carbide or aluminum oxide grinding wheels or sanding belts. Use 80 grit for fast shaping and 120 grit for finishing.

WARNING!

Observe proper respiratory, hazard, and environmental contamination procedures when using a grinder. Dust from tungsten electrode grinding can be hazardous and may contain radioactive elements. Refer to the manufacturer's SDS/MSDS to determine any specific hazards. Safety glasses must be worn when using a grinder.

Follow these steps to point the end of a tungsten electrode using a grinder:

Step 1 With the color band end held farthest from the grinder, hold the tungsten electrode against the grinder so that the tungsten electrode is in the plane of the wheel and makes an angle of about 30° with the wheel or belt surface (*Figure 35*).

WARNING!

When grinding tungsten electrodes, gloves may get snagged in rotating parts. Before grinding a tungsten electrode, be sure to check that the grinding wheel is in safe operating condition with all required guards and shields.

Step 2 Slowly rotate the tungsten electrode as it is ground to keep the point on the center line of the axis. When rotating the tungsten, also move the tungsten side to side so the grinding wheel wears away evenly. Not doing so can create ridges in the grinding wheel, which is a safety hazard. Orient the grind marks toward the point along the long dimension of the tungsten electrode, not around the point. For DCEN electrodes, the length of the taper should be between two and three times the electrode diameter. The point should be truncated to the approximate tip diameter shown in *Table 8*.

Step 3 Check the ground end to make sure the tapered length is correct and that the point is in the center of the tungsten electrode and not toward one side. Correct the shape if necessary.

Figure 36 shows preferred and incorrect grinds for prepared pointed electrode ends.

CAUTION

Always wear clean gloves when handling tungsten electrodes to avoid contaminating the electrode.

CAUTION

Use a grinding wheel designated only for tungsten. Be careful to never use the grinding wheel for any other material, or contaminants will be deposited on the tungsten electrode and the grinding wheel will have to be discarded. Always wear safety glasses when using a grinder.

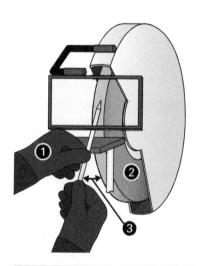

❶ Clean Gloves

❷ Grinding Wheel

❸ Tungsten Electrode Angle to Wheel Surface (Usually 15° to 30°)

Figure 35 Proper grinding angle and tungsten electrode orientation.

TABLE 8 Tip Diameters

Electrode Diameter	Diameter at Tip
0.040" (1.0 mm)	0.005" (0.13 mm)
0.040" (1.0 mm)	0.010" (0.25 mm)
0.062" (1.6 mm)	0.020" (0.51 mm)
0.062" (1.6 mm)	0.030" (0.76 mm)
0.093" (2.4 mm)	0.030" (0.76 mm)
0.093" (2.4 mm)	0.045" (1.14 mm)
0.125" (3.2 mm)	0.045" (1.14 mm)
0.125" (3.2 mm)	0.060" (1.52 mm)

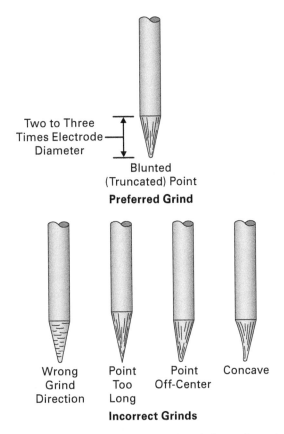

Two to Three Times Electrode Diameter

Blunted (Truncated) Point

Preferred Grind

Wrong Grind Direction   Point Too Long   Point Off-Center   Concave

Incorrect Grinds

Figure 36 Properly and improperly prepared pointed electrode ends.

Diamond Wheel Grinders

Diamond wheel grinders are used to prepare tungsten electrodes. Both stationary and handheld, portable grinders are available. A stationary water-bath diamond wheel grinder (A) can be set to different electrode grind angles as necessary. A water bath captures the tungsten dust that is generated during grinding, so the dust can be disposed of later. A handheld, portable diamond wheel grinder (B), can also be set to different electrode grind angles as necessary. The dust generated during grinding is captured inside the grinder's fully enclosed grind head. The grinder also comes with an attachment that allows the grinder to be connected to a shop vacuum, so accumulated dust can be removed from the grinding chamber.

(B) Handheld Grinder

(A) Stationary Grinder

Sources: Courtesy of Inelco Grinders A/S (A); CK Worldwide, Inc. (B)

3.3.3 Balling the Electrode Tip

Only pure tungsten (EWP) or zirconiated tungsten (EWZr) can be balled, or spherical. The thorium in thoriated tungstens or other rare earth (except cerium) tungstens inhibits the formation of a ball. If welders attempt to ball most rare earth tungstens, the tungsten forms several small protrusions at the tip instead of a ball. These protrusions cause the arc to be unstable.

If the electrode is to be used for lower-amperage DCEP welding, the electrode end should be tapered about two diameters in length and be quite blunt. A ball will form on the blunt end when welding.

For AC welding, the end is balled with or without tapering. Balling is done by arcing the electrode over clean copper or another clean metal. Do not arc over carbon, because it will contaminate the electrode.

AC or DCEP can be used to ball an electrode. When balling an electrode, start off with a low current and gradually increase it until the end of the electrode starts to melt and form a hemispherical (half round) end or slight ball. The ball should be no more than $1\frac{1}{2}$ times the electrode diameter. Using DCEP will cause a ball to form at a much lower amperage setting than would be formed using AC current. Using DCEP, only about 35 A to 45 A of welding current will be required to ball $\frac{3}{32}$" (2.4 mm) to $\frac{1}{8}$" (3.2 mm) diameter electrodes. *Figure 37* shows ideal, acceptable, and unacceptable prepared balled electrode ends.

CAUTION

Using excess amperage will cause the ball to melt and drop off the electrode. Be especially careful when using DCEP as it will generate much more heat on the electrode than the same AC setting.

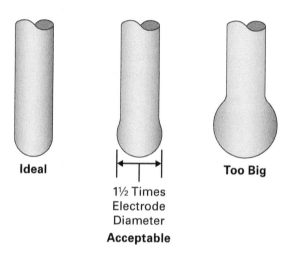

Ideal

$1\frac{1}{2}$ Times Electrode Diameter

Acceptable

Too Big

Figure 37 Properly and improperly prepared balled, nontapered electrode ends for AC welding.

3.4.0   Selecting and Installing the Nozzle

Typically, a range of nozzle sizes and styles are available for a given torch to accommodate different electrode sizes and welding applications. Nozzles are made for close clearances, for long reaches, for reaching deep into narrow places, and for covering wide beads. Discard nozzles with chips, cracks, or metal buildup on the end because they can affect the gas flow pattern and produce weld defects.

The nozzle must be sized for the electrode diameter to be used and for any unique welding requirements, such as reaching into a V-groove. The nozzle diameter must be large enough to cover the entire weld area with shielding gas. A wide, easy flow of shielding gas is generally preferable to a narrow jet, which is apt to create turbulence and draw air into the weld zone. Small-diameter nozzles tend to overheat and break easily. However, a small-diameter nozzle is useful for steadying the torch and arc for the root pass in a V-groove weld.

To select the proper size and shape nozzle, always refer to the manufacturer's nozzle recommendations for the torch being used. In general, the inside diameter of the nozzle should be about three times the diameter of the electrode being used.

3.4.1 Installing the Electrode

The electrode is usually installed after the nozzle because the electrode stickout is measured from the end of the nozzle. The electrode is clamped in place in the torch body by the collet assembly. The collet is a cylindrical clamp that tightens as it is pulled or pushed (depending on the torch design) into a tapered holder as the threaded electrode cap is screwed on and tightened. To adjust the electrode stickout or to remove the electrode, loosen the cap assembly. Collet designs vary with torch manufacturers (*Figure 38*).

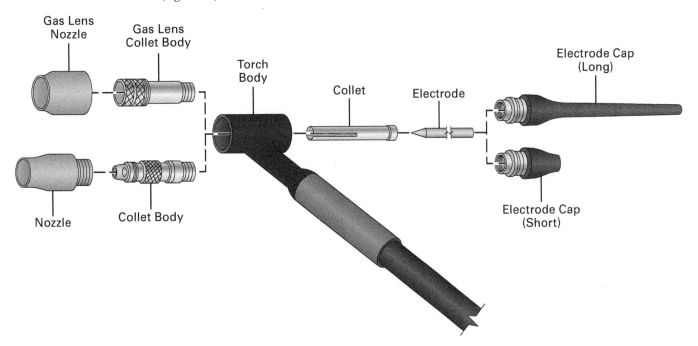

Figure 38 Collet assembly of a typical GTAW torch.

Follow these steps to install the electrode:

Step 1   Remove the electrode cap from the rear of the torch.

Step 2   Check the collet size and collet body. If it is not the correct size for the electrode to be installed, remove and replace it with the correct size.

Step 3   Replace the electrode cap, but do not tighten it. Screw the cap only about one turn to leave the collet clamp loose.

Step 4   Insert the electrode into the collet from the nozzle end of the torch, and adjust it so that it extends from the nozzle about 1 to $1\frac{1}{2}$ times the electrode diameter. Insert the color-coded end first to prevent destroying the electrode markings.

Step 5   Hand tighten the electrode cap to lock the electrode in place.

CAUTION

Handle tungsten electrodes with clean gloves to avoid contamination of the electrode.

CAUTION

If the electrode cap is overtightened, the threads will be damaged as the torch heats with use. If the cap is too loose, the collet and electrode will overheat and damage the torch.

Electrode Stickout

Various manufacturers recommend stickout distances that vary from flush at the end of the nozzle to distances equal to the inside diameter of the nozzle, depending on the application and task. For instance, a root pass in a deep, narrow groove can be accomplished more easily with a narrow nozzle and a stickout that is about equal to the inside diameter of the nozzle. However, the use of a gas lens allows a longer tungsten stickout for welding under the same circumstances.

3.0.0 Section Review

1. Before plugging a welding machine into an electrical outlet, be sure to identify the location of the _____.
   a. supplemental oxygen source
   b. backup generator
   c. alternator back cap
   d. electrical disconnect

2. When connecting the shielding gas for GTAW, if the gas hose is connected to the torch cable and the torch does not contain a gas shutoff valve, install a valve in the line between the torch cable and the _____.
   a. welding machine's gas solenoid
   b. gas cylinder's regulator/flowmeter
   c. gas cylinder's protective cap
   d. remote control foot pedal

3. Welders should select the size of GTAW electrode rated for the _____.
   a. amperage to be used
   b. voltage of the welding machine
   c. temperature at the worksite
   d. required type of shielding gas

4. In general, the inside diameter of the nozzle should be about how many times the diameter of the GTAW electrode that is being used?
   a. One
   b. Two
   c. Three
   d. Four

Module 29207   Review Questions

1. Welders wear safety clothing to reduce the risk of injury when _____.
   a. storing tungsten electrodes
   b. grinding tungsten electrodes
   c. shipping tungsten electrodes
   d. counting tungsten electrodes

2. When making GTAW welds, *always* use safety glasses with a(n) _____.
   a. flame-retardant coating
   b. air-supplied respirator
   c. full face shield or a helmet
   d. OSHA-approved dust mask

3. When welding, the gas that should *never* be released in a large amount as compressed air around flammable materials is _____.
   a. oxygen
   b. argon
   c. carbon dioxide
   d. nitrogen

4. After leaving their sources, vapors and fumes tend to _____.
   a. fall in the air
   b. remain level
   c. be visible
   d. rise in the air

5. When a GTAW arc is struck across a gap, the resistance generates heat in the range of _____.
   a. 1,000°F to 2,000°F (550°C to 1,100°C)
   b. 2,000°F to 5,000°F (1,100°C to 2,750°C)
   c. 6,000°F to 10,000°F (3,300°C to 5,500°C)
   d. 15,000°F to 18,000°F (8,300°C to 10,000°C)

6. The duty cycle of a welding machine is based on a period of how many minutes?
   a. 5
   b. 10
   c. 15
   d. 20

7. What percent duty cycle do most heavy-duty industrial machines used for manual welding have?
   a. 50
   b. 100
   c. 50 or 60
   d. 60 or 100

8. An air-cooled GTAW torch that contains a manual gas control valve is typically used if the power source for the torch does *not* have a(n) _____.
   a. backup generator for the shielding gas supply
   b. electric solenoid valve to automatically control the shielding gas flow
   c. backflow prevention valve in the line leading to the electrode
   d. overload connection to a heat exchanger or cooling fan

9. Which of the following types of electrodes is radioactive?
   a. Lanthanated
   b. Thoriated
   c. Ceriated
   d. Zirconiated

10. How many times heavier is argon shielding gas than helium?
    a. 5
    b. 8
    c. 10
    d. 20

11. When transporting and handling cylinders of shielding gases, installation of the safety cap over the valve is _____.
    a. a site-specific decision
    b. mandatory
    c. optional
    d. not recommended

12. What valve is used to adjust the gas flow to the torch nozzle?
    a. Metering
    b. Pressure
    c. Tapering
    d. Shielding

13. The filler metal used for manual GTAW is generally supplied in diameters of $\frac{1}{16}$" (1.6 mm) to $\frac{1}{4}$" (6.4 mm) and lengths of _____.
    a. 48" (1,219 mm)
    b. 36" (914 mm)
    c. 24" (610 mm)
    d. 12" (305 mm)

14. The aluminum alloy filler metal rod ER4043, widely used in GTAW, also contains _____.
    a. iron
    b. flux
    c. silicon
    d. tungsten

15. When setting up GTAW equipment, the shielding gas supply must be located reasonably close to the welding site because of the _____.
    a. limited length of GTAW torch cables
    b. risk of acetylene dispersion at the worksite
    c. weight and immobility of the gas cylinders
    d. need to refuel the torch frequently

16. When a standard welding machine is used for a GTAW operation, the shielding gas hose connects to the _____.
    a. torch cable
    b. electrode back cap
    c. retaining cup
    d. nozzle gasket

17. The type of GTAW electrode to be used for a welding operation is specified in _____.
    a. OSHA regulations
    b. ASTM and ANSI appendixes
    c. SDS/MSDS and NFPA requirements
    d. WPS or site quality standards

18. When using sanding to taper the point on a tungsten electrode, what size grit should a welder use for finishing?
    a. 60
    b. 80
    c. 120
    d. 200

19. When selecting and installing the nozzle, the inside diameter of the nozzle, compared to the diameter of the electrode, should generally be _____.
    a. 3 times larger
    b. equal
    c. 10 times larger
    d. 15 times larger

20. A GTAW electrode is usually installed after the nozzle because the _____.
    a. nozzle grounds the entire torch assembly
    b. electrode stickout is measured from the end of the nozzle
    c. electrode back cap threads onto the nozzle to form a seal
    d. nozzle determines the maximum outer diameter of the electrode

---

Answers to odd-numbered questions are found in the Review Question Answer Keys at the back of this book.

Answers to Section Review Questions

Answer	Section Reference	Objective
Section 1.0.0		
1. c	1.1.3	1a
2. a	1.2.2	1b
Section 2.0.0		
1. b	2.1.3	2a
2. c	2.2.0	2b
3. d	2.3.1	2c
4. d	2.4.1	2d
5. a	2.5.4	2e
Section 3.0.0		
1. d	3.1.0	3a
2. b	3.2.0	3b
3. a	3.3.0	3c
4. c	3.4.0	3d

Source: Photo courtesy of The Lincoln Electric Company, Cleveland, OH, U.S.A.

GTAW – Plate

Objectives

Successful completion of this module prepares you to do the following:

1. Identify GTAW-related safety practices and explain how to set up for welding.
   a. Identify GTAW-related safety practices.
   b. Explain how to safely set up the equipment and work area for welding.
2. Describe welding techniques for GTAW and explain how to produce basic weld beads.
   a. Describe welding techniques related to GTAW.
   b. Explain how to produce basic GTAW weld beads.
3. Describe the welding techniques needed to produce proper fillet and open V-groove welds using GTAW welding techniques.
   a. Describe the welding techniques needed to produce proper fillet welds using GTAW.
   b. Describe the welding techniques needed to produce proper open V-groove welds using GTAW.

Performance Tasks

Under supervision, you should be able to do the following:

1. Build a pad with stringer beads on carbon steel plate coupons in the flat position using GTAW equipment and carbon steel filler metal.
2. Perform multipass fillet welds on carbon steel plate coupons in all four 1F through 4F positions using GTAW equipment and carbon steel filler metal.
3. Perform multipass open V-groove welds on carbon steel plate coupons in all four 1G through 4G positions using GTAW equipment and carbon steel filler metal.

Overview

Gas tungsten arc welding (GTAW) is one of several types of arc welding processes that uses the heat from an electric arc to join a filler metal and a base metal. The arc is generated between a tungsten electrode and the base metal, and a handheld rod provides the filler metal. This module identifies GTAW-related safety practices, describes common GTAW equipment, and examines different techniques used to produce various types of weld beads.

Digital Resources for Welding

NCCER

SCAN ME

Scan this code using the camera on your phone or mobile device to view the digital resources related to this craft.

1.0.0 GTAW Plate Welding

Performance Tasks

There are no Performance Tasks in this section.

Objective

Identify GTAW-related safety practices and explain how to set up for welding.

  a. Identify GTAW-related safety practices.

  b. Explain how to safely set up the equipment and work area for welding.

In gas tungsten arc welding (GTAW), an arc is generated between a nonconsumable tungsten electrode and the base metal to melt together the base metal and a filler metal (*Figure 1*). The filler metal is a rod of similar composition to the base metal, usually handheld and manually fed into the leading edge of the weld puddle. The GTAW process is often called *tungsten inert gas (TIG) welding* because the weld area is protected from atmospheric contamination by a flow of inert gas from the torch nozzle and a tungsten-based electrode. In GTAW, the electrode and the filler metal are not one and the same.

GTAW produces high-quality welds without slag or oxidation. Since there is no flux, there can be no corrosion due to flux entrapment, so no postweld cleaning is necessary. An exception is the slag left by some flux-cored rods, which must be removed. Flux paste, also called *backup flux*, is another method used to provide root protection for open-root welds when weld quality requirements are less critical. To remove the flux paste after welding, use a wire brush and hot water. Before using a flux paste, ensure it is permitted by the WPS.

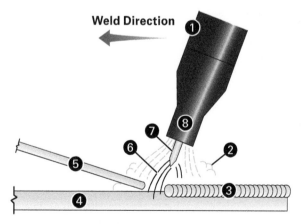

Weld Direction

❶ GTAW Torch	❺ Filler Metal
❷ Shielding Gas Flow	❻ Arc
❸ Weld Bead	❼ Tungsten Electrode
❹ Base Metal	❽ GTAW Torch Nozzle

Figure 1 GTAW process.

1.1.0   Safety Summary

The following sections summarize the safety procedures and practices that welders must observe when cutting or welding. Complete safety coverage is provided in NCCER Module 29101, *Welding Safety*, which should be completed before continuing with this module. The safety procedures and practices are in place to prevent potentially life-threatening injuries from welding accidents. Above all, be sure to wear appropriate protective clothing and equipment when welding or cutting.

1.1.1 Protective Clothing and Equipment

Welding activities can cause injuries unless you wear all the protective clothing and equipment designed specifically for the welding industry. The following safety guidelines about protective clothing and equipment should be followed to prevent injuries:

- Wear a face shield over snug-fitting cutting goggles or safety glasses for gas welding or cutting. The welding hood should be equipped with an approved shade for the application. A welding hood equipped with a properly tinted lens is best for all forms of welding.

- Wear proper protective leather and/or flame-retardant clothing along with welding gloves that protect the welder from flying sparks, molten metal, and heat.

- Wear high-top safety shoes or boots. Make sure the tongue and lace area of the footwear will be covered by a pant leg. If the tongue and lace area is exposed or the footwear must be protected from burn marks, wear leather spats under the pants or chaps over the front and top of the footwear.

- Wear a 100% cotton cap with no mesh material included in its construction. The bill of the cap points to the rear. If a hard hat is required for the environment, use one that allows the attachment of rear deflector material and a face shield. A hard hat with a rear deflector is generally preferred when working overhead and may be required by some employers and jobsites.

- Wear earmuffs, or at least earplugs, to protect your ear canals from sparks.

WARNING!

Using proper personal protective equipment (PPE) for the hands and eyes is particularly important. The most common injuries that welders experience during GTAW operations are injuries to the fingers and eyes. When they are holding and grinding the tungsten electrode by hand, welders who fail to use proper gloves, safety glasses, and face shields can cut or burn their fingers, have splintered tungsten electrode lodge in their hands or fingers, or get small slivers of the electrode stuck in their eyes.

1.1.2 Fire/Explosion Prevention

Welding activities usually involve the use of fire or extreme heat to melt metal. Whenever fire is used in a weld, the fire must be controlled and contained. Welding or cutting activities are often performed on vessels that may have once contained flammable or explosive materials. Residues from those materials can catch fire or explode when a welder begins work on such a vessel. The following fire and explosion prevention guidelines associated with welding contribute to a safe work zone:

- Never carry matches or gas-filled lighters in your pockets. Sparks can cause the matches to ignite or the lighter to explode, causing serious injury.

- Never use oxygen to blow dust or dirt from clothing. The oxygen can remain trapped in the fabric for a time. If a spark hits clothing during this time, the clothing can burn rapidly and violently out of control.

- Make sure any flammable material in the work area is moved or shielded by a fire-resistant covering. Approved fire extinguishers must be available before attempting any heating, welding, or cutting operations. If a hot work permit and a fire watch are required, be sure those items are in place before beginning and that all site requirements have been observed.

- Never release a large amount of oxygen or use oxygen as compressed air. The presence of oxygen around flammable materials or sparks can cause rapid and uncontrolled combustion. Keep oxygen away from oil, grease, and other petroleum products.

- Never release a large amount of fuel gas, especially acetylene. Propane tends to concentrate in and along low areas and can ignite at a considerable distance from the release point. Acetylene is lighter than air but is even more dangerous than propane. When mixed with air or oxygen, acetylene will explode at much lower concentrations than any other fuel.

- To prevent fires, maintain a neat and clean work area, and make sure that any metal scrap or slag is cold before disposing of it.

- Before cutting containers such as tanks or barrels, check to see if they have contained any explosive, hazardous, or flammable materials, including petroleum products, citrus products, or chemicals that decompose into toxic fumes when heated. As a standard practice, always clean and then fill any tanks or barrels with water or purge them with a flow of inert gas such as nitrogen to displace any oxygen. Containers must be cleaned by steam cleaning, flushing with water, or washing with detergent until all traces of the material have been removed.

WARNING!

Welding or cutting must never be performed on drums, barrels, tanks, vessels, or other containers until they have been emptied and cleaned thoroughly, eliminating all flammable materials and all substances (such as detergents, solvents, greases, tars, or acids) that might produce flammable, toxic, or explosive vapors when heated. Clean containers only in well-ventilated areas, as vapors can accumulate during cleaning, causing explosions or injury.

Proper procedures for cutting or welding hazardous containers are described in the *American Welding Society (AWS) F4.1, Safe Practices for the Preparation of Containers and Piping for Welding, Cutting, and Allied Processes.*

1.1.3 Work Area Ventilation

Vapors and fumes tend to rise in the air from their sources. Welders must often work above the welding area where the fumes are being created. Welding fumes can cause personal injuries including long-term respiratory harm. Good work area ventilation helps to remove the vapors and protect the welder. The following is a list of work area ventilation guidelines to consider before and during welding activities:

- Follow confined space procedures before conducting any welding or cutting in a confined space.

- Never use oxygen for ventilation in confined spaces.

- Always perform cutting or welding operations in a well-ventilated area. Cutting or welding operations involving zinc or cadmium materials or coatings result in toxic fumes. For long-term cutting or welding of such materials, always wear an approved full-face, supplied-air respirator that uses breathing air supplied from outside the work area. For occasional, very short-term exposure, a high-efficiency particulate arresting-rated (also called HEPA-rated) or metal fume filter may be used on a standard respirator.

- Make sure confined spaces are properly ventilated for cutting or welding purposes. Use powered extraction systems when available.

WARNING!

Backup flux may contain silica, fluorides, or other toxic materials. Alcohol and acetone used for the mixture are flammable. Weld only in well-ventilated spaces. Failure to provide proper ventilation could result in personal injury or death.

1.1.4 GTAW-Specific Safety

GTAW, also referred to as *Heliarc* or TIG welding, is a different process than those used in SMAW, GMAW, and FCAW. In other forms of welding, the electrode or wire is consumed in the welding process. However, in GTAW, a separate rod of filler metal is consumed rather than the electrode.

Vapors from the melted base metal, filler metal, and shielding gases used in the GTAW process can create respiratory issues. As always, welders performing any kind of work must be aware of the vapors generated from the welding processes. Argon gas is often used for shielding in the GTAW process. A combination of the arc and gas generates substantial levels of ultraviolet (UV) light, which reacts with the oxygen near the weld. UV radiation can ionize oxygen in the atmosphere near the weld, which can lead to a rise in ozone and nitrogen dioxide levels. If the work is being performed in well-ventilated areas, the risks are minimal. If work must be performed in confined spaces, consider using additional ventilation or respirators. Follow your employer's safety policies in these situations.

Because GTAW work is often performed on delicate welds, the welder is frequently positioned very close to the work. The risk of UV radiation being emitted from the electrical arc is higher because the shielding gas used with GTAW is more transparent than gases used with other forms of welding. This transparency is due to the minimal smoke produced from the weld. Because of the brightness of the GTAW process, safety representatives recommend that the welder wear darker clothing to minimize UV reflections under the welding helmet. They also recommend that the walls of the work area be painted with a paint that reduces reflections from the UV light generated by the GTAW process. Because aluminum and stainless steel reflect more light than common carbon steels, welders need to take extra precautions to protect their vision when performing GTAW work on these metals.

The GTAW process requires the base metal and filler metal to be extremely clean. Cleaning solvents are used to prepare the metals. Welders must read all Safety Data Sheet (SDS) information about any cleaning agent before using it. Some cleaning agents are flammable, while others are nonflammable. Make sure you know which kind you are using. Pay special attention to any storage requirements for such cleaning agents.

> **WARNING!**
>
> Chlorinated cleaning solvents can produce toxic gases. For this reason, they should never be used for cleaning metals in a welding setting.

The tungsten electrodes used for GTAW become contaminated over time and with use. When electrodes lose their original shapes, welders must grind the tungsten tips to reshape them and make them suitable for continued use. The tungsten dust and flying particles from these grinding activities can damage the welder's vision and present a respiratory hazard. Read and understand the SDS that comes with each batch of electrodes to protect your health in the workplace.

> **WARNING!**
>
> Observe proper respiratory, hazard, and environmental contamination procedures when using a grinder. Dust from tungsten electrode grinding can be hazardous. Refer to the manufacturer's SDS/MSDS to determine any specific hazards. Safety glasses must be worn when using a grinder.

1.2.0 GTAW Welding Equipment Setup

Before welding can take place, the work area must be made ready, the welding equipment must be set up, and the metal to be welded must be prepared. The following sections explain how to prepare the area and set up the equipment that will be used to perform GTAW welding of carbon steel plate.

1.2.1 Preparing the Welding Area

Practice welding on a welding table, bench, or stand. The welding surface must be nonflammable and electrically conductive, and provisions must be made for mounting practice weld coupons out of position.

Follow these steps to set up an area for welding:

Step 1 Make sure the area is properly ventilated. Make use of doors, windows, and fans.

Step 2 Check the area for fire hazards. Remove any flammable materials before proceeding.

Step 3 Know the location of the nearest fire extinguisher and how to use it. Do not proceed unless the extinguisher is charged.

Step 4 Set up welding curtains around the welding area.

1.2.2 Preparing the Practice Weld Coupons

Use the material thickness as directed by the instructor. If possible, the practice weld coupons should be $\frac{1}{4}$" (6 mm) thick.

Clean the steel plate before welding by using a wire brush or grinder to remove mill scale in the weld zone and a minimum of $\frac{1}{2}$" (13 mm) beyond the weld zone. This is done because the direct current electrode negative (DCEN) used in GTAW has little or no cleaning action.

The following outlines how to prepare weld coupons for various methods of practice:

- *Running beads* — The coupons can be any size or shape that can be easily handled.

- *Overlapping beads* — The coupons can be any size or shape that can be easily handled.

- *Fillet welds* — Cut the metal into 2" × 6" (51 mm × 152 mm) rectangles for the base and 1" × 6" (25 mm × 152 mm) rectangles for the web.

- *Open V-groove welds* — As shown in *Figure 2*, cut the metal into 3" × 7" (about 76 mm × 178 mm) rectangles with one (or both) of the lengths beveled at 30° to 37.5°. Grind a 0" to $\frac{1}{8}$" (0 mm to 3.2 mm) root face on one bevel as directed by your instructor.

NOTE

Reference *Appendix 29208A* for Performance Accreditation Tasks (PATs) designed to evaluate your ability to run fillet and groove welds with GTAW equipment.

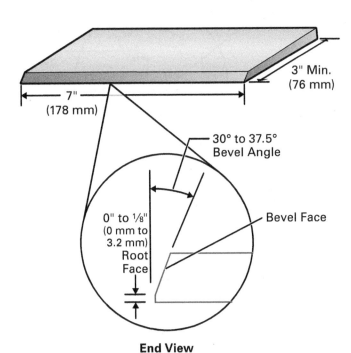

End View
Note: Base Metal Carbon Steel at Least ¼" (6 mm) Thick

Figure 2 Metal cut for open V-groove weld coupons.

Follow these steps to prepare each open V-groove weld coupon:

Step 1 Check the bevel face. There should be no dross and a 0" to $\frac{1}{8}$" (0 mm to 3.2 mm) root face. The bevel angle should be 30° to 37.5°. An example of this is shown in *Figure 3*.

Step 2 Center the beveled strips using a root opening determined by the instructor and tack-weld them in place. Place the tack welds on the ends of the joint.

Going Green

Recycling Material

When metals are no longer usable in the welding shop or on the job, the metals should be collected and sold to a recycling company rather than sent to a landfill. Selling the scrap metal can make money for the company or welding school and protect the environment from resource depletion. Steel is the most recycled material in the world.

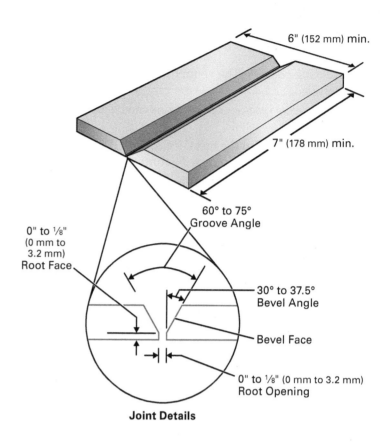

Joint Details

Note: Base Metal Carbon Steel, at Least ⅜" (10 mm) Thick

Figure 3 Open V-groove weld coupon.

1.2.3 The Welding Machine

Identify the proper welding machine for GTAW use and follow these steps to set it up for welding:

Step 1 Verify that the welding machine can be used for GTAW, with or without internal gas shielding control. If desired, identify an optional cooling unit.

Step 2 Identify an air- or water-cooled GTAW torch. Make sure it is compatible with the welding machine and any cooling unit.

Step 3 Verify the location of the primary disconnect.

Step 4 Configure the welding machine and torch for GTAW welding (*Figure 4*) as directed by the instructor. Configure the torch polarity and equip the torch with a tungsten electrode of the correct diameter and properly prepared type for the filler metal used for the application.

Step 5 Connect the proper shielding gas for the application as directed by the instructor, filler metal manufacturer, the Welding Procedure Specification (WPS), or site quality standards.

Step 6 Connect the clamp of the workpiece lead to the table or workpiece.

Step 7 Turn on the welding machine and purge the torch as directed by the manufacturer's instructions.

Step 8 Set the initial welding current for the desired GTAW application.

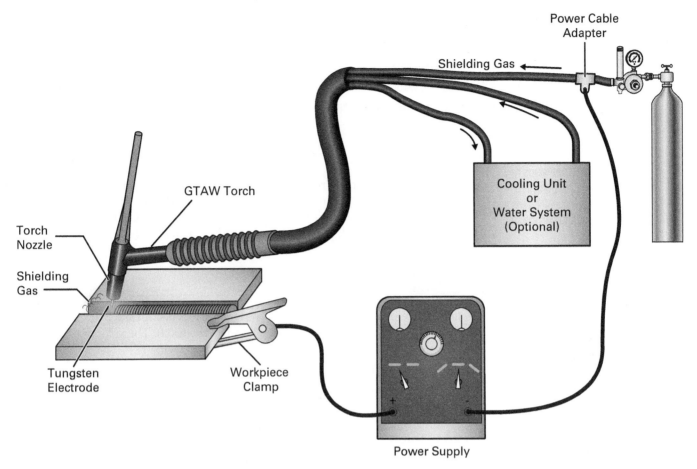

Figure 4 Configuration diagram of typical GTAW welding equipment.

1.0.0 Section Review

1. For long-term cutting or welding of materials that may produce toxic fumes, always wear an approved respirator that uses _____.
   a. a HEPA-rated charcoal vacuum system
   b. breathing air from outside the work area
   c. compressed oxygen from an approved tank
   d. an inert gas that will displace the fumes

2. When configuring a welding machine for GTAW welding, the torch *must* be equipped with a _____.
   a. reel of tungsten wire
   b. polarity limiter
   c. reel of filler wire
   d. tungsten electrode

Performance Tasks

1. Build a pad with stringer beads on carbon steel plate coupons in the flat position using GTAW equipment and carbon steel filler metal.

2. Perform multipass fillet welds on carbon steel plate coupons in all four 1F through 4F positions using GTAW equipment and carbon steel filler metal.

3. Perform multipass open V-groove welds on carbon steel plate coupons in all four 1G through 4G positions using GTAW equipment and carbon steel filler metal.

2.0.0 GTAW Techniques and Basic Beads

Objective

Describe welding techniques for GTAW and explain how to produce basic weld beads.

a. Describe welding techniques related to GTAW.

b. Explain how to produce basic GTAW weld beads.

Effective welding requires the welder to properly control the equipment being used and employ various welding techniques to produce strong and acceptable weld beads. This section examines factors that affect GTAW and how these factors are controlled. The process for producing basic GTAW weld beads is also described.

2.1.0 GTAW Techniques

GTAW weld bead characteristics and quality are affected by several factors that result from the way the welder handles the torch. These factors include the following:

- Torch travel speed and arc length
- Torch angles
- Torch and filler metal handling techniques

2.1.1 Torch Travel Speed and Arc Length

Torch travel speed and arc length affect the GTAW weld puddle and penetration of the weld. A slow travel speed allows more heat to concentrate and forms a larger, more deeply penetrating puddle. Faster travel speeds prevent heat buildup and form smaller, shallower puddles. As the torch is raised, the arc length increases, voltage increases, and the bead width increases.

2.1.2 Torch Angles

The two basic torch angles that must be controlled when performing GTAW are the work angle and the travel angle. The definition of these angles is the same as for all other methods of arc welding.

The torch work angle (*Figure 5*) is an angle less than 90° between a line perpendicular to the major workpiece surface at the point of electrode contact and a plane determined by the electrode axis and weld axis. In other words, the torch work angle is the side-to-side angle in relation to the weld joint. For a T-joint or corner joint, the line is 45° to the non-butting member. For pipe, the plane is determined by the electrode axis and a line tangent to the pipe surface at the same point.

The torch travel angle (*Figure 6*) is an angle less than 90° between the electrode axis and a line perpendicular to the weld axis at the point of electrode contact in a plane determined by the electrode axis and weld axis. A push angle is used for GTAW. A push angle is the travel angle created when the torch is tilted back so the electrode is pointing in the direction of weld progression. In other words, the electrode tip and shielding gas are being directed ahead of the weld bead. Push angles of 15° to 20° are normally used for GTAW. Too great a push angle will tend to draw air from under the back edge of the torch nozzle, where it will mix with the shielding gas stream and contaminate the weld.

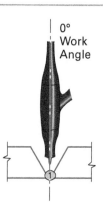

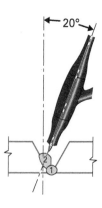

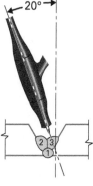

Figure 5 Typical torch work angles.

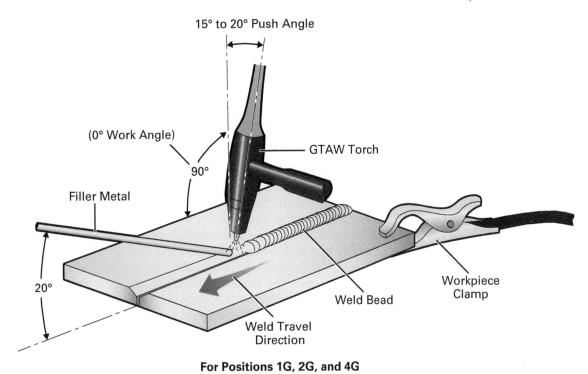

15° to 20° Push Angle

(0° Work Angle)

GTAW Torch

90°

Filler Metal

20°

Weld Bead

Workpiece Clamp

Weld Travel Direction

For Positions 1G, 2G, and 4G

20°

Filler Metal

Weld Travel Direction

15° to 20° Push Angle

90°

(0° Work Angle)

For Position 3G (Vertical)

Figure 6 Torch travel angles.

Walking the cup: A welding method in which the welder rests the edge of the torch nozzle (cup) against the base metal or groove edges to steady the torch and maintain a constant arc length.

2.1.3 Torch and Filler Metal Handling Techniques

The two basic handling techniques used to perform GTAW are freehand and **walking the cup**. Try both techniques and use the one that produces the best results.

In the freehand technique, the torch electrode tip is held just above the weld puddle or base metal. The torch is supported by the welder's hand, which is usually steadied by resting some part of it on or against the base metal to maintain the proper arc length. If required, the welder can move the torch tip in a small circular motion within the molten puddle to maintain its size and advance the puddle. Filler metal is added as needed.

The GTAW filler metal is held in the hand that is not holding the torch. For all positions, the filler rod is held at an angle of about 20° above the base metal surface and in line with the weld. The tip of the filler rod is always kept within the shielding gas envelope to protect the filler rod from atmospheric contamination and to keep it preheated.

Insert the filler metal into the leading edge of the weld puddle, using extreme care not to touch the tungsten electrode with the end of the filler metal. If the tungsten electrode touches the filler metal or weld puddle, it will become contaminated with filler metal. The electrode must then be removed and cleaned by grinding or chemical cleaning before proceeding. Also, do not insert the end of the filler metal rod into the molten puddle under the electrode and then attempt to melt it off. Even if it does not touch the tungsten, this technique can cause hard spots and weld defects. A technique for preventing contamination of the electrode is to coordinate the movement of the electrode toward the back edge of the weld puddle as the filler metal is dabbed into the leading edge of the weld puddle.

In the walking-the-cup method, the welder rests the edge of the torch nozzle (cup) against the base metal or groove edges to steady the torch and maintain a constant arc length. The torch is rocked from side to side on the edge of the nozzle as it is advanced to maintain the puddle size and heat both sides of the groove.

The filler metal is added in the same manner as in the freehand technique, using care not to contaminate the electrode.

2.1.4 Filler Metal Handling Techniques for Continuous Feed

Feeding the filler wire consistently as the weld progresses is important to ensure the weld puddle develops evenly and produces an even bead. Like all welding techniques, the technique described here requires practice to master but is well worth the effort.

Cradle the filler wire between the pointer finger and middle finger, similar to how you might hold a pencil. Then, use the thumb to consistently push the wire forward in small motions. Once the thumb is extended from pushing the wire forward, quickly bend and reposition the thumb to begin a new forward push. Consistent motion is important, so the thumb must be repositioned quickly to maintain the pace. The thumb can also be used to slowly roll the electrode during the pushing motion. This technique is easily practiced without actively welding; simply practice with a wire at home or during free time to become comfortable with the technique before trying it on a weld.

Another technique to consider is referred to as the finger-pull technique. The wire is secured between the pointer and middle fingers, which are used to pull the electrode forward. At the end of the forward stroke, the fingers are slipped backward along the wire to reposition without pulling the wire backward and out of the weld puddle in the process. Once the two fingers are repositioned, the pulling motion begins again.

A Rare Feat

In 2012, a pipe welding team employed by Team Fabricators LLC, a Texas-based subsidiary of Team Industries, accomplished something quite rare. Working on a piping project to do full-penetration, open V-groove pipe welds, the team completed literally thousands of welds. After 847 radiographic examinations of randomly selected welds, done in accordance with the ASME code, not a single pipe weld had to be rejected.

2.2.0 GTAW Bead Types

The following sections explain how to create weld beads using GTAW equipment. The two basic bead types are stringer beads and weave beads.

2.2.1 Stringer Beads

Stringer beads (*Figure 7*) are made with little or no side-to-side movement (oscillation) of the torch. The bead is generally no more than three times the diameter of the filler rod being used.

Practice running stringer beads in the flat position using both the freehand and walking-the-cup techniques. Flat position stringer beads should be run with a 15° to 20° push angle. Practice some stringer beads without filler metal until comfortable with the torch and you get rippled beads that are consecutively uniform.

To run stringer beads with filler metal, hold the filler metal in front of the weld bead, with the end of the filler metal within the torch gas envelope. The filler rod should be held at a 20° angle with the plate surface (*Figure 8*).

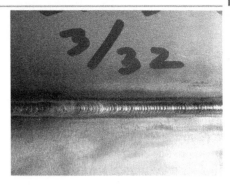

Figure 7 Stringer beads on stainless steel.
Source: Zachry Group

CAUTION

Cooling with water (quenching) is done only on practice coupons. Never cool test coupons or on-the-job welds with water. Cooling with water can cause weld cracks and affect the mechanical properties of the base metal.

WARNING!

Use pliers to handle the hot practice coupons. Wear gloves when quenching the practice coupons in water. Steam will rise off the coupons and can burn or scald unprotected hands and arms.

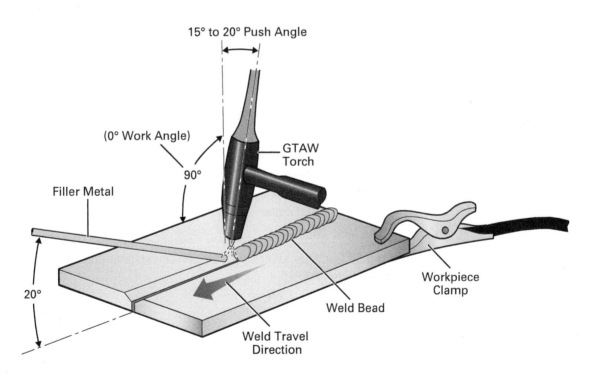

Figure 8 Flat position torch and filler metal angles.

Follow these steps to run flat position stringer beads:

Step 1 Position the torch at a 15° to 20° push angle, with the electrode tip directly over the point where the weld is to begin.

Step 2 Press the remote control to energize the electrode and start the shielding gas flow.

Step 3 Initiate the arc by bringing the electrode close to the base metal (with high frequency) or by lightly touching the electrode to the base metal (no high frequency).

Step 4 Move the electrode tip in a small circular motion until the weld puddle begins to form.

Step 5 While using either the freehand or walking-the-cup technique, slowly advance the torch and add filler metal as needed.

Step 6 Continue to weld until a bead about 2" to 6" (50 mm to 152 mm) long is formed, and then stop the arc without withdrawing the torch.

Step 7 Continue to hold the torch in place so that the postflow gas will protect the hot weld and filler metal until they are cool enough to prevent oxidation damage.

Step 8 Remove the torch, and, if the equipment does not have remote control, close the shielding gas valve.

Step 9 Inspect the bead for the following:

- Straightness of the bead
- Uniform rippled appearance of the bead face
- Smooth, flat transition with complete fusion at the toes of the weld
- No excessive porosity with no pores larger than $3/32$" (2.38 mm)
- No undercut greater than $1/32$" (0.8 mm) deep or 10% of the base metal thickness, whichever is less
- No overlap
- No lack of fusion
- Complete penetration
- No pinholes (fisheyes)
- Filled crater
- No cracks

Step 10 Continue practicing stringer beads until acceptable welds are made every time.

NOTE

If the equipment being used does not have remote control, the electrode will be energized when the power source is turned on. Start the shielding gas flow by opening the manual shielding gas valve on the torch or in the line.

CAUTION

Do not remove the torch or filler metal from the torch-shielding gas flow until the puddle has solidified and cooled. This continued flow of gas will protect the molten metal and the tungsten electrode. Removing the torch and shielding gas too soon can cause crater porosity or cracks.

If the postflow time period is too short, the tungsten electrode may oxidize (turn a bluish-black color), which will require repointing the electrode. If the filler metal becomes contaminated, the contaminated end will have to be discarded.

Going Green

Conserving Material

Steel for practice welding is expensive and difficult to obtain. Every effort should be made to conserve the material available. Reuse welding coupons until all surfaces have been used for welding, then cut the weld coupon apart and reuse the pieces. Use material that cannot be cut into weld coupons to practice running beads.

Stopping the Arc without Remote Control

If the equipment being used does not have remote control, slowly swing the torch up the groove face to break the arc, and then immediately return it to shield the weld and the end of the filler metal.

2.2.2 Weave Beads

Weave beads are made with wide, side-to-side motions of the electrode. The width of a weave bead is determined by the amount of side-to-side motion.

When making weave beads, use care at the toes to be sure there is proper tie-in to the base metal. To ensure proper tie-in at the toes, slow down or pause slightly at the edges. The pause at the edges will also flatten out the weld, giving it the proper profile.

Always check the WPS or site quality standards to determine if stringer or weave beads should be used.

Follow these steps to run flat position weave beads:

Step 1   Position the torch at a 15° to 20° push angle, with the electrode tip directly over the point where the weld is to begin.

Step 2   Press the remote control to energize the electrode and start the shielding gas flow.

Step 3   Initiate the arc by bringing the electrode close to the base metal (with high frequency) or by lightly touching the electrode to the base metal (no high frequency).

Step 4   Move the electrode tip in a small circular motion until the weld puddle begins to form.

Step 5   Slowly advance the torch in a weaving motion and add filler metal as needed.

Step 6   Continue to weld until a bead about 2" to 6" (50 mm to 152 mm) long is formed, and then stop the arc without withdrawing the torch.

Step 7   Continue to hold the torch in place so that the postflow gas will protect the hot weld and filler metal until they are cool enough to prevent oxidation damage.

Step 8   Remove the torch, and, if the equipment does not have remote control, close the shielding gas valve.

Step 9   Inspect the bead for the following:

- Straightness of the bead
- Uniform rippled appearance of the bead face
- Smooth, flat transition with complete fusion at the toes of the weld
- No excessive porosity with no pores larger than $3/32$" (2.38 mm)
- No undercut greater than $1/32$" (0.8 mm) deep or 10% of the base metal thickness, whichever is less
- No overlap
- No lack of fusion
- Complete penetration
- No pinholes (fisheyes)
- Filled crater
- No cracks

Step 10 Continue practicing weave beads until acceptable welds are made every time.

Tacking and Aligning Longer Workpieces

When tacking longer than typical practice workpieces together, both sides of the workpieces are usually tacked with welds about $1/2$" (13 mm) long to position the workpieces and minimize distortion when the final welds are made. After the first tack weld, use a hammer or other tool to align the workpieces side-to-side and end-to-end, and then tack the opposite side. Tack the far ends of the workpieces in the same manner. Intermediate tack welds can be made every 5" to 6" (125 mm to 150 mm) as necessary to minimize lengthwise distortions.

2.2.3 Weld Restarts

A restart is the junction where a new weld connects to and continues the bead of a previous weld. Restarts are important because an improperly made restart may create a discontinuity. A restart must be made so it blends smoothly with the previous weld and does not stand out. If possible, avoid restarts by running a bead the full length of the weld joint that is to be made.

Follow these steps to make a restart:

Step 1 Use a wire brush to clean the restart area.

Step 2 Hold the torch at the proper angle and arc distance, and then restart the arc directly over the center of the crater. Remember that the welding codes do not allow arc strikes outside the area to be welded.

Step 3 Move the electrode tip in a small circular motion over the crater, and when the molten puddle is the same size as the crater, add filler metal.

Step 4 As soon as the puddle fills the crater, advance the puddle slightly and continue to add filler metal as needed.

Step 5 Inspect the restart. A properly made restart will blend into the bead, making it hard or impossible to detect. If the restart has underfill or undercut, not enough filler metal was added. If the restart is higher than the rest of the bead, too much filler metal was added.

Step 6 Continue to practice restarts until they are correct. Use the same techniques for making restarts whenever performing GTAW.

2.2.4 Weld Terminations

A weld termination is made at the end of a weld. A termination normally leaves a crater. When making a termination, the welding codes require that the crater must be filled to the full cross section of the weld to prevent crater cracking. This can be difficult because most terminations are at the edge of a plate, where welding heat tends to build up, making filling the crater more difficult. Filling the crater is much easier when remote control equipment with a potentiometer is used. The potentiometer allows the welding current to be reduced as the end of the weld is approached.

Follow these steps to make a termination using remote control equipment equipped with a potentiometer:

Step 1 As the end of the weld is approached, slowly reduce the welding current. The amount of current reduction is determined by the weld puddle width. If the puddle starts to become wider, there is too much heat. If the puddle starts to become narrower, there is not enough heat.

Step 2 When adding filler metal, continue to reduce the welding current until the crater is filled.

Step 3 Stop the arc and hold the torch in place until the gas postflow cools the weld metal and filler metal.

Follow these steps to make a termination without a potentiometer:

Step 1 As the end of the weld is approached, add filler metal at a faster pace. The filler metal will help absorb excess heat as it melts.

Step 2 Continue to add filler metal until the crater is filled, and then stop the arc.

Step 3 Hold the torch in place until the gas postflow cools the weld metal and filler metal enough to prevent oxidation damage.

CAUTION

Do not use the same brush on different metals.

For either process, inspect the termination for the following:

- Crater filled to the full cross section of the weld (no underfill)
- No crater cracks
- No excessive porosity with no pores larger than $^3/_{32}$" (2.38 mm)
- No undercut greater than $^1/_{32}$" (0.8 mm) deep or 10% of the base metal thickness, whichever is less

Continue to practice terminations until they are correct.

2.2.5 Overlapping Beads

Overlapping beads are made by depositing connective weld beads parallel to one another. The parallel beads overlap, forming a flat surface. This is also called *padding*. Overlapping beads are used to build up a surface and to make multipass welds. Both stringer and weave beads can be overlapped. When viewed from the end, properly overlapped beads will have a relatively flat surface. *Figure 9* shows proper and improper overlapping stringer and weave beads.

Thin Materials

On thin materials, stop about 1" to 2" (25 mm to 50 mm) short of the edge without terminating the weld. Then, back-weld from the edge to the crater, filling the crater.

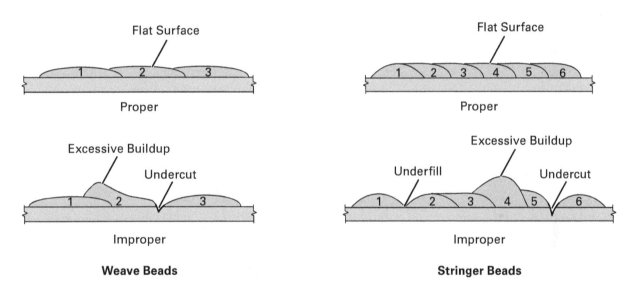

Weave Beads **Stringer Beads**

Completed Welding Coupon

Figure 9 Proper and improper overlapping beads.
Source: Rich Samanich

A Few Suggestions

To improve your technique, practice with both hands in opposing directions when welding in the 1G, 2G, and 4G positions. Also, the first weld bead can be done up the center of the pad, with the remaining beads being worked from the center out. This tactic gives you practice overlapping beads on the pad as you might do in a groove weld.

Follow these steps to make GTAW overlapping stringer beads using $^3/_{32}$" or $^1/_8$" (2.4 mm or 3.2 mm) carbon steel filler metal:

Step 1   Use a 2" × 6" × $^1/_4$" (50 mm × 152 mm × 6 mm) piece of steel.

Step 2   Weld a stringer or weave bead along one edge.

Step 3   Clean the weld.

Step 4   Run a second bead along the previous bead. Be sure to overlap the previous bead to obtain a good tie-in and produce a flat profile.

Step 5   Continue running overlapping stringer or weave beads until the face of the metal square is covered.

Step 6   Continue building layers of stringer or weave beads, one on top of the other, until perfecting the technique.

Build a pad using stringer beads (*Figure 9*). Repeat building the pad using weave beads.

2.0.0 Section Review

1. The two basic torch and filler metal handling techniques used to perform GTAW are _____.
    a. travel angle and walking-the-bead
    b. push angle and freehand
    c. freehand and walking-the-cup
    d. walking-the-cup and push bead

2. When running stringer beads using GTAW, the filler metal rod should be held _____.
    a. at a 20° angle with the plate surface
    b. at the toe padding behind the weld bead
    c. just outside the torch gas envelope
    d. at a 45° angle with the electrode

3.0.0 GTAW Fillet and V-Groove Welds

Performance Tasks

1. Build a pad with stringer beads on carbon steel plate coupons in the flat position using GTAW equipment and carbon steel filler metal.

2. Perform multipass fillet welds on carbon steel plate coupons in all four 1F through 4F positions using GTAW equipment and carbon steel filler metal.

3. Perform multipass open V-groove welds on carbon steel plate coupons in all four 1G through 4G positions using GTAW equipment and carbon steel filler metal.

Objective

Describe the welding techniques needed to produce proper fillet and open V-groove welds using GTAW welding techniques.

a. Describe the welding techniques needed to produce proper fillet welds using GTAW.

b. Describe the welding techniques needed to produce proper open V-groove welds using GTAW.

Fillet welds and V-groove welds are two common types of welds produced with GTAW. This section examines the procedures used to make each of these welds in four different positions: flat, horizontal, vertical, and overhead.

3.1.0   Fillet Welds

Fillet welds require little base metal preparation, except for the cleaning of the weld area and the removal of any excess material from cut surfaces. Any dross from cutting will cause porosity in the weld. For this reason, the codes require that this material is entirely removed prior to welding.

The most common fillet welds made with GTAW are made in lap and T-joints. The weld position for plate is determined by the weld axis and the orientation of the workpiece. The positions for fillet welding on plate are flat (1F, where F stands for fillet), horizontal (2F), vertical (3F), and overhead (4F), as shown in *Figure 10*. In the 1F and 2F positions, the weld axis can be inclined up to 15°. Any weld axis inclination for the other positions varies with the rotational position of the weld face as specified in AWS standards.

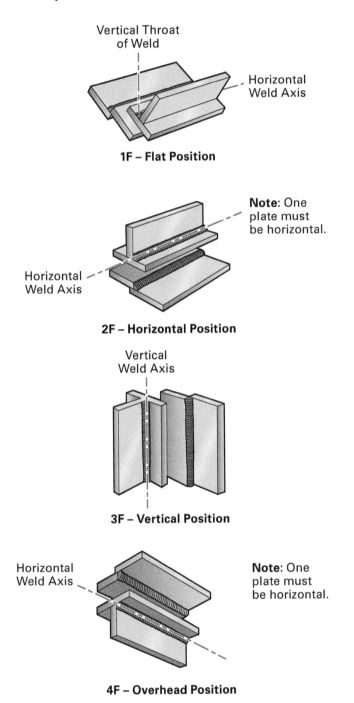

Figure 10 Fillet weld positions.

Fillet welds can be concave or convex, depending on the WPS or site quality standards. The welding codes require a fillet weld to have a uniform concave or convex face, although a slightly nonuniform face is acceptable. Convexity is the distance the weld extends above a line drawn between the toes of the weld. The convexity of a fillet weld or individual surface bead must not exceed that permitted by the applicable code or standard. In single-pass, weave-bead fillet welds where two workpieces are being joined at an angle (not lap joints), flat or slightly convex faces are usually preferred because weld stresses are more uniformly distributed through the fillet weld and workpieces. A fillet weld must be repaired if the profile has defects.

A fillet weld is unacceptable and must be repaired if the profile has insufficient throat, excessive convexity, undercut, overlap, insufficient leg, or incomplete fusion. *Figure 11* shows a variety of ideal, acceptable, and unacceptable fillet weld profiles.

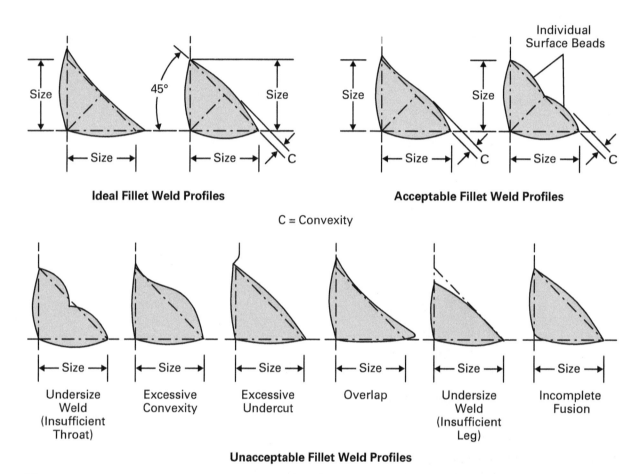

Figure 11 Ideal, acceptable, and unacceptable fillet weld profiles.

T-Joint Heat Dissipation

In T-joints, the welding heat dissipates more rapidly in the thicker or non-butting member. On various bead passes, the arc may have to be concentrated slightly more on the thicker or non-butting member to compensate for the heat loss.

3.1.1 Practicing Flat (1F) Position Fillet Welds

Practice 1F position fillet welds by making multipass (six-pass) slightly convex fillet welds in a T-joint with an appropriate filler metal. When welding in the 1F position, pay close attention to the work angle and travel speed. For the first bead, the work angle is vertical, forming a 45° angle to both plate surfaces. The angle is then adjusted for all subsequent beads. *Figure 12* shows the work angles for 1F position fillet welds.

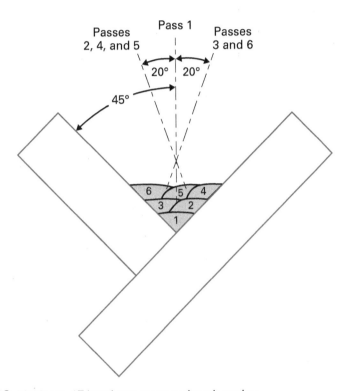

Figure 12 Multipass 1F bead sequence and work angles.

Follow these steps to make a flat fillet weld. Be sure to clean each bead before beginning the next:

Step 1 Tack two plates together to form a T-joint for the fillet weld coupon (*Figure 13*). Clean the tack welds.

Step 2 Clamp or tack-weld the coupon in the flat position.

Step 3 Run the first bead along the root of the joint, using a work angle of 45° with a 10° to 15° push angle. Use a slight side-to-side motion (oscillation).

Step 4 Run the second bead along a toe of the first weld, overlapping about 75% of the first bead. Alter the work angle as shown in *Figure 12* and use a 10° to 15° push angle with a slight oscillation.

Step 5 Run the third bead along the other toe of the first weld, filling the groove created when the second bead was run. Use the work angle shown in *Figure 12* and a 10° to 15° push angle with a slight oscillation.

Step 6 Run the fourth bead along the outside toe of the second weld, overlapping about half the second bead. Use the work angle shown in *Figure 12* and a 10° to 15° push angle with a slight oscillation.

Step 7 Run the fifth bead along the inside toe of the fourth weld, overlapping about half the fourth bead. Use the work angle shown in *Figure 12* and a 10° to 15° push angle with a slight oscillation.

Figure 13 Fillet weld coupon.
Source: Holley Thomas, SME

Raising the Working Height of a GTAW Torch

In some cases, raising the torch height (within limits, of course) can be used to widen second and subsequent weld beads without oscillating the torch.

Step 8 Run the sixth bead along the toe of the fifth weld, filling the groove created when the fifth bead was run. Use the work angle shown in *Figure 12* and a 10° to 15° push angle with a slight oscillation.

Step 9 Inspect the weld. The weld is acceptable if it has the following features:

- Uniform rippled appearance on the bead face
- Craters and restarts filled to the full cross section of the weld
- Uniform weld size $\pm^1/_{16}$" ($\pm$1.6 mm)
- Acceptable weld profile in accordance with the applicable code or standard
- Smooth, flat transition with complete fusion at the toes of the weld
- No excessive porosity with no pores larger than $^3/_{32}$" (2.38 mm)
- No undercut greater than $^1/_{32}$" (0.8 mm) deep or 10% of the base metal thickness, whichever is less
- No overlap
- No cracks

3.1.2 Practicing Horizontal (2F) Position Fillet Welds

Practice horizontal (2F) fillet welding by placing multipass fillet welds in a T-joint using an appropriate filler metal as directed by the instructor. When making horizontal fillet welds, pay close attention to the work angles. For the first bead, the electrode work angle is 45°. The work angle is adjusted for all other welds.

Follow these steps to make a horizontal fillet weld:

Step 1 Tack two plates together to form a T-joint for the fillet weld coupon. Clean the tack welds.

Step 2 Clamp or tack-weld the coupon in the horizontal position.

Step 3 Run the first bead along the root of the joint using a work angle of approximately 45° with a 10° to 15° push angle and a slight side-to-side oscillation (*Figure 14*).

Step 4 Clean the weld.

Step 5 Run the remaining passes using a 10° to 15° push angle and a slight oscillation at the appropriate work angles (refer to *Figure 14*). Overlap each previous pass. Clean the weld after each pass.

Step 6 Have the instructor inspect the weld. The weld is acceptable if it has these features:

- Uniform rippled appearance of the bead face
- Craters and restarts filled to the full cross section of the weld
- Uniform weld size $\pm^1/_{16}$" ($\pm$1.6 mm)
- Acceptable weld profile in accordance with the applicable code or standard
- Smooth, flat transition with complete fusion at the toes of the weld
- No excessive porosity with no pores larger than $^3/_{32}$" (2.38 mm)
- No undercut greater than $^1/_{32}$" (0.8 mm) deep or 10% of the base metal thickness, whichever is less
- No overlap
- No cracks

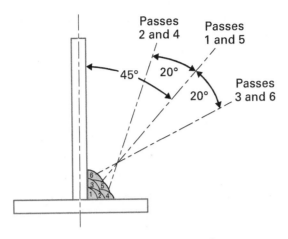

Passes 2 and 4

Passes 1 and 5

45°   20°

20°

Passes 3 and 6

Figure 14 Multipass 2F bead sequence and work angles.

3.1.3 Practicing Vertical (3F) Position Fillet Welds

Practice vertical (3F) fillet welding by placing multipass fillet welds in a T-joint using an appropriate filler metal as directed by the instructor. Normally, vertical welds are made welding uphill (also called *vertical-up welding*) from the bottom to the top, using a torch push angle (up-angle). When vertical welding, either stringer or weave beads can usually be used. Check with the instructor to see if stringer or weave beads, or both techniques, should be practiced. However, weave beads are avoided on stainless steel or in downhill welding. On the job, the site WPS or quality standard will specify which technique to use.

Follow these steps to make an uphill fillet weld with weave beads:

Step 1 Tack two plates together to form a T-joint for the fillet weld coupon (*Figure 15*).

Step 2 Clamp or tack-weld the coupon in the vertical position.

Step 3 Starting at the bottom, run the first bead along the root of the joint using a work angle of approximately 45° with a 10° to 15° push angle. Pause in the weld puddle to fill the crater. Ensure it's filled to the full cross section at the end of the pass.

Step 4 Clean the weld.

Step 5 Run the remaining passes using a 10° to 15° push angle and a side-to-side weave technique with a 45° work angle (*Figure 15*). Use a slow motion across the face of the weld, pausing at each toe for penetration and to fill the crater. Clean the weld after each pass.

Step 6 Have the instructor inspect the weld. The weld is acceptable if it has these features:

- Uniform rippled appearance of the bead face
- Craters and restarts filled to the full cross section of the weld
- Uniform weld size $\pm^1/_{16}$" (±1.6 mm)
- Acceptable weld profile in accordance with the applicable code or standard
- Smooth, flat transition with complete fusion at the toes of the weld
- No excessive porosity with no pores larger than $^3/_{32}$" (2.38 mm)
- No undercut greater than $^1/_{32}$" (0.8 mm) deep or 10% of the base metal thickness, whichever is less
- No overlap
- No cracks

Repeat vertical fillet (3F) welding using stringer beads. Use a slight oscillation and side-to-side motion, pausing slightly at each toe to prevent undercut. For stringer beads, use a 10° to 15° push angle and the work angles shown in *Figure 15* as required.

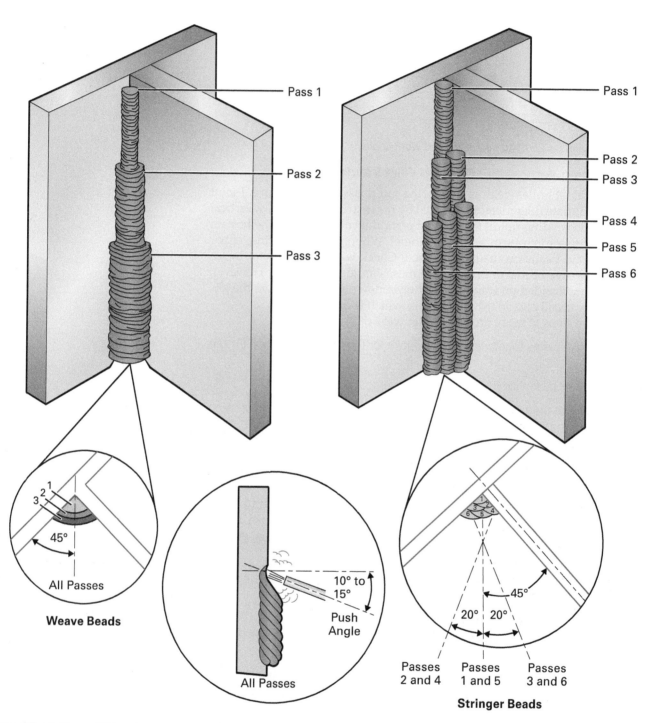

Figure 15 Multipass 3F bead sequences and work angles for stringer and weave beads.

3.1.4 Practicing Overhead (4F) Position Fillet Welds

Practice overhead (4F) fillet welding by welding multipass fillet welds in a T-joint using an appropriate filler metal as directed by the instructor. When making overhead fillet welds, pay close attention to the work angles. For the first bead, the work angle is approximately 45°. The work angle is adjusted for all other welds.

Follow these steps to make an overhead fillet weld:

Step 1 Tack two plates together to form a T-joint for the fillet weld coupon.

Step 2 Clamp or tack-weld the coupon so it is in the overhead position.

Step 3 Run the first bead along the root of the joint using a work angle of approximately 45° with a 10° to 15° push angle.

Step 4 Clean the weld.

Step 5 Using a slight oscillation, run the remaining passes using a 10° to 15° push angle and the work angles shown in *Figure 16*. Overlap each previous pass. Clean the weld after each pass.

Step 6 Have the instructor inspect the weld. The weld is acceptable if it has these features:

- Uniform rippled appearance of the bead face
- Craters and restarts filled to the full cross section of the weld
- Uniform weld size $\pm^1/_{16}$" (±1.6 mm)
- Acceptable weld profile in accordance with the applicable code or standard
- Smooth, flat transition with complete fusion at the toes of the weld
- No excessive porosity with no pores larger than $^3/_{32}$" (2.38 mm)
- No undercut greater than $^1/_{32}$" (0.8 mm) deep or 10% of the base metal thickness, whichever is less
- No overlap
- No cracks

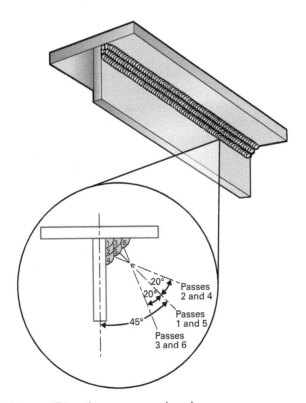

Figure 16 Multipass 4F bead sequence and work.

3.2.0 Open V-Groove Welds

The open V-groove weld is a common groove weld normally made on plate and pipe. Practicing the open V-groove weld on plate will help with making the more difficult pipe welds covered in *Welding Level Three*.

Open groove welds can be made in all positions (*Figure 17*). The weld position is determined by the axis of the weld. Groove weld positions are flat (1G), horizontal (2G), vertical (3G), and overhead (4G).

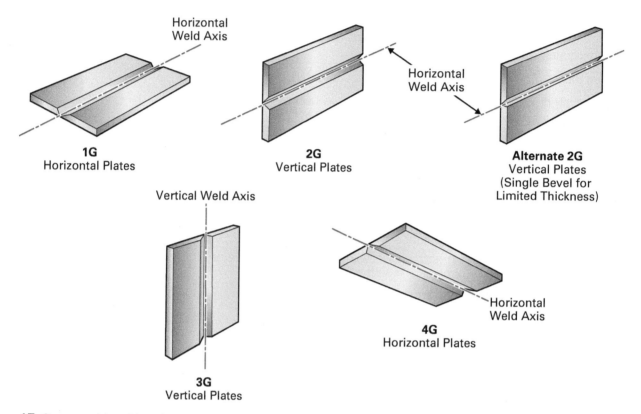

Figure 17 Groove weld positions for plate.

3.2.1 Open V-Groove Root Pass Techniques

The most difficult part of welding an open V-groove butt joint is the root pass. The root pass must have complete penetration but not an excessive amount of penetration or melt-through root reinforcement. The penetration is controlled by using the correct amperage, filler metal application, and torch travel speed for the root opening and root face being used. Root reinforcement should be flush to $1/8$" (3.2 mm), depending on the site specifications.

The two techniques for running the root pass are **lay wire**, also called *on the wire*, and keyhole. Regardless of the technique used, always start the root pass from a tack weld.

To perform the lay-wire root pass technique, secure or hold the filler metal in the bottom of the root opening of the V-groove parallel to the plates. Fuse the filler metal in place by moving the torch along the groove using either the freehand or walking-the-cup torch technique. Advance the torch slowly enough to thoroughly fuse the filler metal with the plate edges, but not so slowly that it can burn through. This technique is used with a filler metal the same size as or slightly larger than the width of the root opening. The method is similar to pipe welding using a consumable insert. With this technique, to ensure a uniform bead thickness, the filler metal should not be moved into or out of the weld puddle. *Figure 18* shows the preparation for the lay-wire root pass technique.

Lay wire: A root pass technique, also called *on the wire*, in which the filler metal is held in the bottom of the groove parallel to the plates and then melted and fused in place by moving the torch along the groove.

Figure 18 Lay-wire root pass technique.
Source: Rich Samanich

Backing and/or Backing Gas

Normally, backing and/or backing gas is not required for GTAW of carbon steel. However, it is required for stainless steel, most low-alloy steel, and most nonferrous metals.

To perform the keyhole root pass technique (*Figure 19*), hold the filler metal at an angle to the plate surfaces and feed it into the groove root from the weld side. This technique uses a filler metal rod diameter slightly larger than the root opening so that the rod can rest on the groove.

When making a root pass, start on a tack weld. When a keyhole weld puddle has been established by melting the root faces of both plates together, move the arc to the back of the puddle and insert the end of the filler metal into the leading edge of the weld puddle. Then, move the filler metal away from the weld puddle. Advance the torch, using either the walking-the-cup or freehand technique, maintaining the keyhole. Insert the filler metal back into the leading edge of the molten puddle and forward out of the molten puddle to add filler metal as needed. Be sure the filler metal is always kept in the shielding gas stream to keep the filler metal preheated and protected from atmospheric contamination.

After the root pass has been completed, clean and inspect it for the following:

- Uniform and smooth face and root
- Excessive root reinforcement
- Excessive face buildup
- Undercut
- Porosity
- Cracks

Cutting Filler Rod

Snip the rod at a 45° angle. Then, insert the rod into the puddle with the angled tip at the top of the feathered tack. This results in better penetration of the tie-in and a smoother transition from the previous weld into the new weld.

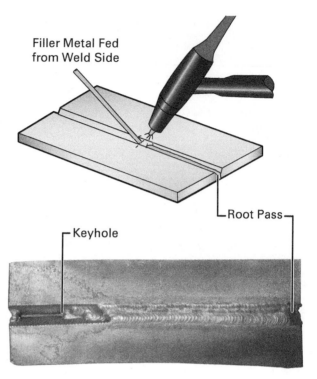

Figure 19 Root pass and keyhole.
Source: Rich Samanich

3.2.2 Acceptable and Unacceptable Groove Weld Profiles

Groove welds should be made with slight reinforcement (not exceeding ⅛" or 3.2 mm) and a gradual transition to the base metal at each toe. Groove welds must not have excessive reinforcement, undercut, underfill, or overlap (*Figure 20*). If a groove weld has any of these defects, it must be repaired.

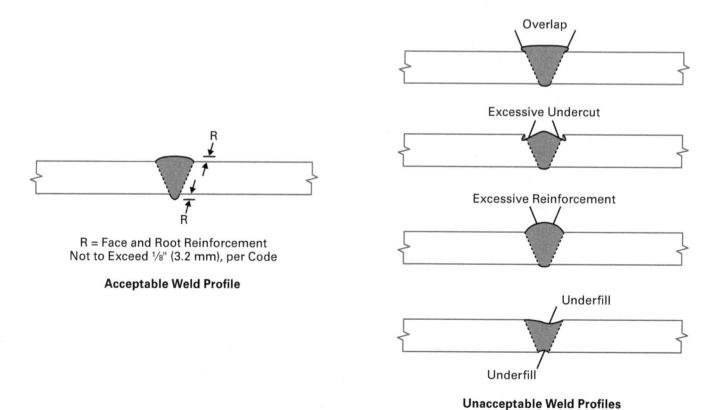

R = Face and Root Reinforcement
Not to Exceed ⅛" (3.2 mm), per Code

Acceptable Weld Profile

Unacceptable Weld Profiles

Figure 20 Acceptable and unacceptable groove weld profiles.

3.2.3 Practicing Flat (1G) Position Open V-Groove Welds

Practice flat (1G) open V-groove welds, as shown in *Figure 21* using GTAW as directed by the instructor. Use a stringer bead for the root pass, keeping the torch angle at 90° to the plate surface (0° work angle) using a 15° to 20° push angle. For fill and cover passes, stringer or weave beads may be used. When using stringer beads, adjust the work angle to tie in and fill the bead. For weave beads, keep the torch angle at 90° to the plate surface (0° work angle). Pay special attention at the termination of the weld to fill the crater.

Follow these steps to make open V-groove welds in the flat (1G) position:

Step 1 Tack-weld the practice coupon together, as explained earlier.

Step 2 Clamp or tack-weld the welding coupon in a flat position above the welding table surface.

Step 3 Run the root pass using an appropriate diameter of carbon steel filler metal and either the lay-wire or the keyhole technique. Use a 15° to 20° push angle, a 20° filler metal angle to the plate surface, and the work angle for Pass 1 (*Figure 21*).

Step 4 Clean the root pass.

Step 5 Run the remaining passes using either $^3/_{32}$" or $^1/_8$" (2.4 mm or 3.2 mm) carbon steel filler metal, a 15° to 20° push angle, and the work angles shown in *Figure 21*. Hold the filler metal about 20° above the plate surface. Clean the weld between each pass.

Step 6 Inspect the weld. The weld is acceptable if it has the following:

- Uniform rippled appearance of the bead face
- Craters and restarts filled to the full cross section of the weld
- Uniform weld width ±$^1/_{16}$" (1.6 mm)
- Acceptable weld profile in accordance with the applicable code or standard
- Smooth, flat transition with complete fusion at the toes of the weld
- Complete uniform root reinforcement at least flush with the base metal to a maximum buildup of $^1/_8$" (3.2 mm)
- No excessive porosity with no pores larger than $^3/_{32}$" (2.38 mm)
- No undercut greater than $^1/_{32}$" (0.8 mm) deep or 10% of the base metal thickness, whichever is less
- No cracks

Open V-groove welds can be run with a root and a single-layer hot pass before beginning the split layer technique shown in *Figure 21*.

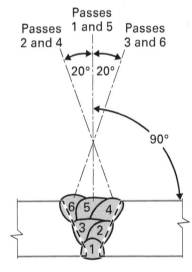

Stringer Bead Sequence

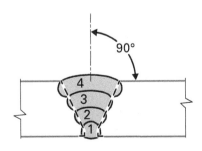

Weave Bead Sequence

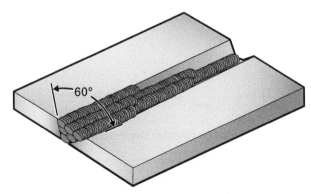

Note: The actual number of weld beads will vary depending on the plate thickness.

Figure 21 Multipass 1G bead sequence and work angles.

3.2.4 Practicing Horizontal (2G) Position Open V-Groove Welds

Practice horizontal (2G) open V-groove welds, as shown in *Figure 22,* using an appropriate filler metal as directed by the instructor. Run the root bead at a work angle of 0° and a push angle of 15° to 20°. The work angle is adjusted for all other welds. Pay particular attention to filling the crater at the termination of the weld.

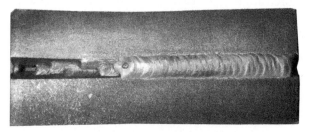

Note: The actual number of weld beads will vary depending on the plate thickness.

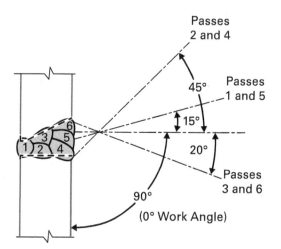

Alternate Joint Representation

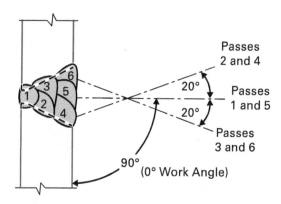

Standard Joint Representation

Figure 22 Multipass 2G bead sequences and work.
Source: Rich Samanich

Follow these steps to practice open V-groove welds in the horizontal position:

Step 1 Tack-weld the practice coupon together, as explained earlier. Use the standard or alternate weld coupon as directed by the instructor.

Step 2 Clamp or tack-weld the coupon in the horizontal position.

Step 3 Run the root pass with either the lay-wire or the keyhole welding technique and use an appropriate diameter of carbon steel filler metal. Use a 15° to 20° push angle, a 20° filler metal angle with the plate surface, and the appropriate work angle for Pass 1 (*Figure 22*).

Step 4 Clean the weld.

Step 5 Run the remaining passes using either $^3/_{32}$" or $^1/_8$" (2.4 mm or 3.2 mm) carbon steel filler metal, a 15° to 20° push angle, and the work angles shown in *Figure 22*. Clean the weld between each pass.

Step 6 Inspect the weld. The weld is acceptable if it has the following:

- Uniform rippled appearance of the bead face
- Craters and restarts filled to the full cross section of the weld
- Uniform weld width $\pm^1/_{16}$" (1.6 mm)
- Acceptable weld profile in accordance with the applicable code or standard
- Smooth, flat transition with complete fusion at the toes of the weld
- Complete uniform root reinforcement at least flush with the base metal to a maximum buildup of $^1/_8$" (3.2 mm)
- No excessive porosity with no pores larger than $^3/_{32}$" (2.38 mm)
- No undercut greater than $^1/_{32}$" (0.8 mm) deep or 10% of the base metal thickness, whichever is less
- No cracks

3.2.5 Practicing Vertical (3G) Position Open V-Groove Welds

Practice vertical (3G) open V-groove welds, as shown in *Figure 23*, using an appropriate filler metal as directed by the instructor. When welding vertically, either stringer or weave beads can be used. However, weave beads are not normally used on stainless steel. On the job, the WPS or site quality standards will specify which technique to use. Check with the instructor to see if stringer or weave beads, or both techniques, should be practiced. The root, fill, and cover beads can be run either uphill or downhill. However, downhill weave beads are not normally used. Use the direction preferred at the site or as specified in the WPS or site quality standards.

Run the root bead using a work angle of 0° and a push angle of 15° to 20°. The work angle is adjusted for all other welds.

Follow these steps to practice open V-groove welds in the vertical (3G) position:

Step 1 Tack-weld the practice coupon together, as explained earlier.

Step 2 Clamp or tack-weld the coupon in the vertical position.

Step 3 Run the root pass uphill with either the lay-wire or keyhole welding technique and use an appropriate diameter of carbon steel filler metal. Use a push angle of 15° to 20°, a filler metal angle of 20° with the plate surface, and the Pass 1 work angle (*Figure 23*).

Step 4 Clean the weld.

Step 5 Run the remaining passes uphill using either stringer or weave beads with $^3/_{32}$" or $^1/_8$" (2.4 mm or 3.2 mm) carbon steel filler metal, a 15° to 20° push angle, and the work angles shown in *Figure 23*. Clean the weld between each pass.

Step 6   Inspect the weld. The weld is acceptable if it has the following:

- Uniform rippled appearance of the bead face
- Craters and restarts filled to the full cross section of the weld
- Uniform weld width $\pm\frac{1}{16}$" (1.6 mm)
- Acceptable weld profile in accordance with the applicable code or standard
- Smooth, flat transition with complete fusion at the toes of the weld
- Complete uniform root reinforcement at least flush with the base metal to a maximum buildup of $\frac{1}{8}$" (3.2 mm)
- No excessive porosity with no pores larger than $\frac{3}{32}$" (2.38 mm)
- No undercut greater than $\frac{1}{32}$" (0.8 mm) deep or 10% of the base metal thickness, whichever is less
- No cracks

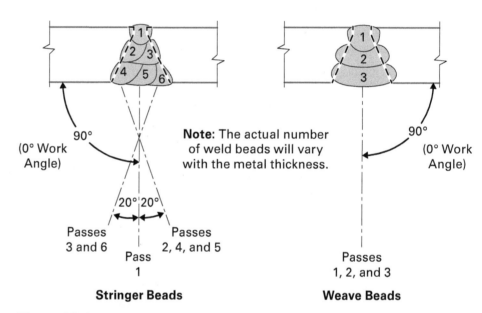

Figure 23   Multipass 3G bead sequence and work angles.

3.2.6 Practicing Overhead (4G) Position Open V-Groove Welds

Practice overhead (4G) open V-groove welds, as shown in *Figure 24*, using an appropriate filler metal as directed by the instructor. Run the root bead using a work angle of 0° and a push angle of 15° to 20°. The work angle is adjusted for all other welds.

Follow these steps to practice open V-groove welds in the overhead (4G) position:

Step 1   Tack-weld the practice coupon together, as explained earlier.

Step 2   Clamp or tack-weld the coupon in the overhead position.

Step 3   Run the root pass with either the lay-wire or the keyhole welding technique, using an appropriate diameter carbon steel filler metal. Use a 15° to 20° push angle, a 20° filler metal angle to the plate surface, and the Pass 1 work angle (*Figure 24*).

Step 4   Clean the weld.

Step 5   Run the remaining passes using either $\frac{3}{32}$" or $\frac{1}{8}$" (2.4 mm or 3.2 mm) carbon steel filler metal, a 15° to 20° push angle, and the work angles shown in *Figure 24*. Clean the weld between each pass.

Step 6 Inspect the weld. The weld is acceptable if it has the following:

- Uniform rippled appearance of the bead face
- Craters and restarts filled to the full cross section of the weld
- Uniform weld width $\pm^1/_{16}$" (1.6 mm)
- Acceptable weld profile in accordance with the applicable code or standard
- Smooth, flat transition with complete fusion at the toes of the weld
- Complete uniform root reinforcement at least flush with the base metal to a maximum buildup of $^1/_8$" (3.2 mm)
- No excessive porosity with no pores larger than $^3/_{32}$" (2.38 mm)
- No undercut greater than $^1/_{32}$" (0.8 mm) deep or 10% of the base metal thickness, whichever is less
- No cracks

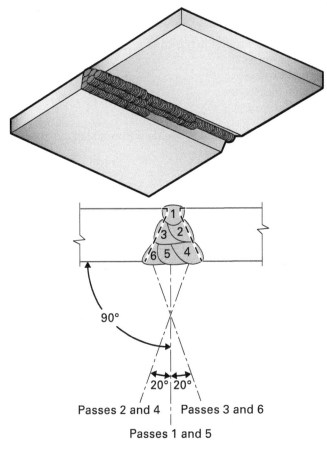

Stringer Bead Sequence

Figure 24 Multipass 4G bead sequence and work angles.

3.0.0 Section Review

1. The type of GTAW fillet weld that is normally made by welding uphill using a torch push angle is a(n) _____.
   a. horizontal position fillet weld
   b. flat position fillet weld
   c. vertical position fillet weld
   d. overhead position fillet weld

2. The most difficult part of welding an open V-groove butt joint is the _____.
   a. oscillation
   b. root pass
   c. up-cut
   d. termination

Module 29208   Review Questions

1. In GTAW, the filler metal is typically a _____.
   a. flux-cored wire
   b. handheld rod
   c. tungsten electrode
   d. base metal flux

2. GTAW produces high-quality welds without slag, corrosion, or _____.
   a. penetration
   b. ventilation
   c. termination
   d. oxidation

3. Any safety glasses, face shields, or helmet lenses used during GTAW activities *must* have the proper _____.
   a. light-reducing shade
   b. mirroring capabilities
   c. temperature tolerance
   d. electrical resistance

4. Before a tank or barrel is cut or welded, it should be cleaned and _____.
   a. sealed shut to prevent leaks or fumes
   b. filled with the proper shielding gas
   c. neutralized with compressed oxygen
   d. filled with water or purged with inert gas

5. When performing GTAW, open V-groove welds generally have a bevel angle of _____.
   a. 0° to 15°
   b. 20° to 25.5°
   c. 30° to 37.5°
   d. 45° to 55°

6. Torch travel speed affects the size of the GTAW weld puddle and the weld _____.
   a. convexity
   b. temperature
   c. appearance
   d. penetration

7. The two basic torch angles that *must* be controlled when performing GTAW are the _____.
   a. work angle and travel angle
   b. freehand angle and pull angle
   c. arc angle and work angle
   d. travel angle and keyhole angle

8. In the freehand technique, the torch electrode tip is manually held _____.
   a. against the base metal
   b. just above the weld puddle or base metal
   c. just below the weld puddle or base metal
   d. against the groove edges

9. Flat position stringer beads should be run with a push angle of _____.
   a. 15° to 20°
   b. 20° to 30°
   c. 30° to 40°
   d. 60° to 70°

10. Compared to the puddle width used to produce it, a weave bead is much _____.
    a. deeper
    b. higher
    c. wider
    d. hotter

11. Refer to *Figure RQ01*. An example of an *acceptable* fillet weld is represented by _____.
    a. Letter B
    b. Letter D
    c. Letter E
    d. Letter F

12. Refer to *Figure RQ01*. An example of an *ideal* fillet weld is represented by _____.
    a. Letter A
    b. Letter C
    c. Letter E
    d. Letter F

13. Refer to *Figure RQ01*. An example of an *unacceptable* overlap fillet weld is represented by _____.
    a. Letter B
    b. Letter C
    c. Letter D
    d. Letter E

14. If a groove weld has defects, it must be _____.
    a. thrown away
    b. recycled
    c. melted
    d. repaired

15. When practicing flat (1G) position open V-groove welds, run the root pass using either the lay-wire or _____.
    a. side-to-side movement
    b. keyhole technique
    c. walking-the-cup angle
    d. freehand technique

Figure RQ01

Answers to odd-numbered questions are found in the Review Question Answer Keys at the back of this book.

Answers to Section Review Questions

Answer	Section Reference	Objective
Section 1.0.0		
1. b	1.1.3	1a
2. d	1.2.3	1b
Section 2.0.0		
1. c	2.1.3	2a
2. a	2.2.1	2b
Section 3.0.0		
1. c	3.1.3	3a
2. b	3.2.1	3b

User Update

Did you find an error? Submit a correction by visiting **https://www.nccer.org/olf** or by scanning the QR code using your mobile device.

APPENDIX 29202A

Example of a Three-View Detail Drawing

Figure A01 is an example of an object that may be encountered in welding. *Figure A02* is an example of a three-view detail drawing representing that same object.

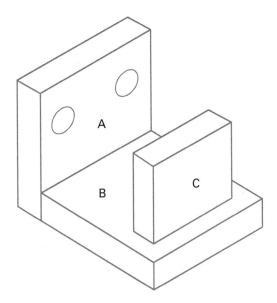

Details			
Part	Thickness	Length	Width
A	¾"	3½" (Tall)	3½"
B	¾"	4" (Long)	3½"
C	¾"	2½" (Tall)	2½"

Hole Dimensions and Location
- Hole diameter is ¾"
- Hole centers are 1¼" from the top edge of block "A"
- Hole centers are ¾" from the outside edges of block "A"

Figure A01

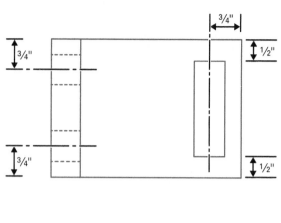

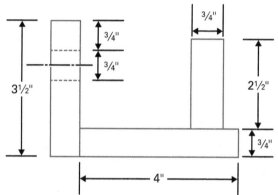

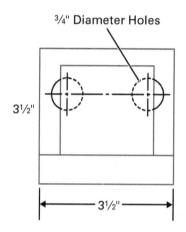

**Result of Redrawing the Object Using
Three Orthographic Views**

Figure A02

APPENDIX 29209A

Performance Accreditation Tasks

The Performance Accreditation Tasks (PATs) correspond to and support the learning objectives in *AWS EG2.0, Guide for the Training of Welding Personnel: Level 1—Entry Welder.*

PATs provide specific acceptable criteria for performance and help to ensure a true competency-based welding program for students.

The following tasks are designed to evaluate your ability to run fillet and groove welds with GMAW equipment. Perform each task when you are instructed to do so by your instructor. As you complete each task, show it to your instructor for evaluation. Do not proceed to the next task until instructed to do so by your instructor.

PATs 1 through 4 correspond to *AWS EG2.0, Module 5—Gas Metal Arc Welding (GMAW-S, GMAW Spray Transfer)*, Key Indicators 5 and 10.

PATs 5 through 8 correspond to *AWS EG2.0, Module 5—Gas Metal Arc Welding (GMAW-S, GMAW Spray Transfer)*, Key Indicators 6 and 11.

Make a Fillet Weld in the (1F) Flat Position

As directed by the instructor, use GMAW with carbon steel solid electrodes and appropriate shielding gas to make a six-pass fillet weld on carbon steel plate, as shown.

Note: Base metal thickness is at the instructor's discretion.

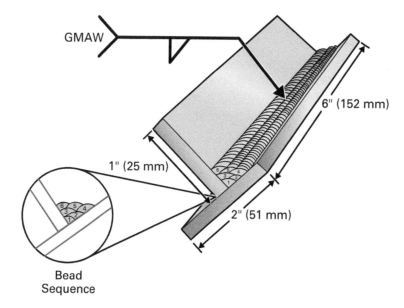

Bead
Sequence

Criteria for Acceptance:

- Uniform rippled appearance of the bead face
- Craters and restarts filled to the full cross section of the weld
- Uniform weld size ±$\frac{1}{16}$" (1.6 mm)
- Acceptable weld profile in accordance with the application code or standard
- Smooth, flat transition with complete fusion at the toes of the welds
- No excessive porosity with no pores larger than $\frac{3}{32}$" (2.38 mm)
- No undercut greater than $\frac{1}{32}$" (0.8 mm) deep or 10% of the base metal thickness, whichever is less
- No cracks

Make a Fillet Weld in the (2F) Horizontal Position

As directed by the instructor, use GMAW with carbon steel solid electrodes and appropriate shielding gas to make a six-pass fillet weld on carbon steel plate, as shown.

Note: Base metal thickness is at the instructor's discretion.

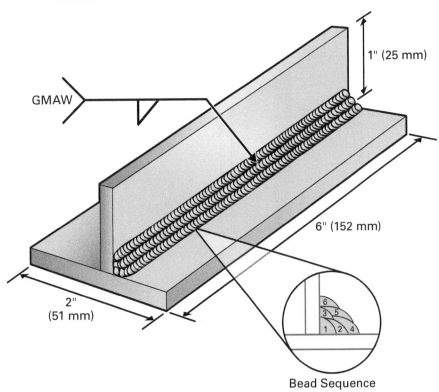

GMAW

1" (25 mm)

6" (152 mm)

2" (51 mm)

Bead Sequence

Criteria for Acceptance:

- Uniform rippled appearance of the bead face
- Craters and restarts filled to the full cross section of the weld
- Uniform weld size ±$\frac{1}{16}$" (1.6 mm)
- Acceptable weld profile in accordance with the application code or standard
- Smooth, flat transition with complete fusion at the toes of the welds
- No excessive porosity with no pores larger than $\frac{3}{32}$" (2.38 mm)
- No undercut greater than $\frac{1}{32}$" (0.8 mm) deep or 10% of the base metal thickness, whichever is less
- No cracks

Make a Fillet Weld in the (3F) Vertical Position

As directed by the instructor, use GMAW with carbon steel solid electrodes and appropriate shielding gas to make a vertical fillet weld on carbon steel plate, as shown.

Note: Base metal thickness is at the instructor's discretion.

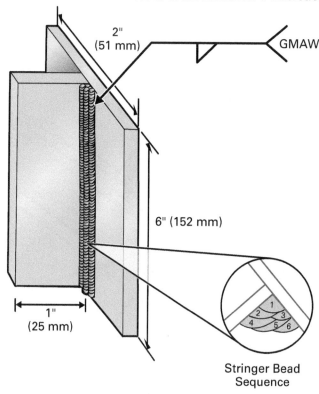

2" (51 mm)

GMAW

6" (152 mm)

1" (25 mm)

Stringer Bead Sequence

Criteria for Acceptance:

- Uniform rippled appearance of the bead face   _____
- Craters and restarts filled to the full cross section of the weld   _____
- Uniform weld size ±$\frac{1}{16}$" (1.6 mm)   _____
- Acceptable weld profile in accordance with the application code or standard   _____
- Smooth, flat transition with complete fusion at the toes of the welds   _____
- No excessive porosity with no pores larger than $\frac{3}{32}$" (2.38 mm)   _____
- No undercut greater than $\frac{1}{32}$" (0.8 mm) deep or 10% of the base metal thickness, whichever is less   _____
- No cracks   _____

Make a Fillet Weld in the (4F) Overhead Position

As directed by the instructor, use GMAW with carbon steel solid electrodes and appropriate shielding gas to make a six-pass fillet weld on carbon steel plate, as shown.

Note: Base metal thickness is at the instructor's discretion.

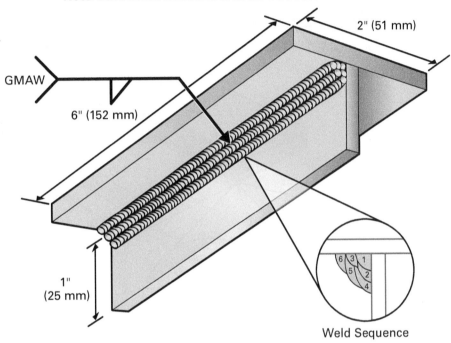

Weld Sequence

Criteria for Acceptance:

- Uniform rippled appearance of the bead face _____
- Craters and restarts filled to the full cross section of the weld _____
- Uniform weld size ±$\frac{1}{16}$" (1.6 mm) _____
- Acceptable weld profile in accordance with the application code or standard _____
- Smooth, flat transition with complete fusion at the toes of the welds _____
- No excessive porosity with no pores larger than $\frac{3}{32}$" (2.38 mm) _____
- No undercut greater than $\frac{1}{32}$" (0.8 mm) deep or 10% of the base metal thickness, whichever is less _____
- No cracks _____

Make a Groove Weld, With or Without Backing, in the (1G) Flat Position

As directed by the instructor, use GMAW with carbon steel solid electrodes and appropriate shielding gas to make a multipass groove weld on carbon steel plate, with or without backing, as shown.

Note: Base metal thickness is at the instructor's discretion.

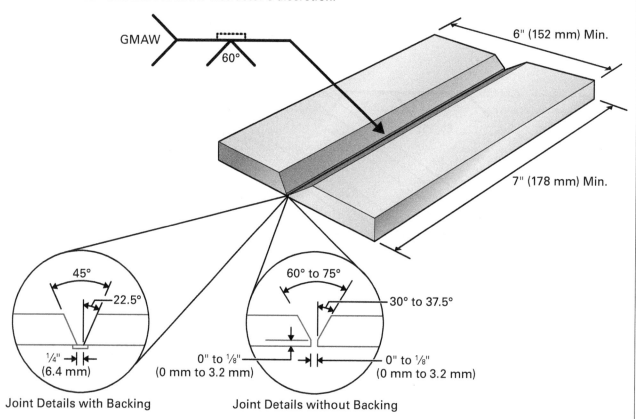

Joint Details with Backing   Joint Details without Backing

Criteria for Acceptance:

- Uniform rippled appearance of the bead face   _____
- Craters and restarts filled to the full cross section of the weld   _____
- Uniform weld size ±$\frac{1}{16}$" (1.6 mm)   _____
- Acceptable weld profile in accordance with the application code or standard   _____
- Smooth, flat transition with complete fusion at the toes of the welds   _____
- Complete joint penetration with uniform root and face reinforcement at least flush with the base metal to a maximum buildup of $\frac{1}{8}$" (3.2 mm), if applicable   _____
- No excessive porosity with no pores larger than $\frac{3}{32}$" (2.38 mm)   _____
- No undercut greater than $\frac{1}{32}$" (0.8 mm) deep or 10% of the base metal thickness, whichever is less   _____
- No cracks   _____

Performance Accreditation Tasks

Make a Groove Weld, With or Without Backing, in the (2G) Horizontal Position

As directed by the instructor, use GMAW with carbon steel solid electrodes and appropriate shielding gas to make a multipass groove weld on carbon steel plate, with or without backing, as shown.

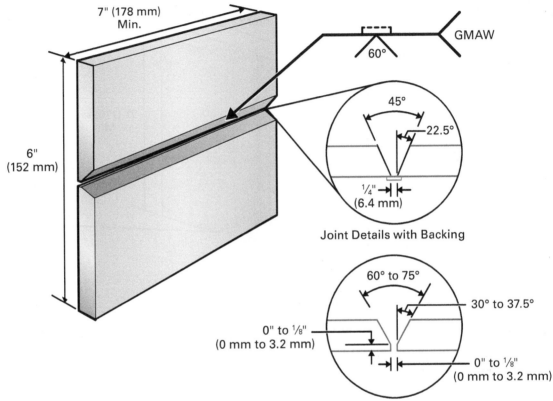

Joint Details with Backing

Joint Details without Backing

Note: Base metal thickness is at the instructor's discretion.

Criteria for Acceptance:

- Uniform rippled appearance of the bead face _____
- Craters and restarts filled to the full cross section of the weld _____
- Uniform weld size $\pm\frac{1}{16}$" (1.6 mm) _____
- Acceptable weld profile in accordance with the application code or standard _____
- Smooth, flat transition with complete fusion at the toes of the welds _____
- Complete joint penetration with uniform root and face reinforcement at least flush with the base metal to a maximum buildup of $\frac{1}{8}$" (3.2 mm), if applicable _____
- No excessive porosity with no pores larger than $\frac{3}{32}$" (2.38 mm) _____
- No undercut greater than $\frac{1}{32}$" (0.8 mm) deep or 10% of the base metal thickness, whichever is less _____
- No cracks _____

Performance Accreditation Tasks   **Module 29209**

Make a Groove Weld, With or Without Backing, in the (3G) Vertical Position

As directed by the instructor, use GMAW with carbon steel solid electrodes and appropriate shielding gas to make a multipass groove weld on carbon steel plate, with or without backing, as shown.

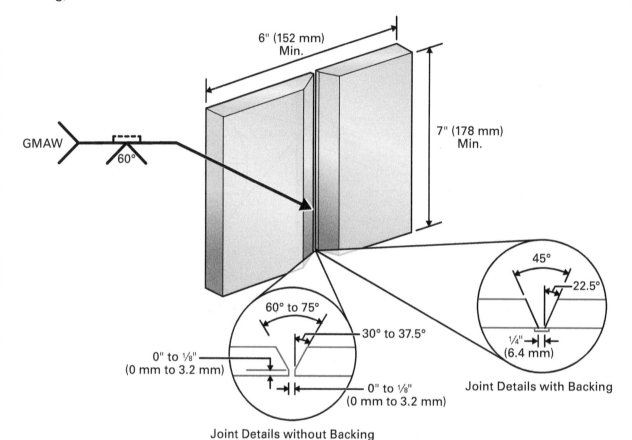

Joint Details without Backing

Joint Details with Backing

Note: Base metal thickness is at the instructor's discretion.

Criteria for Acceptance:

- Uniform rippled appearance of the bead face   _____
- Craters and restarts filled to the full cross section of the weld   _____
- Uniform weld size ±$\frac{1}{16}$" (1.6 mm)   _____
- Acceptable weld profile in accordance with the application code or standard   _____
- Smooth, flat transition with complete fusion at the toes of the welds   _____
- Complete joint penetration with uniform root and face reinforcement at least flush with the base metal to a maximum buildup of $\frac{1}{8}$" (3.2 mm), if applicable   _____
- No excessive porosity with no pores larger than $\frac{3}{32}$" (2.38 mm)   _____
- No undercut greater than $\frac{1}{32}$" (0.8 mm) deep or 10% of the base metal thickness, whichever is less   _____
- No cracks   _____

Performance Accreditation Tasks Module 29209

Make a Groove Weld, With or Without Backing, in the (4G) Overhead Position

As directed by the instructor, use GMAW with carbon steel solid electrodes and appropriate shielding gas to make a multipass groove weld on carbon steel plate, with or without backing, as shown.

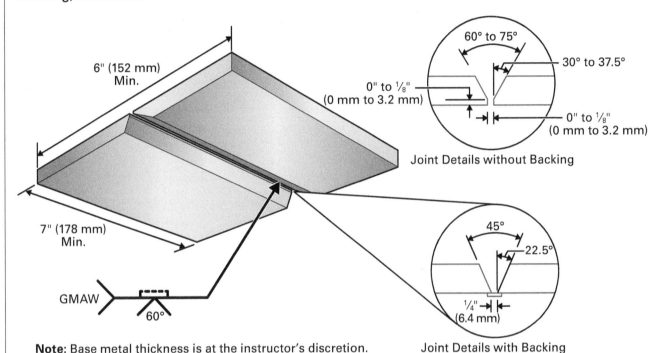

6" (152 mm) Min.

7" (178 mm) Min.

GMAW 60°

Note: Base metal thickness is at the instructor's discretion.

60° to 75°
30° to 37.5°
0" to 1/8" (0 mm to 3.2 mm)
0" to 1/8" (0 mm to 3.2 mm)
Joint Details without Backing

45°
22.5°
1/4" (6.4 mm)
Joint Details with Backing

Criteria for Acceptance:

- Uniform rippled appearance of the bead face _____
- Craters and restarts filled to the full cross section of the weld _____
- Uniform weld size $\pm\frac{1}{16}$" (1.6 mm) _____
- Acceptable weld profile in accordance with the application code or standard _____
- Smooth, flat transition with complete fusion at the toes of the welds _____
- Complete joint penetration with uniform root and face reinforcement at least flush with the base metal to a maximum buildup of $\frac{1}{8}$" (3.2 mm), if applicable _____
- No excessive porosity with no pores larger than $\frac{3}{32}$" (2.38 mm) _____
- No undercut greater than $\frac{1}{32}$" (0.8 mm) deep or 10% of the base metal thickness, whichever is less _____
- No cracks _____

APPENDIX 29210A

Performance Accreditation Tasks

The Performance Accreditation Tasks (PATs) correspond to and support the learning objectives in *AWS EG2.0, Guide for the Training of Welding Personnel: Level 1—Entry Welder*.

PATs provide specific acceptable criteria for performance and help to ensure a true competency-based welding program for students.

The following tasks are designed to evaluate your ability to run fillet and groove welds with FCAW equipment. Perform each task when you are instructed to do so by your instructor. As you complete each task, show it to your instructor for evaluation. Do not proceed to the next task until instructed to do so by your instructor.

As a reminder, open-root passes are not typically done with FCAW. Short-circuit GMAW, GTAW, or SMAW processes are typically used for root passes on open V-groove welds. For this module, you have the option of using other processes for the root pass of an open V-groove weld or the FCAW process with a backing bar.

PATs 1 through 4 correspond to AWS EG2.0, Module 6—Flux Core Arc Welding (FCAW-G/GM, FCAW-S), Key Indicators 5 and 10.

PATs 5 through 8 correspond to AWS EG2.0, Module 6—Flux Core Arc Welding (FCAW-G/GM, FCAW-S), Key Indicators 6 and 11.

Performance Accreditation Tasks   Module 29210

Make a Fillet Weld in the (1F) Flat Position

As directed by the instructor, use FCAW with flux-cored electrodes and, if required, the appropriate shielding gas to make a six-pass fillet weld using stringer beads on carbon steel plate, as shown.

Note: Base metal thickness is at the instructor's discretion.

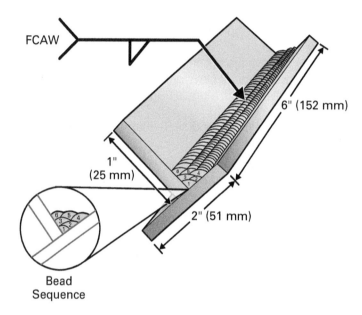

FCAW

6" (152 mm)

1" (25 mm)

2" (51 mm)

Bead Sequence

Criteria for Acceptance:

- Uniform rippled appearance of the bead face   _____
- Craters and restarts filled to the full cross section of the weld   _____
- Uniform weld size $\pm\frac{1}{16}$" (1.6 mm)   _____
- Acceptable weld profile in accordance with the application code or standard   _____
- Smooth, flat transition with complete fusion at the toes of the welds   _____
- No excessive porosity with no pores larger than $\frac{3}{32}$" (2.38 mm)   _____
- No undercut greater than $\frac{1}{32}$" (0.8 mm) deep or 10% of the base metal thickness, whichever is less   _____
- No cracks   _____
- No inclusions   _____

Performance Accreditation Tasks   Module 29210

Make a Fillet Weld in the (2F) Horizontal Position

As directed by the instructor, use FCAW with flux-cored electrodes and, if required, the appropriate shielding gas to make a six-pass fillet weld on carbon steel plate, as shown.

Note: Base metal thickness is at the instructor's discretion.

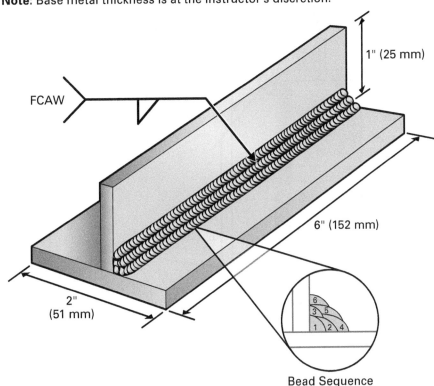

Bead Sequence

Criteria for Acceptance:

- Uniform rippled appearance of the bead face   _____
- Craters and restarts filled to the full cross section of the weld   _____
- Uniform weld size ±$\frac{1}{16}$" (1.6 mm)   _____
- Acceptable weld profile in accordance with the application code or   _____
  standard
- Smooth, flat transition with complete fusion at the toes of the welds   _____
- No excessive porosity with no pores larger than $\frac{3}{32}$" (2.38 mm)   _____
- No undercut greater than $\frac{1}{32}$" (0.8 mm) deep or 10% of the base metal   _____
  thickness, whichever is less
- No cracks   _____
- No inclusions   _____

Make a Fillet Weld in the (3F) Vertical Position

As directed by the instructor, use FCAW with flux-cored electrodes and, if required, the appropriate shielding gas to make a vertical fillet weld on carbon steel plate, as shown.

Note: Base metal thickness is at the instructor's discretion.

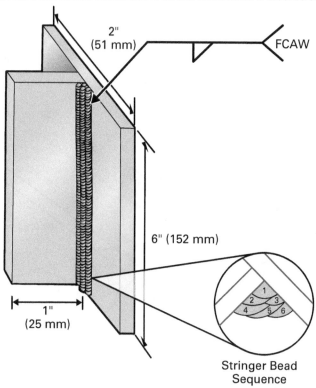

Stringer Bead Sequence

Criteria for Acceptance:

- Uniform rippled appearance of the bead face
- Craters and restarts filled to the full cross section of the weld
- Uniform weld size $\pm\frac{1}{16}$" (1.6 mm)
- Acceptable weld profile in accordance with the application code or standard
- Smooth, flat transition with complete fusion at the toes of the welds
- No excessive porosity with no pores larger than $\frac{3}{32}$" (2.38 mm)
- No undercut greater than $\frac{1}{32}$" (0.8 mm) deep or 10% of the base metal thickness, whichever is less.
- No cracks
- No inclusions

Performance Accreditation Tasks   **Module 29210**

Make a Fillet Weld in the (4F) Overhead Position

As directed by the instructor, use FCAW with flux-cored electrodes and, if required, the appropriate shielding gas to make a six-pass fillet weld on carbon steel plate, as shown.

Note: Base metal thickness is at the instructor's discretion.

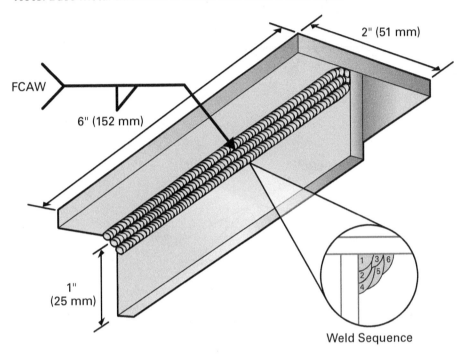

FCAW

2" (51 mm)

6" (152 mm)

1" (25 mm)

Weld Sequence

Criteria for Acceptance:

- Uniform rippled appearance of the bead face   _____
- Craters and restarts filled to the full cross section of the weld   _____
- Uniform weld size ±$\frac{1}{16}$" (1.6 mm)   _____
- Smooth, flat transition with complete fusion at the toes of the weld   _____
- Acceptable weld profile in accordance with the application code or standard   _____
- No excessive porosity with no pores larger than $\frac{3}{32}$" (2.38 mm)   _____
- No undercut greater than $\frac{1}{32}$" (0.8 mm) deep or 10% of the base metal thickness, whichever is less   _____
- No cracks   _____
- No inclusions   _____

Make a Groove Weld, With or Without Backing, in the (1G) Flat Position

As directed by the instructor, use FCAW with flux-cored electrodes and, if required, the appropriate shielding gas to make a multipass groove weld on carbon steel plate, with or without backing, as shown. If using the open-root joint design, perform root pass with GMAW, GTAW, or SMAW as directed, then fill and cap with FCAW.

Note: Base metal thickness is at the instructor's discretion.

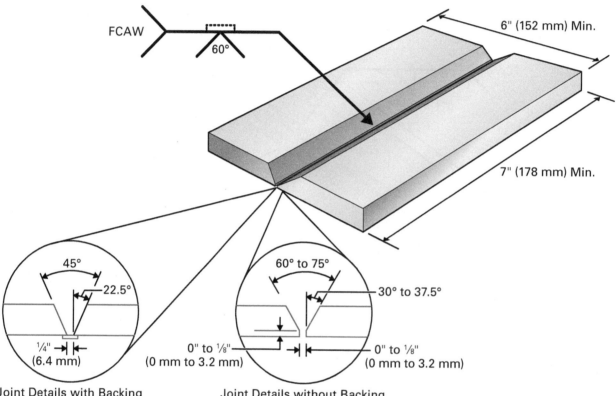

Joint Details with Backing   Joint Details without Backing

Criteria for Acceptance:

- Uniform rippled appearance of the bead face   _____
- Craters and restarts filled to the full cross section of the weld   _____
- Uniform weld size ±$\frac{1}{16}$" (1.6 mm)   _____
- Acceptable weld profile in accordance with the application code or standard   _____
- Smooth, flat transition with complete fusion at the toes of the welds   _____
- Complete joint penetration with uniform root and face reinforcement at least flush with the base metal to a maximum buildup of $\frac{1}{8}$" (3.2 mm), if applicable   _____
- No excessive porosity with no pores larger than $\frac{3}{32}$" (2.38 mm)   _____
- No undercut greater than $\frac{1}{32}$" (0.8 mm) deep or 10% of the base metal thickness, whichever is less   _____
- No cracks   _____
- No inclusions   _____

Performance Accreditation Tasks   **Module 29210**

Make a Groove Weld, With or Without Backing, in the (2G) Horizontal Position

As directed by the instructor, use FCAW with flux-cored electrodes and, if required, the appropriate shielding gas to make a multipass groove weld on carbon steel plate, with or without backing, as shown. If using the open-root joint design, perform root pass with GMAW, GTAW, or SMAW as directed, then fill and cap with FCAW.

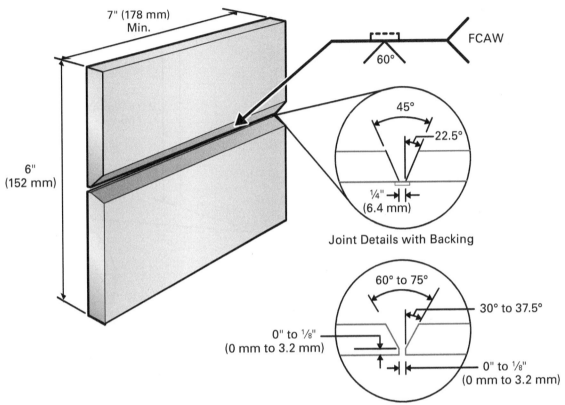

7" (178 mm) Min.

6" (152 mm)

FCAW
60°

45°
22.5°
¼" (6.4 mm)
Joint Details with Backing

60° to 75°
30° to 37.5°
0" to ⅛" (0 mm to 3.2 mm)
0" to ⅛" (0 mm to 3.2 mm)
Joint Details without Backing

Note: Base metal thickness is at the instructor's discretion.

Criteria for Acceptance:

- Uniform rippled appearance of the bead face _____
- Craters and restarts filled to the full cross section of the weld _____
- Uniform weld size ±¹⁄₁₆" (1.6 mm) _____
- Acceptable weld profile in accordance with the application code or standard _____
- Smooth, flat transition with complete fusion at the toes of the welds _____
- Complete joint penetration with uniform root and face reinforcement at least flush with the base metal to a maximum buildup of ⅛" (3.2 mm), if applicable _____
- No excessive porosity with no pores larger than ³⁄₃₂" (2.38 mm) _____
- No undercut greater than ¹⁄₃₂" (0.8 mm) deep or 10% of the base metal thickness, whichever is less _____
- No cracks _____
- No inclusions _____

Make a Groove Weld, With or Without Backing, in the (3G) Vertical Position

As directed by the instructor, use FCAW with flux-cored electrodes and, if required, the appropriate shielding gas to make a multipass groove weld on carbon steel plate, with or without backing, as shown. If using the open-root joint design, perform root pass with GMAW, GTAW, or SMAW as directed, then fill and cap with FCAW.

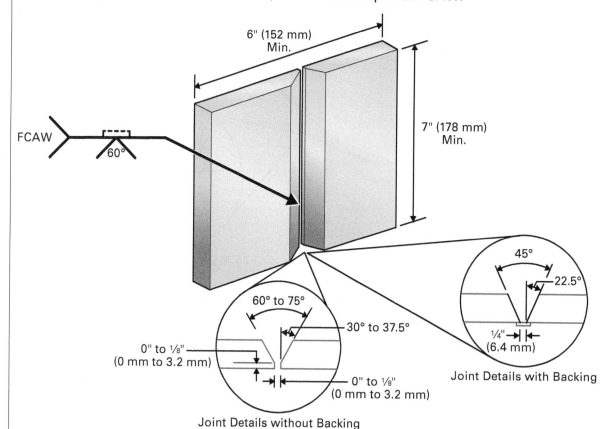

6" (152 mm) Min.

7" (178 mm) Min.

FCAW 60°

60° to 75°

30° to 37.5°

0" to ⅛" (0 mm to 3.2 mm)

0" to ⅛" (0 mm to 3.2 mm)

Joint Details without Backing

45°

22.5°

¼" (6.4 mm)

Joint Details with Backing

Note: Base metal thickness is at the instructor's discretion.

Criteria for Acceptance:

- Uniform rippled appearance of the bead face _____
- Craters and restarts filled to the full cross section of the weld _____
- Uniform weld size ±$\frac{1}{16}$" (1.6 mm) _____
- Acceptable weld profile in accordance with the application code or standard _____
- Smooth, flat transition with complete fusion at the toes of the welds _____
- Complete joint penetration with uniform root and face reinforcement at least flush with the base metal to a maximum buildup of ⅛" (3.2 mm), if applicable _____
- No excessive porosity with no pores larger than $\frac{3}{32}$" (2.38 mm) _____
- No undercut greater than $\frac{1}{32}$" (0.8 mm) deep or 10% of the base metal thickness, whichever is less _____
- No cracks _____
- No inclusions _____

Performance Accreditation Tasks   **Module 29210**

Make a Groove Weld, With or Without Backing, in the (4G) Overhead Position

As directed by the instructor, use FCAW with flux-cored electrodes and, if required, the appropriate shielding gas to make a multipass groove weld on carbon steel plate, with or without backing, as shown. If using the open-root joint design, perform root pass with GMAW, GTAW, or SMAW as directed, then fill and cap with FCAW.

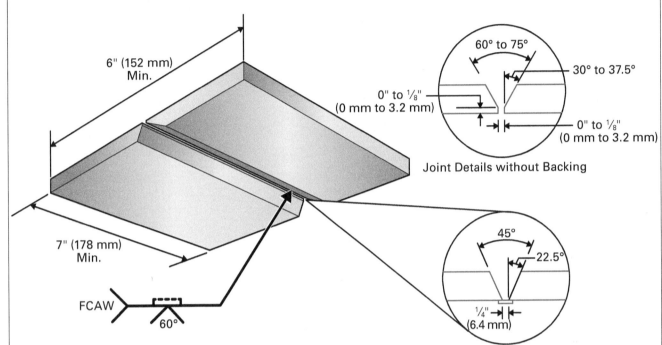

6" (152 mm) Min.

7" (178 mm) Min.

FCAW 60°

Note: Base metal thickness is at the instructor's discretion.

60° to 75°
30° to 37.5°
0" to ⅛" (0 mm to 3.2 mm)
0" to ⅛" (0 mm to 3.2 mm)

Joint Details without Backing

45°
22.5°
¼" (6.4 mm)

Joint Details with Backing

Criteria for Acceptance:

- Uniform rippled appearance of the bead face   _____
- Craters and restarts filled to the full cross section of the weld   _____
- Uniform weld size ±$\frac{1}{16}$" (1.6 mm)   _____
- Acceptable weld profile in accordance with the application code or standard   _____
- Smooth, flat transition with complete fusion at the toes of the welds   _____
- Complete joint penetration with uniform root and face reinforcement at least flush with the base metal to a maximum buildup of ⅛" (3.2 mm), if applicable   _____
- No excessive porosity with no pores larger than $\frac{3}{32}$" (2.38 mm)   _____
- No undercut greater than $\frac{1}{32}$" (0.8 mm) deep or 10% of the base metal thickness, whichever is less   _____
- No cracks   _____
- No inclusions   _____

APPENDIX 29208A

Performance Accreditation Tasks

The Performance Accreditation Tasks (PATs) correspond to and support the learning objectives in *AWS EG2.0, Guide for the Training of Welding Personnel: Level 1—Entry Welder.*

PATs provide specific acceptable criteria for performance and help to ensure a true competency-based welding program for students.

The following tasks are designed to evaluate your ability to run fillet and groove welds with GTAW equipment. Perform each task when you are instructed to do so by your instructor. As you complete each task, show it to your instructor for evaluation. Do not proceed to the next task until instructed to do so by your instructor.

PATs 1 through 4 correspond to *AWS EG2.0, Module 7—Gas Tungsten Arc Welding (GTAW)*, Key Indicator 5.

PATs 5 through 8 correspond to *AWS EG2.0, Module 7—Gas Tungsten Arc Welding (GTAW)*, Key Indicator 6.

Make Multipass Fillet Welds on Carbon Steel Plate in the (1F) Flat Position

As directed by the instructor, use GTAW process with carbon steel filler metal to make a six-pass fillet weld on carbon steel plate, as shown.

Note: Base metal thickness is at the instructor's discretion.

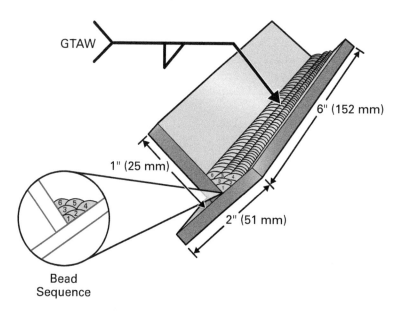

Bead Sequence

Criteria for Acceptance:

- Uniform rippled appearance of the bead face   _____
- Craters and restarts filled to the full cross section of the weld   _____
- Uniform weld size ±$\frac{1}{16}$" (1.6 mm)   _____
- Acceptable weld profile in accordance with the applicable code or standard   _____
- Smooth, flat transition with complete fusion at the toes of the welds   _____
- No excessive porosity with no pores larger than $\frac{3}{32}$" (2.38 mm)   _____
- No undercut greater than $\frac{1}{32}$" (0.8 mm) deep or 10% of the base metal thickness, whichever is less   _____
- No cracks   _____

Make Multipass Fillet Welds on Carbon Steel Plate in the (2F) Horizontal Position

As directed by the instructor, use GTAW process with carbon steel solid filler metal to make a six-pass fillet weld on carbon steel plate, as shown.

Note: Base metal thickness is at the instructor's discretion.

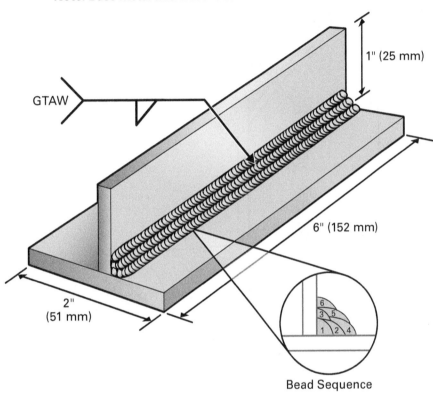

1" (25 mm)

GTAW

6" (152 mm)

2" (51 mm)

Bead Sequence

Criteria for Acceptance:

- Uniform rippled appearance of the bead face   _____
- Craters and restarts filled to the full cross section of the weld   _____
- Uniform weld size ±$\frac{1}{16}$" (1.6 mm)   _____
- Acceptable weld profile in accordance with the applicable code or standard   _____
- Smooth, flat transition with complete fusion at the toes of the welds   _____
- No excessive porosity with no pores larger than $\frac{3}{32}$" (2.38 mm)   _____
- No undercut greater than $\frac{1}{32}$" (0.8 mm) deep or 10% of the base metal thickness, whichever is less   _____
- No cracks   _____

Performance Accreditation Tasks   Module 29208

Make Multipass Fillet Welds on Carbon Steel Plate in the (3F) Vertical Position

As directed by the instructor, use GTAW process with carbon steel filler metal to make a vertical fillet weld on carbon steel plate, as shown.

Note: Base metal thickness is at the instructor's discretion.

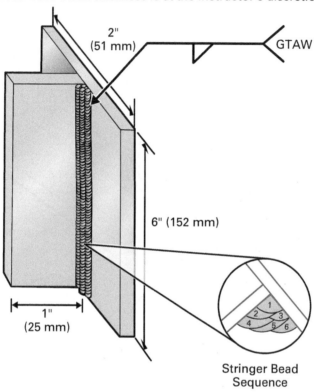

2" (51 mm)

GTAW

6" (152 mm)

1" (25 mm)

Stringer Bead Sequence

Criteria for Acceptance:

- Uniform rippled appearance of the bead face   _____
- Craters and restarts filled to the full cross section of the weld   _____
- Uniform weld size $\pm\frac{1}{16}$" (1.6 mm)   _____
- Acceptable weld profile in accordance with the applicable code or standard   _____
- Smooth, flat transition with complete fusion at the toes of the welds   _____
- No excessive porosity with no pores larger than $\frac{3}{32}$" (2.38 mm)   _____
- No undercut greater than $\frac{1}{32}$" (0.8 mm) deep or 10% of the base metal thickness, whichever is less   _____
- No cracks   _____

Make Multipass Fillet Welds on Carbon Steel Plate in the (4F) Overhead Position

As directed by the instructor, use GTAW process with carbon steel filler metal to make a six-pass fillet weld on carbon steel plate, as shown.

Note: Base metal thickness is at the instructor's discretion.

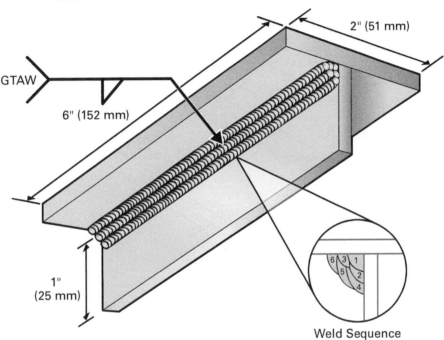

GTAW

2" (51 mm)

6" (152 mm)

1" (25 mm)

6 3 1
5 2
4

Weld Sequence

Criteria for Acceptance:

- Uniform rippled appearance of the bead face

- Craters and restarts filled to the full cross section of the weld

- Uniform weld size ±$\frac{1}{16}$" (1.6 mm)

- Acceptable weld profile in accordance with the applicable code or standard

- Smooth, flat transition with complete fusion at the toes of the welds

- No excessive porosity with no pores larger than $\frac{3}{32}$" (2.38 mm)

- No undercut greater than $\frac{1}{32}$" (0.8 mm) deep or 10% of the base metal thickness, whichever is less

- No cracks

Performance Accreditation Tasks

Module 29208

Make Multipass Open V-Groove Welds on Carbon Steel Plate in the Flat (1G) Position

As directed by the instructor, use GTAW process with carbon steel filler metal to make multipass open V-groove welds on carbon steel plate in the flat (1G) position, as shown.

Note: Base metal thickness is at the instructor's discretion.

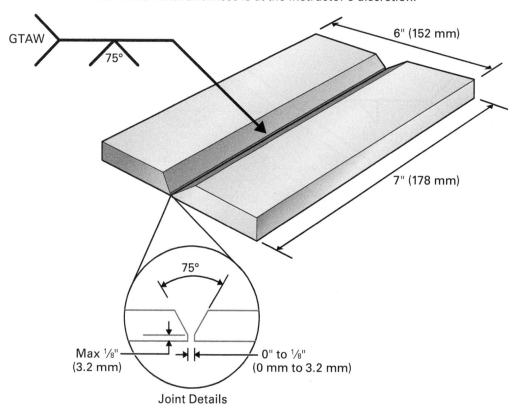

Joint Details

Criteria for Acceptance:

- Uniform rippled appearance of the bead face _____
- Craters and restarts filled to the full cross section of the weld _____
- Uniform weld size $\pm\frac{1}{16}$" (1.6 mm) _____
- Acceptable weld profile in accordance with the applicable code or standard _____
- Smooth, flat transition with complete fusion at the toes of the welds _____
- Complete uniform root and face reinforcement at least flush with the base metal to a maximum buildup of $\frac{1}{8}$" (3.2 mm) _____
- No excessive porosity with no pores larger than $\frac{3}{32}$" (2.38 mm) _____
- No undercut greater than $\frac{1}{32}$" (0.8 mm) deep or 10% of the base metal thickness, whichever is less _____
- No cracks _____
- No pinholes _____

Make Multipass Open V-Groove Welds on Carbon Steel Plate in the Horizontal (2G) Position

As directed by the instructor, use GTAW process with carbon steel filler metal to make multipass open V-groove welds on carbon steel plate in the horizontal (2G) position, as shown.

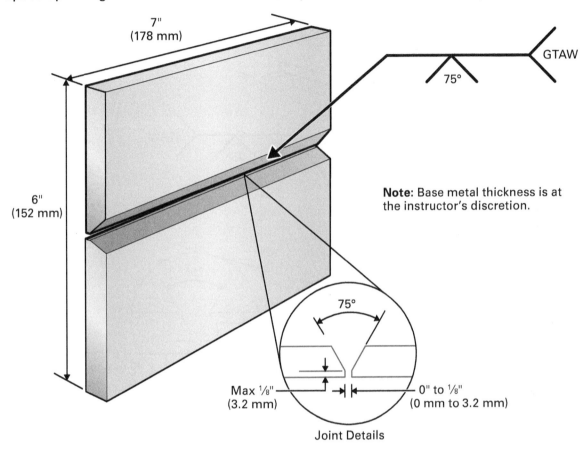

7"
(178 mm)

GTAW

75°

6"
(152 mm)

Note: Base metal thickness is at the instructor's discretion.

75°

Max 1/8"
(3.2 mm)

0" to 1/8"
(0 mm to 3.2 mm)

Joint Details

Criteria for Acceptance:

- Uniform rippled appearance of the bead face   _____
- Craters and restarts filled to the full cross section of the weld   _____
- Uniform weld size ±$\frac{1}{16}$" (1.6 mm)   _____
- Acceptable weld profile in accordance with the applicable code or standard   _____
- Smooth, flat transition with complete fusion at the toes of the welds   _____
- Complete uniform reinforcement at least flush with the base metal to a maximum buildup of $\frac{1}{8}$" (3.2 mm)   _____
- No excessive porosity with no pores larger than $\frac{3}{32}$" (2.38 mm)   _____
- No undercut greater than $\frac{1}{32}$" (0.8 mm) deep or 10% of the base metal thickness, whichever is less   _____
- No cracks   _____
- No pinholes   _____

Make Multipass Open V-Groove Welds on Carbon Steel Plate in the Vertical (3G) Position

As directed by the instructor, use GTAW process with carbon steel filler metal to make multipass open V-groove welds on carbon steel plate in the vertical (3G) position, as shown.

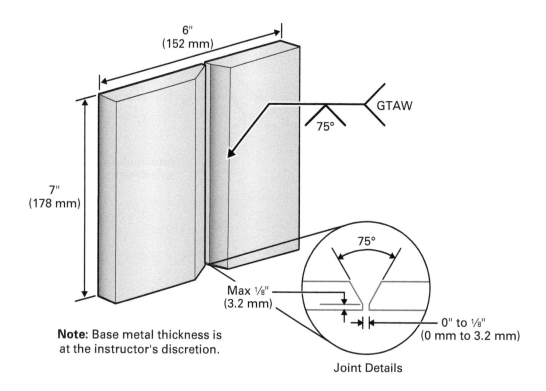

Note: Base metal thickness is at the instructor's discretion.

Joint Details

Criteria for Acceptance:

- Uniform rippled appearance of the bead face   _____
- Craters and restarts filled to the full cross section of the weld   _____
- Uniform weld size $\pm\frac{1}{16}$" (1.6 mm)   _____
- Acceptable weld profile in accordance with the applicable code or standard   _____
- Smooth, flat transition with complete fusion at the toes of the welds   _____
- Complete uniform root reinforcement at least flush with the base metal to a maximum buildup of $\frac{1}{8}$" (3.2 mm)   _____
- No excessive porosity with no pores larger than $\frac{3}{32}$" (2.38 mm)   _____
- No undercut greater than $\frac{1}{32}$" (0.8 mm) deep or 10% of the base metal thickness, whichever is less   _____
- No cracks   _____
- No pinholes   _____

Performance Accreditation Tasks Module 29208

Make Multipass Open V-Groove Welds on Carbon Steel Plate in the Overhead (4G) Position

As directed by the instructor, use GTAW process with carbon steel filler metal to make multipass open V-groove welds on carbon steel plate in the overhead (4G) position, as shown.

Note: Base metal thickness is at the instructor's discretion.

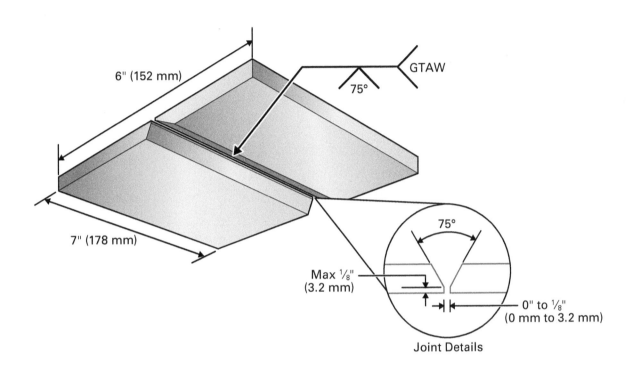

6" (152 mm)

GTAW

75°

7" (178 mm)

75°

Max 1/8"
(3.2 mm)

0" to 1/8"
(0 mm to 3.2 mm)

Joint Details

Criteria for Acceptance:

- Uniform rippled appearance of the bead face   _____
- Craters and restarts filled to the full cross section of the weld   _____
- Uniform weld size ±$\frac{1}{16}$" (1.6 mm)   _____
- Acceptable weld profile in accordance with the applicable code or standard   _____
- Smooth transition with complete fusion at the toes of the welds   _____
- Complete uniform root reinforcement at least flush with the base metal to a maximum buildup of $\frac{1}{8}$" (3.2 mm)   _____
- No excessive porosity with no pores larger than $\frac{3}{32}$" (2.38 mm)   _____
- No undercut greater than $\frac{1}{32}$" (0.8 mm) deep or 10% of the base metal thickness, whichever is less   _____
- No cracks   _____
- No pinholes   _____

REVIEW QUESTION ANSWER KEYS

MODULE 29201

Answer	Section Reference
1. a	1.1.0
3. c	1.1.0
5. a	1.2.0; *Figure 13*
7. a	1.2.2
9. d	1.3.2
11. b	1.3.3; *Table 1*
13. a	1.4.1; *Figure 30*
15. a	1.4.6

MODULE 29202

Answer	Section Reference
1. d	1.0.0
3. a	1.2.1
5. b	2.1.3
7. a	2.1.6
9. c	2.1.0

MODULE 29202 CALCULATIONS

Question 9

Part 1B is 6" tall, and parts 1A and 1B are a total of 12" tall. Subtract part 1B from the total parts to get part 1A.

1B = 6"

1A + 1B = 12"

(1A + 1B) − 1B = 1A

12" − 6" = **6"**

MODULE 29203

Answer	Section Reference
1. d	1.0.0
3. b	1.2.0
5. b	1.3.5
7. b	1.4.5
9. c	2.2.2
11. d	2.3.3
13. b	2.4.1
15. a	3.1.0
17. c	3.2.0
19. b	3.4.1

MODULE 29204

Answer	Section Reference
1. d	1.0.0
3. b	1.1.3
5. c	1.2.0
7. c	1.3.2
9. c	2.2.1

MODULE 29205

Answer	Section Reference
1. a	1.1.2
3. c	1.2.0
5. b	1.3.3
7. c	2.1.1
9. d	2.1.3
11. c	2.2.1
13. a	2.5.0
15. d	2.6.1
17. b	2.7.4
19. a	3.1.1

MODULE 29209

Answer	Section Reference
1. b	1.1.1
3. a	1.2.2
5. d	1.3.3
7. c	2.2.2
9. d	2.2.4
11. c	3.1.1
13. d	3.2.2
15. a	3.2.7

MODULE 29210

Answer	Section Reference
1. b	1.2.1
3. a	1.1.2
5. d	1.3.3
7. c	2.2.2
9. d	2.2.4
11. c	3.1.0
13. d	3.2.2
15. c	3.2.6

MODULE 29207

Answer	Section Reference
1. b	1.1.1
3. a	1.1.2
5. c	1.2.0
7. d	2.1.6
9. b	2.3.2
11. b	2.4.4
13. b	2.5.0
15. a	3.1.0
17. d	3.3.0
19. a	3.4.0

MODULE 29208

Answer	Section Reference
1. b	1.0.0
3. a	1.1.1
5. c	1.2.2
7. a	2.1.2
9. a	2.2.1
11. d	3.1.0; *Figure 11*
13. c	3.1.0; *Figure 11*
15. b	3.2.3

ADDITIONAL RESOURCES

29201 Welding Symbols

ASTM A325, Standard Specification for Structural Bolts, Steel, Heat Treated, 120/105 ksi Minimum Tensile Strength. West Conshohocken, PA: ASTM International.

AWS A2.4, Standard Symbols for Welding, Brazing, and Nondestructive Examination. Miami, FL: American Welding Society.

AWS A3.0, Standard Definitions; Including Terms for Adhesive Bonding, Brazing, Soldering, Thermal Cutting, and Thermal Spraying. Miami, FL: American Welding Society.

AWS D1.1/D1.1M, Structural Welding Code Steel. Miami, FL: American Welding Society.

How to Read Shop Drawings. Cleveland, OH: The James F. Lincoln Arc Welding Foundation, 2012.

ISO 2553, Welded, Brazed, and Soldered Joints — Symbolic Representation on Drawings. Geneva, Switzerland: International Organization for Standardization (ISO).

Steel Construction Manual. 14th Edition. Chicago, IL: American Institute of Steel Construction, 2011.

29202 Reading Welding Detail Drawings

ASTM A325, Standard Specification for Structural Bolts, Steel, Heat Treated, 120/105 ksi Minimum Tensile Strength. West Conshohocken, PA: ASTM International.

AWS A2.4, Standard Symbols for Welding, Brazing, and Nondestructive Examination. Miami, FL: American Welding Society.

AWS A3.0, Standard Welding Terms and Definitions. Miami, FL: American Welding Society.

AWS D1.1/D1.1M, Structural Welding Code Steel. Miami, FL: American Welding Society.

How to Read Shop Drawings. Cleveland, OH: The James F. Lincoln Arc Welding Foundation, 2012.

Steel Construction Manual. 14th Edition. Chicago, IL: American Institute of Steel Construction, 2011.

Welding Symbols on Drawings. E. N. Gregory and A. A. Armstrong. Boca Raton, FL: CRC Press, 2005.

29203 Physical Characteristics and Mechanical Properties of Metals

AWS B1.10M/B1.10, Guide for the Nondestructive Examination of Welds. Miami, FL: American Welding Society.

AWS B1.11, Guide for the Visual Examination of Welds. Miami, FL: American Welding Society.

AWS D3.5-93R, Guide for Steel Hull Welding. Miami, FL: American Welding Society.

AWS D3.6M, Underwater Welding Code. Miami, FL: American Welding Society.

AWS D3.7, Guide for Aluminum Hull Welding. Miami, FL: American Welding Society.

Machinery's Handbook. Erik Oberg, Franklin D. Jones, and Christopher J. McCauley. New York, NY: Industrial Press, Inc., 2004.

Metals and How to Weld Them. T. B. Jefferson. Cleveland, OH: The James F. Lincoln Arc Welding Foundation.

OSHA 1926.351, Arc Welding and Cutting. Latest Edition. Washington, DC: Occupational Safety and Health Administration (OSHA).

Stick Electrode Product Catalog. Cleveland, OH: The Lincoln Electric Company, 2008.

Stick Electrode Welding Guide. Cleveland, OH: The Lincoln Electric Company, 2004.

The Procedure Handbook of Arc Welding. 14th Edition. Cleveland, OH: The James F. Lincoln Arc Welding Foundation, 2000.

Welding Handbook, Volume 4: Materials and Applications. 9th Edition. Miami, FL: American Welding Society, 2011.

Lincoln Electric website offers sources for products and training **http://www.lincolnelectric.com.**

29204 Preheating and Postheating of Metals

The Procedure Handbook of Arc Welding, 14th Edition. Cleveland, OH: The James F. Lincoln Arc Welding Foundation, 2000.

Welding Essentials: Questions and Answers. 2nd Edition. William L. Galvery, Jr. and Frank M. Marlow. New York, NY: Industrial Press, Inc., 2007.

Welding Handbook. Volume 2, Part 1: Welding Processes. 9th Edition. Miami, FL: American Welding Society, 2004.

Welding Handbook. Volume 3, Part 2: Welding Processes. 9th Edition. Miami, FL: American Welding Society, 2004.

29205 GMAW and FCAW – Equipment and Filler Metals

AWS B1.10M/B1.10, Guide for the Nondestructive Examination of Welds. Miami, FL: American Welding Society.

AWS B1.11, Guide for the Visual Examination of Welds. Miami, FL: American Welding Society.

AWS A5.20/A5.20M, Specification for Carbon Steel Electrodes for Flux Cored Arc Welding. Miami, FL: American Welding Society.

AWS A5.22, Specification for Stainless Steel Electrodes for Flux Cored Arc Welding and Stainless Steel Flux Cored Rods for Gas Tungsten Arc Welding. Miami, FL: American Welding Society.

AWS A5.29/A5.29M, Specification for Low-Alloy Steel Electrodes for Flux Cored Arc Welding. Miami, FL: American Welding Society.

AWS A5.34/A5.34M, Specification for Nickel Based Flux Cored and Metal Cored Electrodes. Miami, FL: American Welding Society.

Modern Welding Technology. 6th Edition. Howard B. Cary. Englewood Cliffs, NJ: Prentice Hall, Inc., 2004.

OSHA 1926.351, Arc Welding and Cutting. Latest Edition. Washington, DC: Occupational Safety and Health Administration (OSHA).

Welding Handbook. Volume 1, Welding Science & Technology. 9th Edition. Miami, FL: American Welding Society, 2001.

Welding Handbook. Volume 2, Part 1: Welding Processes. 9th Edition. Miami, FL: American Welding Society, 2004.

29209 GMAW – Plate

AWS B1.10M/B1.10, Guide for the Nondestructive Examination of Welds. Miami, FL: American Welding Society.

AWS B1.11, Guide for the Visual Examination of Welds. Miami, FL: American Welding Society.

AWS C5.6-89R, Recommended Practices for Gas Metal Arc Welding. Miami, FL: American Welding Society.

AWS D1.1 3G/4G, Certification. Miami, FL: American Welding Society.

AWS D3.5-93R, Guide for Steel Hull Welding. Miami, FL: American Welding Society.

GMAW Welding Guide. Latest Edition. Cleveland, OH: The Lincoln Electric Company.

OSHA 1910.269, Appendix C, Protection from Hazardous Differences in Electric Potential. Latest Edition. Washington, DC: Occupational Safety and Health Administration (OSHA).

OSHA 1926.351, Arc Welding and Cutting. Latest Edition. Washington, DC: Occupational Safety and Health Administration (OSHA).

The Procedure Handbook of Arc Welding. 14th Edition. Cleveland, OH: The James F. Lincoln Arc Welding Foundation, 2000.

Welding Handbook. Volume 1, Welding Science & Technology. 9th Edition. Miami, FL: American Welding Society, 2001.

Welding Handbook. Volume 2, Part 1: Welding Processes. 9th Edition. Miami, FL: American Welding Society, 2004.

29210 FCAW – Plate

AWS B1.10M/B1.10, Guide for the Nondestructive Examination of Welds. Miami, FL: American Welding Society.

AWS B1.11, Guide for the Visual Examination of Welds. Miami, FL: American Welding Society.

AWS D3.5-93R, Guide for Steel Hull Welding. Miami, FL: American Welding Society.

OSHA 1910.269, Appendix C, Protection from Hazardous Differences in Electric Potential. Latest Edition. Washington, DC: Occupational Safety and Health Administration (OSHA).

OSHA 1926.351, Arc Welding and Cutting. Latest Edition. Washington, DC: Occupational Safety and Health Administration (OSHA).

The Procedure Handbook of Arc Welding. 14th Edition. Cleveland, OH: The James F. Lincoln Arc Welding Foundation, 2000.

Welding Handbook. Volume 1, Welding Science & Technology. 9th Edition. Miami, FL: American Welding Society, 2001.

Welding Handbook. Volume 2, Part 1: Welding Processes. 9th Edition. Miami, FL: American Welding Society, 2004.

29207 GTAW – Equipment and Filler Metals

AWS B1.10, Guide for the Nondestructive Examination of Welds. Miami, FL: American Welding Society.

AWS B1.11, Guide for the Visual Examination of Welds. Miami, FL: American Welding Society.

AWS C5.5, Recommended Practices for Gas Tungsten Arc Welding. Miami, FL: American Welding Society.

AWS D3.5, Guide for Steel Hull Welding. Miami, FL: American Welding Society.

AWS D3.7, Guide for Aluminum Hull Welding. Miami, FL: American Welding Society.

AWS Safety and Health Fact Sheet 27. Miami, FL: American Welding Society.

Modern Welding Technology. 6th Edition. Howard B. Cary. Englewood Cliffs, NJ: Prentice Hall, Inc., 2004.

OSHA 1910.269, Appendix C, Protection from Hazardous Differences in Electric Potential. Latest Edition. Washington, DC: Occupational Safety and Health Administration (OSHA).

OSHA 1926.351, Arc Welding and Cutting. Latest Edition. Washington, DC: Occupational Safety and Health Administration (OSHA).

Tungsten Electrodes: an Arc-Zone.com® Technical Focus Paper (pdf). Carlsbad, CA: Arc-Zone.com, Inc. **https://www.arc-zone.com/azc-scip/TungstenElectrodes.pdf**.

Tungsten Guidebook (pdf). Newbury Park, CA: Diamond Ground Products, Inc. **http://www.diamondground.com/TungstenGuidebook2013.pdf**.

Welding Handbook. Volume 1, Welding Science & Technology. 9th Edition. Miami, FL: American Welding Society, 2001.

Welding Handbook. Volume 2, Part 1: Welding Processes. 9th Edition. Miami, FL: American Welding Society, 2004.

29208 GTAW – Plate

OSHA 1926.351, Arc Welding and Cutting. Latest Edition. Washington, DC: Occupational Safety and Health Administration (OSHA).

The Procedure Handbook of Arc Welding. 14th Edition. Cleveland, OH: The James F. Lincoln Arc Welding Foundation, 2000.

Welding Handbook. Volume 1, Welding Science & Technology. 9th Edition. Miami, FL: American Welding Society, 2001.

Welding Handbook. Volume 2, Part 1: Welding Processes. 9th Edition. Miami, FL: American Welding Society, 2004.

GLOSSARY

Alloy: A metal that has had other elements added to it that substantially change its mechanical properties.

American Iron and Steel Institute (AISI): An industry organization responsible for preparing standards for steel and steel alloys that are based upon the code system of SAE International.

American Society of Mechanical Engineers (ASME): An educational and technical organization founded for the practice of mechanical and multidisciplinary engineering.

Annealing: A postweld heat treatment for stress-relieving weldments that is done at temperatures approximately 100°F (38°C) above the critical temperature with a prolonged holding period.

Arrow line: The line drawn at an angle from the reference line (either end or both ends) to an arrowhead at the location of the weld.

ASTM International: Formerly known as American Society for Testing and Materials, an organization that developed a code system for identifying and labeling steels.

Austenitizing: Heating a steel up to or above the transformation temperature range to achieve partial or complete transformation to austenite grain structure.

Axially: In a straight line along an axis.

Axis: The centerline of the weld or prepared base metal.

Blind hole: A drilled hole that does not go all the way through the object.

Blowholes: The vents that form when air or gases escape.

Casting: The metal object produced by pouring molten metal into a mold.

Coefficient: A numerical measure of a physical or chemical property that is constant for a system under specified conditions, such as the coefficient of friction.

Composite metal-cored wire electrodes: Filler wire electrodes with hollow cores containing powdered materials, primarily metals.

Constituents: The elements and compounds, such as metal oxides, that make up a mixture or alloy.

Contact tip to work distance (CTWD): The distance between the contact tip of the electrode and the base metal to be welded.

Counterbore: To enlarge a portion of a drilled hole so that a socket head fastener will seat flush or below the surface.

Countersink: A hole with tapered sides and a wider opening that allows a flat-head fastener to seat flush to the surface of the material in which the hole is made.

Critical temperature: The temperature at which iron crystals in a ferrous-based metal transform from being face-centered to body-centered. This dramatically changes the strength, hardness, and ductility of the metal.

Crystalline: Like a crystal; having a uniform atomic structure throughout the entire material.

Degassing: The escape of gases formed during the welding process that can cause imperfections in the weld.

Discontinuity: The deformation, or interruption, of the crystal structure inside a metal.

Ductile: Able to be bent or shaped without breaking.

Ductility: The characteristic of metal that allows it to be stretched, drawn, or hammered without breaking.

Elevation: A height measurement above sea level or other identified point; also, the vertical sides of an object in an orthographic view.

Ferrous: Relating to iron or an alloy that is mostly iron; a metal containing iron.

Forgeability: The ability to be heated and then beaten or hammered into shape.

Globular transfer: A GMAW process in which the electrode wire does not touch the base metal, and the filler metal is transferred randomly in the arc as large, irregular globules.

Hardenability: A characteristic of a metal that enables it to become hard, usually through heat treatment.

Heat-affected zone (HAZ): The part of the base metal that has not been melted but has been altered structurally by welding or other heat-intensive operation such as oxyfuel cutting.

Hot shortness: The condition of metal when it proves to be very brittle and unbendable at red heat but can be bent without showing signs of brittleness when cold or at white heat. This condition is often a result of high sulfur and phosphorus content.

Inductance: An electrical circuit property, adjustable on some FCAW power supplies, that influences the pinch effect.

Interpass temperature: In a multipass weld, the temperature of the base metal at the time the next weld pass is started; it is usually the same temperature that is used for preheating.

Interpass temperature control: The process of maintaining the temperature of the weld zone within a given range during welding.

Isometric view: A drawing depicting a three-dimensional view of an object.

Lay wire: A root pass technique, also called *on the wire*, in which the filler metal is held in the bottom of the groove parallel to the plates and then melted and fused in place by moving the torch along the groove.

Lift-arc technology: A type of GTAW technology in which the tungsten electrode may be touched to the work, but the arc will not strike until the tip is lifted away to an appropriate distance.

Malleable: Able to be hammered or pressed into another shape without breaking.

Mark: A system of letters and numbers used with structural steel to identify the type of member and where it is placed.

Martensite: A solid solution of carbon in alpha-iron that is formed when steel is cooled so rapidly that the change from austenite to pearlite is suppressed; responsible for the hardness of quenched steel.

Mechanical properties: The characteristics or traits that indicate how flexible and strong a metal will be under stress.

Metallic: Having characteristics of metal, such as ductility, malleability, luster, and heat and electrical conductivity.

Mill test report (MTR): A quality assurance report provided to a customer by a metal manufacturer that shows the chemical content and testing results of the metal being purchased; also called a *certified mill test report (CMTR)*.

Nonferrous: Relating to a metal, such as aluminum, copper, or brass, that lacks sufficient quantities of iron to be affected by its properties.

Notch toughness: The impact load required to break a test piece of weld metal measured by a notch toughness test (Charpy V-notch test). This test may be performed on metal below room temperature, at which point it may be more brittle.

Orthographic projection: A method of developing multiple flat views of an object by projecting lines out from the object at right angles from the face.

Pinch effect: The tendency of the end of the wire electrode to melt into a droplet during short-circuiting GMAW welding.

Pitch: The center-to-center distance between welds.

Postweld heat treatment (PWHT): The process of heat-treating the base metal and weld after the weld has been made.

Potentiometer: A three-terminal resistor with an adjustable center connection; used for variable voltage control.

Preheat: The process of heating the base metal before making a weld.

Pulsed transfer: A GMAW welding process in which the welding power is cycled from a low level (background level) to a higher level (welding level), resulting in lower average voltage and current; this is a variation of the spray-transfer process.

Quench: To rapidly cool a hot metal using air, water, or oil.

Reference line: The horizontal line in the center of the welding symbol from which all elements of the welding symbol are referenced. The reference line is one of the most important elements of the welding symbol.

Residual stresses: Strains that occur in a welded joint after the welding has been completed, as a result of mechanical action, thermal action, or both.

SAE International: Formerly named the Society of Automotive Engineers, an automotive society responsible for creating an early system for classifying carbon steels.

Scale: A method of drawing an object larger or smaller than its actual size while keeping the proportions the same.

Short-circuiting transfer: A GMAW welding process in which the electrode wire shorts against the base metal. The welding current repeatedly and rapidly melts the wire and establishes an arc, which is extinguished when the wire again touches the base metal.

Sintered: A process of making parts from a powdered metal in which the metal is molded and then heated in a furnace without melting to fuse it into a solid metal.

Spray transfer: A GMAW welding process in which the electrode wire does not touch the base metal, and the filler metal is transferred axially (in a straight line) in the arc as fine droplets.

Stickout: The length of electrode that projects from the contact tube, measured from the arc tip to the contact tube.

Strain hardening: A method, also called *work hardening*, used to increase the strength of a metal through permanent deformation.

Stress-relief heat treatment: A technique used to reduce welding stresses and to increase the flexibility and strength of the weld metal and the heat-affected zone (HAZ).

Synergic: Welding technology that controls voltage, current, and wire feed using a program optimized for each welding situation.

Tempering: The process of reheating quench-hardened or normalized steel to a temperature below the transformation range and then cooling it at any rate desired.

Tensile strength: The maximum stress or force that a material can withstand without breaking.

Through hole: A drilled hole that goes completely through an object.

Tolerances: The allowances over or under the specified dimension, value, or other parameter.

Trim: A control found on synergic welding machines similar to the voltage control on regular machines that acts as a kind of override to the program.

Truncated: Having the pointed top (vertex) of a conical shape cut off, leaving only a plane section parallel to the base of the cone.

Tungsten: A nonstandard common name for the tungsten electrode used in GTAW; also, the principal metal from which the electrodes are constructed.

Underbead cracking: Subsurface cracking in the base metal under or near the weld.

Unified numbering system (UNS): An industry-accepted numbering system for identifying ferrous metals and alloys.

Walking the cup: A welding method in which the welder rests the edge of the torch nozzle (cup) against the base metal or groove edges to steady the torch and maintain a constant arc length.

Web: The plate joining the flanges of a girder, joist, or rail.

Weld symbols: Graphic characters connected to the reference line of a welding symbol specifying the weld type.

Welding symbol: A graphical representation of the specifications for producing a welded joint; includes a reference line and arrow line and many also include a weld symbol.

Wide flange (W) beam: A metal I-beam that has an extra wide flange at the top and bottom.

Wrought: Formed or shaped by hammering or rolling.